Nicolas Chuquet, Renaissance Mathematician

Nicolas Chuquet, Renaissance Mathematician

A study with extensive translation of Chuquet's mathematical manuscript completed in 1484

edited by

Graham Flegg

Cynthia Hay

Barbara Moss

D. REIDEL PUBLISHING COMPANY

A MEMBER OF THE KLUWER ACADEMIC PUBLISHERS GROUP

DORDRECHT / BOSTON / LANCASTER

Library of Congress Cataloging in Publication Data

CIP

Flegg, Graham,
Nicolas Chuquet, Renaissance mathematician.

Includes selections of the translation of Chuquet's Triparty en la science des nombres.
Bibliogrpahy: p.
1. Mathematics-Early works to 1800. 2. Mathematics-France-History-15th century. 3. Chuquet, Nicolas, d. 1488. I. Hay, Cynthia. II. Moss, Barbara. III. Chuquet, Nicolas, d. 1488, Triparty en la science des nombres. English. Selections. 1984. IV. Title.
QA32.F57 1984 510 84–18068
ISBN-13: 978-94-009-6504-1 e-ISBN-13: 978-94-009-6502-7
DOI: 10.1007/978-94-009-6502-7

Published by D. Reidel Publishing Company,
P.O. Box 17, 3300 AA Dordrecht, Holland.

Sold and distributed in the U.S.A. and Canada
by Kluwer Academic Publishers,
190 Old Derby Street, Hingham, MA 02043, U.S.A.

In all other countries, sold and distributed
by Kluwer Academic Publishers Group,
P.O. Box 322, 3300 AH Dordrecht, Holland.

CONTENTS

PREFACE

My attention was first drawn to Chuquet's mathematical manuscript whilst undertaking the necessary research for the preparation of the Open University's *History of Mathematics* course, presented initially in 1974. It was whilst editing the English edition of *Mathématiques et Mathématiciens* (P. Dedron and J. Itard, trans. J. Field) that I noted that it was stated that "the whole manuscript ... comprises 324 folios, i.e. 648 pages", and that, in addition to the *Triparty* (by which the work is generally known) the manuscript includes sections on problems, on the application of algebraic methods to geometry, and on commercial arithmetic. It was clear from references to Chuquet in most of the secondary sources that even the existence of the last three parts of the manuscript was very largely unknown. It therefore became clear that a comprehensive presentation of the contents of the entire manuscript was a much needed resource for researches into the mathematics of the Renaissance period.

For the past six years, it has been my privilege to work with colleagues both within and outside the Open University in the preparation of a transcription of those parts of Chuquet's work never previously published and the translation of the whole work into English, and also in making our combined assessment of the role of the manuscript in the history of mathematics and Chuquet's importance as a Renaissance mathematician.

It has not been possible to publish here the complete translation; this would have involved a volume nearly twice the length of this present work. My colleagues and I have therefore had to be selective. The most important and interesting part of the manuscript, the *Triparty*, has been very extensively presented in English translation. The remaining parts have been presented sufficiently extensively to enable their nature and general content to be well appreciated. Our assessment has, however, been based on the total work.

In making the translation, we have attempted to retain the flavour and something of the structure of Chuquet's

French. We have had as a first priority, however, the task of conveying the mathematical content of the work. Sometimes this has involved difficult decisions for us, when our various aims have come into conflict. Because of this, we have deliberately not always been strictly consistent in our translation of specific words, nor have we been by any means entirely faithful to Chuquet's punctuation. We hope, however, that what has resulted is a 'faithful' translation if not always an 'exact' one. We have, for example, often ignored Chuquet's tenses and also his use of singular and plural forms where an 'exact' translation would have made the English excessively awkward. On the other hand we have often retained his order of words, when this enabled us to keep the contemporary 'flavour' of his writing without destroying the sense of the English.

I would wish here to pay especial tribute to my two academic colleagues who have collaborated with me on this venture, Dr. Cynthia Hay of the Open University and Ms. Barbara Moss, formerly of Salford University but now a micro-computing consultant to *the Economist* group. We would wish to record our gratitude to Dr. M. S. Sidarski and Mr. J. L. Taylor who assisted with the initial transcription of Chuquet's manuscript, and to Mrs. Doreen Tucker who undertook the somewhat difficult typing which preparation of this book has demanded. Our gratitude is also due to the Open University for providing funds to support this research, to the Bibliotheque Nationale for granting permission to photograph the entire Chuquet manuscript, to the President and Fellows of Queens College, Cambridge, for permission to reproduce an extract from de la Roche, and especially to our Publishers, who have been extremely patient and helpful when we encountered unavoidable delays in preparing this work for publication.

This book is presented to the public in the quincentenary year of the completion of Chuquet's manuscript in 1484. We offer it as a small tribute to that remarkable French Renaissance mathematician.

Graham Flegg
The Open University
July 1984.

Chapter 1

ANTECEDENTS

Chuquet completed his mathematical manuscript in 1484, at the end of what is usually known as the Middle Ages, that is, the thousand years after the end of the Roman Empire in the West in about AD 500. Chuquet's mathematical achievements can best be appreciated against the general background of medieval mathematics. In this chapter, we can only briefly sketch some of the aspects of the history of mathematics in this period. For a general introduction, the reader may refer to Boyer (1968) or Mahoney (1978); for a study in greater depth, the reader may consult Juschkewitsch (1964).

The main historical factors which should be taken into consideration include the indigenous traditions in medieval mathematics in Western Europe, Indian mathematicians and their contributions to arithmetic and algebra from the 5th to the 10th century, and the Arabic mathematicians active from the 8th century, whose translations and adaptations of mathematical works were seminal for the later development of mathematics in Western Europe. Arabic texts, both original works and translations, were translated into Latin in 12th-century Spain by such figures as Robert of Chester and Adelard of Bath, and these Latin versions stimulated the development of mathematics in Western Europe by making available hitherto unknown concepts and approaches. Chuquet's most eminent predecessors, such as Leonardo of Pisa, Jordanus de Nemore, and Nicole Oresme were all beneficiaries of this scholarly activity.

The mathematical manuscripts of Boethius provide a benchmark for the sparse indigenous mathematical activities in Western Europe for the period prior to the 12th century. Boethius lived from about 480 to 525, during the period of the final disintegration and collapse of the Roman Empire in Western Europe. The historian Edward Gibbon described him as the last of the true Romans. He figures more widely in the history of Western culture and thought as the author of the *Consolations of Philosophy*, written while he was imprisoned towards the end of his life.

The contribution of Boethius to medieval mathematics was to copy and preserve in a much truncated form some of the mathematical manuscripts of the Greeks. What Boethius transmitted provided for a long time an upper limit at a low level to mathematics in the Latin West. His mathematical manuscripts were extensively copied and in the process corrupted; many later interpolations were made, and it is difficult to determine what was due to Boethius himself and what was subsequently added.

Boethius wrote text books for each of the mathematical subjects which formed the second part of the arts curriculum, the *quadrivium*, in medieval schools: arithmetic, geometry, astronomy and music. These works were little more than elementary paraphrases of more impressive works of earlier Greek and Alexandrian mathematicians. For example, his *Arithmetic* was an abbreviated version of the *Introductio* written by Nicomachus who flourished around AD 100. His geometry provided statements without proofs of the simpler parts of the first four books of Euclid's *Elements*. The mathematical knowledge obtainable from the works of Boethius is to our eyes extremely limited, but it was adequate for the purposes for which it was used, such as calculating the religious calendar. Moreover, from the 5th to the 12th century there was little else available in the monastic schools of Western Europe.

Mathematics in the Byzantine Empire was not as limited as in the Latin West, as can be seen from a brief consideration of the works of some commentators. Although none of these commentators can be described as a 'significant mathematician', four of them deserve mention because of the later influence of their commentaries. Metrodorus, who flourished *c* AD 500, demonstrated a basic knowledge of algebra which is absent from the works of Boethius. He compiled the *Greek Anthology*, consisting of some forty mathematical problems, most of which give rise to simple linear equations. Similar problems are found in some later western manuscripts. Simplicius, who flourished around the second decade of the 6th century, was a wide-ranging mathematical commentator, especially concerned with geometry. His verbatim quotations preserved ancient fragments on the calculation of the areas of lunes. His commentary on Euclid was later translated into Arabic. Eutocius of Ascalon, who flourished *c* 510, commentated extensively on the works of

Archimedes and Apollonius. His commentaries were edited by Isodorus of Miletus, who flourished in Constantinople in the 6th century.

Developments which were later to infuse new life into Western mathematics took place in India from the 5th century and in the Arabic world from the 8th century. Indian mathematics of this period contributed the number system which is often called the 'Hindu-Arabic numerals', as well as developing certain standard techniques for solving problems in arithmetic and algebra. For example, solution by the rule of false position is to be found in Indian writings prior to its appearance in Arabic works. This technique, which Chuquet called the 'rule of one position', consists of setting a value for the unknown, working out the corresponding discrepancy between the two sides of the equation, and using this to obtain the correct result by linear interpolation.

Indian mathematics developed as part of a flourishing culture and was closely linked with astronomy, The best-known Indian mathematical works are embedded in astronomical texts. Another aspect of Indian mathematics of this time, which is striking to western eyes, is that mathematical works were written in verse, a feature which translations from Sanskrit cannot reproduce. Lastly, despite the development of some general algorithms for arithmetic and algebra, it is fair to say that Indian mathematics was more empirical than the mathematics of classical Greece. It lacked the logical structure of presentation of which Euclid's *Elements* is the supreme example.

One of the earliest major Indian mathematical works of this period was the *Aryabhatiya*, written in AD 499. Commentators have differed in their assessment of this work. Some have considered its importance for the history of Indian mathematics to be comparable to that of the *Elements* for the mathematics of Ancient Greece; it has also been described as obscurely phrased. The topics covered in this brief work give some indication of the concerns of Indian mathematicians. It introduces names for powers of 10 up to the tenth place, rules for calculating square and cube roots, various methods for determining areas and volumes (not all of which are correct), a value for π, rules for progressions and proportions, and some commercial problems which include compound interest.

A later Indian mathematical work, the *Siddhânta* of Brahmagupta, appears to have been the crucial intermediary in the transmission of the Hindu numerals to Arabic mathematics, and subsequently to the Latin West. This work, written in the 8th century, was translated into Arabic towards the end of that century, thus making the Hindu decimal system and associated methods of calculation available to the Arabic mathematicans of the time.

The development of Arabic mathematics was in part a consequence of the Arab conquests; the Arabs swiftly assimilated the superior cultures of the countries they conquered. Commercial needs also played a part in the development of Arabic science. Their mathematical activities were centred around the school of Baghdad. The most significant period of development spanned the two hundred years from the end of the 8th century. Juschkewitsch (1964) comments on the insecure and difficult position of Arabic scholars, whose activities were dependent on their remaining in favour with their patrons, the Moslem rulers: for example, they could not afford to come into conflict with the official religion. Nevertheless, Arabic mathematicians played a crucial role in transmitting the mathematical knowledge of both the Greeks and the Indians to Western Europe.

The best-known of the Arabic mathematicians of this period was Mohammed ibn-Musa al-Khowarizmi, who lived from *c* 780 to *c* 850. His importance for the history of mathematics is twofold. He had a major role in disseminating the Hindu numerals, and he composed an elementary algebra text, whose importance in this field can be compared to that of Euclid's *Elements* in geometry. He gave a full account of the Hindu numeral system and Hindu methods of calculation in a work, *De numero indorum*, which is known to us only through the Latin translation, as the Arabic original has been lost.

Al-Khowarizmi's algebra, *Al-jabr w'al mugābalah*, is a key work in the history of mathematics. The very word 'algebra' derives from the first part of its Arabic title. The later Latin translation of the first part of this work by Robert of Chester in 1145 can be said to mark the start of European algebra. Al-Khowarizmi's achievement was to synthesize and communicate a substantial body of mathematical knowledge from both Hindu and Greek sources.

Al-Khowarizmi's algebra contrasts in two respects with that of Diophantus, his illustrious Greek predecessor, who flourished in the 3rd century AD. Al-Khowarizmi's work was written to aid the resolution of problems in everyday life, and was thus a much more practical work than that of Diophantus. A second contrast is that Al-Khowarizmi's algebra was a rhetorical one, in which equations were expressed in words alone, whereas Diophantus wrote a syncopated algebra, using a mixture of words and symbols. This latter point illustrates the intermittent historical development of a partial use of symbolism in algebra.

Al-Khowarizmi's algebra followed the example of his work on arithmetic in using the Hindu place-value system of numbers. He considered the solution of the various standard types of equations made up from *jidhr* or 'root', *mal* or 'square', and numbers. In modern terms, these are equations involving x, x^2 and constant terms. He was able to give general formulations of problems in his rhetorical algebra, because he used general terms for the unknown and its square. His methods of solution for quadratic equations are equivalent to completing the square, and can all be found in earlier Hindu works. Proofs of the methods are, however, geometrical, and taken from the Greeks. The problems which Al-Khowarizmi included illustrate the application of quadratic equations to a wide variety of situations, including inheritance questions - which could be extremely complicated under Moslem law. These problems have provided a means of tracing sources of Al-Khowarizmi's mathematics.

Among other Arab mathematicians, one in particular, Al-Khayyami (1048?-1131?) is better known as Omar Khayyam and the author of the *Rubâiyât;* his contributions to algebra include a systematic treatment of cubic equations. Overall, the Arabic works represent a blending together of Greek and Hindu learning with some original contributions, but also with some significant omissions such as the solution of indeterminate problems found in Diophantus.

The introduction of Hindu-Arabic numerals into Europe is attributed to Gerbert, later Pope Sylvester II, in the latter part of the 10th century. His mathematical writings on both arithmetic and geometry were in the Boethian tradition. Although he used Hindu-Arabic numerals on counters used for arithmetic computation on counting boards, the methods of calculation associated with them in Hindu and Arabic works did not appear in his writings.

In the later medieval period there were two main channels for the transmission of mathematics to the West: trade and translations. Spain and Italy, countries on the main Mediterranean trade routes, mediated this transmission. Spain in particular had close links with Arabic culture because it had been part of the Moslem Empire through most of the Middle Ages. In the 12th century, centres of translation in Spain were a major factor in making available mathematical works to scholars in the Latin West. Greek and other mathematical works became known through the medium of Latin translations, often from Arabic. Scholars travelled to Spain in order to make translations of the manuscripts available there; one such translator was the Englishman Robert of Chester, (*fl* 1141) who translated the algebra of Al-Khowarizmi. Italian merchants trading with the Arab world also contributed to the transmission of Arabic mathematics. Leading mathematicians of the 13th century were thus familiar with Al-Khowarizmi's work through one or both of these channels.

Two main traditions of mathematics developed in the Latin West as a result of this dissemination. In church schools and most medieval universities, the knowledge of a small number of classical texts was emphasized, such as Boethius' *Arithmetic*, parts of Euclid's *Elements*, and the treatise of the 12th-century English writer, Sacrobosco, on the sphere. The practical tradition of commercial arithmetic that developed among merchants and traders made use of the Hindu-Arabic numerals. There was a certain amount of rivalry between exponents of these numerals, who became known as algorists, and the abacists, who came from the older practical tradition that relied on the abacus (counting board) and Roman numerals.

The mathematician who did more than any other to popularize the new mathematical methods associated with Hindu-Arabic numerals was Leonardo of Pisa (*c* 1175-1250), also known as Fibonacci. Leonardo was in a unique position to disseminate the ideas of Arab mathematics in Western Europe. He was born in Pisa and brought up in that part of North Africa which is now known as Algeria. His father was a prosperous customs officer at Bougie, the most important port on the North African coast. Leonardo travelled extensively in Egypt, Syria and Greece. He returned to Italy in AD 1200 with a comprehensive knowledge and understanding of the art of algorism and its use in commerical transactions of every kind.

Leonardo's best-known work, the *Liber Abaci*, first appeared in 1202 and subsequently in a revised version in 1228. Its title was one of the first uses of the word 'abacus' in a broader sense to refer to mathematical computations rather than to a counting board. It was a broad compendium of arithmetical and algebraic methods, and included the solution of problems involving the new numerals. One problem in this collection which is particularly well-known has to do with the breeding of rabbits, starting with one pair of rabbits kept surrounded by a wall. This problem gives rise to the Fibonacci series: 1, 1, 2, 3, 5, 8, 13, ... (where $u_n = u_{n-1} + u_{n-2}$), which is perhaps the most widely known of Leonardo's contributions to mathematics.

The *Liber Abaci* reflected Leonardo's indebtedness to Arabic mathematicians, and to Al-Khowarizmi in particular. The algebra in this work was rhetorical, as in the works of Arabic mathematicians. Some geometry was included in the spirit of the later Boethian tradition, that is, less for its own sake than for the sake of underlining the claim that arithmetic and geometry are linked and can be used in each other's mutual support. Leonardo also wrote a geometrical work, *Practica Geometriae* (1220) as well as *Flos* and *Liber Quadratorum* (both *c* 1225).

As with most works in medieval mathematics, the assessment of the influence of the *Liber Abaci* is a complicated matter. There was considerable resistance to Leonardo's recommendation that the Hindu-Arabic numerals and the associated algorithmic methods should be adopted. The new numerals were first adopted for recording numbers in almanacs and calendars, on coins, and for numbering pages in books. In scientific works, the numerals, together with common fractions and algorithmic methods, were adopted comparatively quickly, although astronomers retained the sexagesimal fractions handed down from Babylonian times. The calculation methods were slow to be adopted outside scientific circles for several reasons. Confidence in the existing use of the abacus for commerical calculations was not shaken by Leonardo's recommendation of the Hindu-Arabic alternative. Methods of calculation associated with the abacus had been in use for many centuries and could be carried out mechanically with little or no mathematical thinking. The algorithmic methods for addition and subtraction were more difficult.

For the new calculation methods to spread through Western Europe it was essential that they should be widely adopted by the large mercantile community. Only gradually, over a period of some two centuries, were these methods adopted in the thriving Italian and Spanish centres of trade with the Arab world. Italian merchants, in such cities as Genoa, Venice, Florence and Pisa were among the first to make use of the new algorithmic methods, because the bulk of their trade was with people already using them. Florentine merchants in particular developed and improved some aspects of these methods, and also invented principles of accountancy and book-keeping using the new numerals. The double-entry ledger still used today appears to have originated in Florence.

However, there was still opposition to the algorithmic methods as late as the mid-14th century in edicts of some of the civic authorities. It was considered that forged alterations of the numerals could be made all too easily, for example by changing a 0 to a 6 or a 9. Consequently, as a safeguard, sums of money were often recorded in both words and numerals, a practice which has continued in the writing of cheques to the present day.

Leonardo's *Liber Abaci*, although it was written in Latin, has been seen as the father of the vernacular abacus manuscripts: these were works in which abacus is understood in the broader sense of mathematical computation. They stand in a close relation to the commerical activities and the popularization of the Hindu-Arabic numerals and methods, discussed above. According to a recent study of the Italian abacus manuscripts by Warren van Egmond (1976), the earliest known instances of these manuscripts came from the last decades of the 13th century, some ninety years after Leonardo's *Liber Abaci*. These popular works, written in Italian, used Hindu-Arabic numerals and methods to solve practical and more or less realistic mathematical problems of everyday life, chiefly of business and commerce. Their mathematical contents included arithmetic, algebra, and geometry, as well as material in merchants' handbooks, and on calendars and coinage. The earliest Italian printed arithmetics were similar to the abacus manuscripts. Indeed, there were still abacus manuscripts being written in the 16th century, after the first printed arithmetics had come into existence.

Van Egmond has argued that these manuscripts are of major importance to the history of mathematics. The history of mathematics has tended to concentrate on those persons who have been credited with mathematical innovations, thus contributing to the idea that the history of mathematics is relatively independent of social context. Van Egmond argues that this view is a serious distortion. His alternative view might be described as a case for history of mathematics 'from below'. He argues that the practical mathematics of the Italian abacus manuscripts made a major but neglected contribution to the development of algebra in the 16th century and thus, indirectly, to the mathematical 'revolution' in the 17th century.

As with other fields of medieval mathematics, the story of medieval geometry can be told in terms of before and after: before and after the translations of Euclid's *Elements* from the Arabic versions. Prior to the spreading of Latin translations from Arabic, medieval geometry was characterized by interests in geometry for reasons that would now be considered to be external to it. It was centred on the exegesis of Aristotle and on the understanding of theological references to geometry. The Boethian tradition in geometry, as has been noted, gave only a very limited account of Euclid's work. A tradition of practical geometry had also developed in connection with architecture and surveying, but the geometrical requirements of this tradition were at a fairly low level. The third translation of Euclid by Adelard of Bath (*fl* 1116-1142) in particular had a considerable influence on the subsequent development of medieval geometry. Leonardo's *Practica geometriae* (1220) incorporated both the rediscovery of a substantial geometrical theory and also the practical tradition in geometry. He provided two paths through this work: one for theoretically inclined readers, and one for the more practically inclined. Later 14th-century geometers, such as Johannes de Muris and

Dominicus de Clavasio, followed his lead in including both practical and theoretical elements in their geometrical writings, although not in terms of alternate paths for readers with different interests and capabilities.

In the history of mathematics in the later medieval period, there is both a high road and a low road. The high road is marked by the achievements of notable mathematicians such as Leonardo of Pisa, Jordanus de Nemore, Oresme, and Regiomontanus. Those who will be briefly mentioned may have influenced Chuquet, or may be compared with him. On the low road we find the development of the vernacular and practical traditions. Chuquet drew on these traditions as well as on the achievements of earlier major mathematicans. Both roads meet and cross in his writings and indeed elsewhere.

Jordanus de Nemore, who flourished *c* 1220, was a contemporary of Leonardo of Pisa. His *De numeris datis* has recently been published in a modern edition with an English translation (Hughes, 1983). The work was unusual, not only because it used a rudimentary symbolism, in contrast with the rhetorical algebra of Leonardo, but also because it was more theoretical than most other medieval works, in that Jordanus set his material out in terms of statements of general problems. Hughes describes it as the first textbook of advanced algebra.

A 14th-century figure who made major contributions to mathematics was Nicole Oresme (*c* 1320-1382). Oresme was born in Normandy, became a Parisian academic, and subsequently entered the Church, in which he rose to become Bishop of Lisieux. As well as translating some of Aristotle, he wrote five mathematical works, of which two were largely devoted to the theory of proportions, a complex topic of considerable interest to many medieval mathematicians. Some of his works have been edited and published in recent years . He wrote algebra in the rhetorical mode, without algebraic notation. Both integral and fractional exponents are to be found in the works of Oresme, as well as a suggestion, if not more, of the possibility of irrational proportions. In his writings on motion there is one of the earliest known examples of the representation of a simple function by what has become known as a 'straight-line graph'.

Oresme wrote in Latin and in the tradition of the scholastics. He was more concerned to expound and glorify the works of his Creator than to instruct and guide merchants and their children. His immediate influence is somewhat hard to establish. His works passed without citation in the more pragmatic writings of the 15th century, although L'Huillier (1979) has shown that Chuquet almost certainly had access to at least one Oresme manuscript.

One of the most influential mathematicians in the 15th century was Johann Müller (1426-1476), born in Königsberg and usually known as Regiomontanus, from the Latin name of his birthplace. His work on triangles led to the revival of trigonometry. He was the first major publisher of astronomical and mathematical literature. In publishing mathematical works, he endeavoured to provide texts free from the errors and interpolations of the manuscript tradition, an aim he also pursued in the Latin translations he made of scientific writings. Publication of scientific works was later to be of major importance in the development and diffusion of science. As scientific publishing developed after the death of Regiomontanus, the availability of books, in contrast to the difficulties of studying manuscripts, meant that scientists could cover much more in the way of scientific literature. The contributions of Regiomontanus to both trigonometry and to scientific publishing were cut short by his premature death.

The academic and vernacular traditions came together in the work of Luca Pacioli, whose *Summa de arithmetica, geometrica et proportionalita*, printed in 1494, has an assured place in the history of mathematics. Written in Italian, it can be seen as the culmination of the flourishing tradition of practical mathematics in 15th-century Italy, and has been described as a landmark of 15th-century mathematics. It encompasses a wide range of topics, ranging from algebra and the geometry of Euclid to mercantile tables and double-entry book-keeping. In composing this work, Pacioli acknowledged that he had drawn freely on the works of other mathemticians, and on Leonardo of Pisa in particular: the concept of literary property, as we know it, had not yet developed. The *Summa* was well-known to 16th-century mathematicians; it bears careful comparison with Chuquet's *Triparty*.

Although Pacioli himself had a university background, his book has more affinities with the abacus manuscripts and their successors among early printed books than with academic tradition. These manuscripts have been described above. Most of them are in Italian, but by the late fifteenth century a number had also appeared in French, German and English. Two examples in French will be mentioned in this book: an anonymous manuscript from Pamiers which contains several problems also found in the *Triparty* and an arithmetic by Jehan Adam who refers, as Chuquet does, to Berthelemy des Romans.

The vernacular tradition of arithmetic developed in response to the needs of the commercial world. The merchants of the Renaissance were not interested in the rediscovery of classical mathematics, but rather in a large supply of teachers and texts to explain in the language of the market-place the arithmetical techniques that would enable increase-ingly complicated commercial calculations and transactions to be carried out. As the great Italian banking families established houses throughout Europe, so the vernacular arithmetics which has been popular in Italy for so long spread to other countries. Chuquet's work, covering commercial arithmetic and practical geometry well as more advanced material, can be linked with this tradition, but his originality and his contributions to mathematics go well beyond its usual scope. Also, although it is a milestone in the history of mathematics, it is not an isolated text, for it is part of the much more extensive development of mathematics in the later medieval period which has been briefly sketched above. The work can be assessed with reference to the topics that had been discussed. His algebra can be compared with the works of other important figures in the history of algebra, such as Leonardo of Pisa, Jordanus de Nemore, and Pacioli. His geometry should be assessed in the context of a tradition of practical geometry. His commercial arithmetic was written to teach youths entering the world of business. Thus the manuscript incorporates several major strands of late medieval mathematics.

Chuquet has been described as so advanced for his time that his contemporaries could not understand him and consequently neglected him. According to J.W.N. Sullivan (1925), "Chuquet ... was in advance of his time, particularly in France, which lagged behind Italy in the art of calculation, so that his influence was slight." This assessment has been given currency by Howard Eves' textbook in the history of mathematics (1976): "Chuquet's work was too advanced for the time, to exert much influence on his contemporaries." Sullivan and Eves have oversimplified the matter of Chuquet's reputation. They are right to draw attention to his originality. However, to attribute the neglect of his work solely to the sophistication of his mathematics for his time overlooks the chequered history of his mathematical manuscript - to be recounted in the next chapter. The manuscript, in the form in which Chuquet had written it, was not readily available to the mathematicians who succeeded him. Just as world history might have changed had Cleopatra's nose been longer, so might the history of mathematics have been changed had Chuquet's manuscript circulated more widely in the sixteenth century, or had it been printed in the form in which Chuquet had written it. Estienne de la Roche, possibly one of Chuquet's pupils, incorporated much of Chuquet's manuscript in his *Larismethique nouellement composee*, which first appeared in 1520. However, de la Roche omitted or misunderstood much of the more interesting mathematical material in Chuquet's manuscript. Thus Chuquet's contributions to mathematics were obscured rather than illuminated by de la Roche's treatment.

Chuquet's neglect by his contemporaries was indeed their misfortune, although, to be fair, it should be said that they appear to have been ignorant of his work rather than baffled by it. The justice of the claims that have been made for Chuquet's originality can be fully appreciated only by reading and assessing his manuscript in the context of late medieval mathematics.

Chapter 2

NICOLAS CHUQUET - THE MAN AND HIS MANUSCRIPT

Who was Nicolas Chuquet? It is extremely difficult to answer this question in detail. Chuquet tells us very little about himself. What he does tell us is to be found at the end of the *Triparty*, where he writes:

And thus to the honour of the glorious Trinity ends this book, which on account of these three principal parts I call Triparty. And also, because it has been made by Nicolas Chuquet, Parisian, Bachelor of Medicine, I call it the Triparty of Nicolas in the science of numbers, which was begun, continued and finished at Lyon on the Rhone, the year of salvation 1484.

Although researches into the tax registers of Lyon have thrown a little more light onto Chuquet's later years, no evidence has so far been discovered which would enable us to give a year for his birth with any degree of confidence. At best, we can say that he was probably born in the early 1440s. We cannot be certain that he was born in Paris, since his claim to be a Parisian could refer merely to a period of residence.

In the tax registers of Lyon over the years 1480 to 1484 there is to be found a certain *Maitre Nicolas, escripvain.* He is described as having lived "from the gate of the Little Friars along by the *rue de la Grenette* towards *le Muton*". We can find *l'Eglise des Cordeliers* (the Franciscans' Church) still standing there today. In the registers of 1485 and 1487, Chuquet is given his full name *Nicolas Chuequet* and is described as being an *algoriste.* In the register of 1488 there is reference to his heirs. From this evidence we can reasonably infer that he arrived in Lyon in 1480, or possibly earlier, and that he died in 1487 or 1488.

We see therefore that Chuquet earned his living in Lyon initially as an *escripvain.* In those days this title would mean that he was a copyist or writing master. We can still find the title *ecrivain* on office signs, though today it has the specific connotation of public letter-writer. In the later fifteenth century Lyon was a cosmopolitan city with a highly active commercial community and a flourishing spice market. The fairs, of which the spice market was a significant part, had been established in 1464 under the royal protection of Louis XI. These fairs stimulated the growth of Lyon as an important centre of banking, and the city was also beginning to develop a printing industry. There was a need for the copying of legal and commercial documents, especially on the part of those who were less familiar with the French language, now increasingly being used in place of Latin. As the various commercial activities increased, there was a further need for masters skilled in the art of computation using the Hindu-Arabic numerals and capable of communicating that art to others.

It would appear therefore that Chuquet initially set up in business in Lyon around 1480 to supply the need for copying, translating and drawing up commercial and legal documents. As his skill as a master of algorithm became known, he was increasingly in demand for his mathematical abilities, so that by 1485 he could be described solely as an algorist.

Chuquet describes himself as a bachelor of medecine. This suggests that he abandoned his medical education without obtaining the full qualifications required to practice as a doctor. If he studied in Paris, he would first have obtained the degree of Master of Arts, which ostensibly required the study of mathematical material to be found in the work of Boethius and the first six books of Euclid. However, there is no evidence that Chuquet studied at the Sorbonne. His name does not appear in any of the student registers, kept sporadically from 1458 onwards. There were many other medical schools in France where he could have studied, including Montpellier, where Rabelais was granted his doctorate after a mere six weeks of residence. There is thus no external evidence to support Chuquet's claim to a medical education, neither can we say anything about his formal mathematical training. Many respected doctors of the late fifteenth century had a considerable interest in astrology, but there is no evidence to suggest that Chuquet shared this interest.

Another important question which remains unresolved is whether or not Chuquet visited Italy. Internal evidence from his manuscript suggests strong Italian links, but it is possible to overestimate the significance of this evidence because there can be difficulties in distinguishing between Italianisms and Latinisms. The evidence is strongest in the geometrical part of the manuscript where there are a number of striking similarities with contemporary Italian works. L'Huillier has found an earlier manuscript of the *Geometry*, written on Italian paper and dating from the early 1470s.

If Chuquet did indeed visit Italy, we may well wonder if he met Luca Pacioli, with whose *Summa de arithmetica, geometrica, proportioni et proportionalita* of 1494 his manuscript must inevitably be compared. However, even if we accept that there are many Italianisms in Chuquet's work, this can be explained by pointing to the considerable Italian colony existing at that time in Lyon. He may well have had close dealings with its members both as an *escripvain* and, later, as a teacher of algorism. Many fifteenth-century Italian mathematical manuscripts survive today, some of which have been carefully studied by modern scholars such as Warren van Egmond. A number of them have been published for the first time by the Centre for the Study of Medieval Mathematics at the University of Siena. Nothing is known however, about the extent to which such manuscripts could have circulated outside Italy, and particularly in Lyon. It is possible that Chuquet could have had access to Italian mathematical works through an association with the Italian colony in Lyon. We cannot therefore infer from internal evidence that Chuquet visited Italy.

A further interesting question relates to the nature of Chuquet's relationship, if any, with Estienne de la Roche. De la Roche incorporated substantial portions of Chuquet's manuscript in his *Larismethique nouellement composee* published in Lyon in 1520, and subsequently in 1538 under the revision of Gilles Huguetan. It seems certain from de la Roche's work that Chuquet's manuscript came into his possession. This is confirmed by marginal annotations in de la Roche's hand on a number of folios of the manuscript. The tax registers at Lyon provide us with information about the de la Roche family. Thus, we know that the family owned a house in Lyon on the Rue Neuve and also property situated above Villefranche. Indeed, in the tax register of 1493, Estienne de la Roche is said to be "called Villefranche". He is described as a qualified master of *argorisme*, a spelling to be found also in his *Larismethique*. It seems likely that de la Roche and Chuquet were acquainted, possibly in the respective roles of pupil and master. The former does acknowledge a debt to Chuquet in two passages of *Larismethique*, though referring also in the same terms to a certain Philippe Frisco Baldi of Florence and to "Brother Luke of Burgo San Sepulchro of the order of Friars Minor" (Luca Pacioli). However, the substantial extracts from Chuquet's manuscript incorporated in *Larismethique* show that his main debt was to Chuquet. These extracts include a number of transcription errors, and de la Roche's work is generally of a more conservative nature. He is often unwilling or unable to accept Chuquet's innovations and reverts to earlier and less satisfactory terminology.

The history of Chuquet's manuscript, which can be easily traced through internal evidence from the document itself, is carefully documented in Aristide Marre's introduction to his edition of the *Triparty* (1880). After it had been in the possession of de la Roche, it was bought by an Italian, Leonardo de Villa, according to a note written in Latin on a protecting sheet at the beginning of the work. It then passed into the famous library of Jean Baptiste Colbert (1619-1683), the finance minister of King Louis XIV who was also founder of the French Academy of Science and a noted patron of scholarship and the arts. It was catalogued no. 2170 in Colbert's Library. In 1732, some eight thousand volumes, formely in Colbert's possession, passed into the royal library of King Louis XV (1710-1774), among which was Chuquet's manuscript. Its accession into the King's manuscript collection was recorded on 11th September of that year, and it was catalogued as no. 7482[3-5]. Today it is no. 1346 of the *Fonds francais* in the Bibliothèque Nationale.

From time to time, catalogues of manuscripts in the vaults of the Bibliothèque Nationale have been published, and we can find Chuquet's work listed, for example, in a manuscript list published in 1868. Here it is described as follows:

1. The Triparty of Nicolas Chuquet, a treatise of arithmetic beginning:
   "This book, to the honour of the glorious and sacred Trinity, is divided into three parts...", and concluding: "begun, continued and finished at Lyon on the Rhone, the year of salvation 1484."
2. "Several other devices with numbers in general", a collection of problems in arithmetic, beginning (fol. 148): "First, of 10 I wish to make three parts...",

and concluding: "...are sons of their sons and brothers of their husbands."

3. "How the science of numbers can be applied to the measures of geometry", beginning (fol. 211): "Here begins a little treatise on the practice of geometry...", and concluding: "...and so much of measures contains the measured vessel".
4. "How the science of numbers can be applied to the matters of merchandise", beginning (fol. 264): "One owes a man all the parts which follow...", and concluding: "...concerning merchandise and also all this book".

Neither in this description of the manuscript, nor in other (and much briefer) descriptions included in various earlier catalogues of manuscripts is there any indication that Chuquet's work includes algebra as well as arithmetic. Nor do we have evidence that it had attracted any significant interest on the part of informed mathematicians from the time of de la Roche until its rediscovery by Marre in the late nineteenth century. De la Roche's *Larismethique* also was comparatively little known. In papers presented to the French Academy of Sciences in 1841 and 1842, Michel Chasles remarks that it "was known neither to historians nor to bibliographers, although the name of the author had been cited by two sixteenth-century algebraists, Butéon and Gosselin, and after Buéton by Wallis," and that it was cited also by Heilbronner and Panzer. He notes de la Roche's reference to Chuquet's "treatise on algebra" and expresses the hope that this treatise "has not been completely lost".

The publication of the first of the four sections of Chuquet's work by Aristide Marre in 1880 therefore created something of a sensation for historians of mathematics. It appeared in Volume XIII of Boncompagni's *Bullettino di Bibliografia e di Storica della Scienze Matematiche i Fisiche*, pp. 593-814, together with a substantial introduction by Marre, pp. 555-592, including a table of contents mainly though not entirely based on Chuquet's own words. In the following year, in Volume XIV of the *Bullettino* (pp. 417-460), Marre published an abbreviated version of the second section of Chuquet's manuscript, prefaced by a short *Avertissement* of four pages. This section consists of 166 problems with fully worked solutions, but only the statement of the problems and their 'answers' were published, and their working remains unpublished to this day.

In publishing the *Triparty*, Marre rightly draws attention to its great importance for the history of European algebra. It antedates by ten years the much better known *Summa* of Luca Pacioli. He is, however, largely content to leave the work to speak for itself. We quote: "I leave to the reader the task and the pleasure of finding for himself, in the text before his own eyes, not only the cartesian notation for exponents and the elementary principles of their manipulation, but also the ingenious use of these same exponents in the solution of equations; then the "rule of intermediate terms", invented by Nicolas Chuquet, and so called by him because it serves to "find as many intermediate numbers as one wishes between two neighbouring numbers"; the use of the words "plus" and "minus" with their signs $\bar{p}$. and $\bar{m}$., which a little later become + and -, and also the statement of the "rule of signs" as it is formulated in our nineteenth-century treatises of algebra; and furthermore the germ, I might also say the very idea of "Logarithms", the discovery which, 130 years later, was to immortalize Sir John Napier."

Marre is highly critical of de La Roche and his *Larismethique*. After side-by-side comparison of extracts from the *Triparty* and *Larismethique*, he says of de la Roche, "master of argorisme" that: "...without being accused of injustice or exaggeration one could say that he appropriated the work of Nicolas Chuquet, that he purely and simply copied the *Triparty* in a host of places, that he suppressed some of the most important passages, especially in the algebra, that he shortened or lengthened others, in order to compose his *Arismethique*, vastly inferior to the *Triparty*, and finally that, if for four centuries Nicolas Chuquet and his work have remained in the shadows, it is above all to him [de la Roche] that we must attribute the prime cause."

Many of the commentators after Marre have been unstinting in their praise of Chuquet's work, though in some cases their enthusiasm has carried them well beyond what the text of the *Triparty* would support. Fr. B. Lefebvre's *Aperçu historique sur les origines des mathématiques elementaires* (Namur, 1897) provides an extreme example, including the claim that Chuquet was the inventor of logarithms. The extent to which such claims on Chuquet's behalf can be justified are discussed in Chapter 9. They are, however, best assessed from a study of the actual text of the *Triparty*. A more balanced account of the work appears in the *Dictionary of Scientific Biography*, where J. Itard draws attention to some of Chuquet's limitations as well as his notable achievements.

Itard's article refers to all four sections of the manuscript. This is in contrast to most references in modern works on the history of mathematics, which refer only to the *Triparty*. Indeed, it would seem that even today the existence of the other sections is known to only a few. This is not altogether surprising as only the *Triparty* was available in a complete form until the publication of the third section - that on geometry - by H. l'Huillier in 1979. The fourth and last section is still unpublished, though transcriptions exist both in the United Kingdom (at the Open University) and in France. The second section is, as we have seen published only in a highly abridged form. However, enough has been published for more than a century both of the text itself and of its history to call into question the somewhat superficial assessments which have been repeated in some general histories of mathematics. The examples of J.W.N. Sullivan and Howard Eves have been mentioned in Chapter 1. Superficial explanations of the failure of Chuquet to exert much influence on his contemporaries do not give adequate recognition to other contemporary documents, especially of the Italian vernacular school of mathematics, and neglect the more obvious reasons for his lack of influence such as his comparative isolation in Lyon and the general inaccessibility of the manuscript over several centuries.

Marre discovered Chuquet's manuscript in very much the same state as it is to be found today, though there has been some slight deterioration over the past hundred years. It consists of 342 folios of thick paper. There are three protecting sheets, added by the binder, at the beginning followed by five further such sheets of earlier origin which have become yellowed in the course of centuries. It is on the last of these that we find the reference to Leonardo de Villa. The 325 folios of main text then follow and, at the end, six of the older yellowed protecting sheets, followed by three more recent protecting sheets (corresponding to those at the beginning). The sheets are numbered in black ink at the top of the right on the recto and only from

1 to 20 and from 20bis to 327, the number 1 being on the last guard leaf at the beginning of the work so that the main text begins at page 2. The last page of main text carries the number 324 under the number 320 in a different hand deleted with two horizontal strokes. The numbers 325 to 327 are on protecting leaves. Four folios are entirely without text (82, 94, 153, 263), as are also three of the verso pages (fols. 147, 152, 210). This numbering was described by Marre as being of "a fairly recent date". Originally the manuscript had neither page nor folio numbers, though there are inscriptions at the bottom right-hand corners of the recto pages indicating that an earlier system of labelling the sheets had once existed. A number of these inscriptions are missing, on the assumption that the system was at one time complete. The system involved letters of the alphabet followed by numbers 1 to 8 from a.1 on the first page of text (now erased) to x.4 on the last page of text (fol. 324 r).

The overall size of each folio is 310 mm by 210 mm, and there are generally 33 lines to the page. There are large margins, especially on one side and at the bottom, and these are frequently used for examples of calculations, for geometric figures, and for various marginal annotations. The whole is bound with hard covers of a somewhat faded red colour with a red morocco spine divided into six sections. On the spine there are engraved gold letters describing the work as the "Triparty de Nicolas" and decorations described by Marre as follows: "The ... segments are adorned with the cypher of King Louis XV, two capital Ls in cursive script interlaced, with a fleur-de-lys between their arms, are surmounted with the royal crown of France, surrounded by stars and accompanied by flowers to the right and to the left, the whole being engraved in gold on the red morocco spine." A label indicates that the manuscript has the catalogue reference FR 1346.

The writing throughout the main text is in a very firm and uniform hand which is quite beautiful to the eye. Itard's article on Chuquet claims that it is "the work of a firm of copyists", but it has been established by l'Huillier that it is Chuquet's own hand. There are large and superbly decorated capitals of red and blue on the first page of each of the main parts of the Triparty and at the beginning of the third and fourth sections. Smaller decorated capitals are to be found at the beginning of 'chapters'. There are five pages (fols. 321 v. - 323 v.) of numerical tables in black and red relating to precious metals, of which the last is a table of conversion of "fine silver" to "King's silver".

Chuquet's style of writing is clear and for the most part concise, and is thus particularly well-suited to conveying mathematical information. It is something of a surprise to find how little Chuquet's language differs from modern French. The manuscript can thus be read without great difficulty. Marre provides a glossary of words in the *Triparty* whose sense or spelling has changed significantly over the centuries. This runs to only three double-column pages. One of the most striking differences from a modern text lies with the punctuation, or rather, the lack of it. Often several sentences will be strung together without punctuation though with the occasional capital to suggest the start of a new principal clause. Subsidiary clauses are almost never indicated by punctuation. Neither hyphen nor apostrophe are used, nor does Chuquet employ accents or the cedilla. Words of Latin origin which commonly end with "-tion" are usually terminated in the manuscript by "-cion", for example *multiplicacion*. A number of words seem to be very close to those found in contemporary Italian manuscripts, for example *abreuier* (in modern French *abrêger*) is close to Italian *abbreviare*, and *indaguer* (in modern French *rechercher*) is close to Italian *indagare*. Marre makes considerable play with what he sees as the comparatively large number of Italianisms in the manuscript, but in many cases it is not clear whether or not such so-called Italianisms may in fact be Latinisms. Thus, for example, *indagare* (said by Marre to be Italian) is also perfectly good Latin, and there are other examples of this. In many cases we find "z" in the manuscript where today we would have "s": *foiz* is a typical case. We also find phrases with the words run together, for example *alafoiz*, *alapart*, etc.. As we might expect, Chuquet uses a number of contemporary French words and phrases which we do not find in modern French texts; some examples are *cuyder* (mod. *penser*), icelle (mod. *cette*), *orendroit* (mod. *directement*), and so on. Some of these are specifically mathematical terms: *equipollence* (mod. *equivalence*), *partiteur* (mod. *diviseur*), etc... The entries included in the glossary are sufficient, according to Marre, for everyone to be able to read Chuquet's work. This is a somewhat over-optimistic statement. Whilst a very general appreciation of the content of the manuscript can no doubt be gleaned from a cursory reading assisted by Marre's glossary, a detailed understanding and more particularly a translation, involves recourse to specialist dictionaries of medieval and Renaissance French, especial care being required to ensure that the contemporary meaning of words is preserved. As one of the earliest writers of mathematical material in vernacular French, Chuquet would inevitably have had to face the problem of expressing mathematical technicalities in a language which he could expect his intended readership to understand. The extent of his adoption of Latin or Italian mathematical terms suggests that the *Triparty* at least was intended either for advanced students or for other teachers. In our translation we have not attempted to follow precisely Chuquet's changes from active to passive, nor have we hesitated to change the order of his words in lengthy sentences. Often the order of words in the original text gives rise to debates on interpretation; we have elected to make our own choice of interpretation in the light of the general context and to avoid ambiguities in the translated passages. Also, we have not followed Chuquet's practice of enclosing numbers within periods thus:

.2067.

and omitting a period if it occurs at the beginning or end of a line of text. We have chosen to ignore these periods. In general, we have preserved Chuquet's special notation exactly as in the manuscript only when it has a particular significance.

Unfortunately Chuquet reveals very little about his sources. In the *Triparty* we find Boethius and Campanus mentioned, and also a reference to Archimedes as " a certain wise man". There are references to Euclid and Lull in the section on geometry, and in the set of problems we find criticism of the work of a certain "Berthelemy de Rōmans". In addition to the very few names quoted, likely sources can to some extent be inferred from internal evidence, though here we must move with caution. We can also consider sources of other similar roughly contemporary works with a view to inferring possible common sources shared by Chuquet with his mathematical compatriots. This is, however, a dangerous path opening the way to speculation rather than serious historical analysis. The internal evidence does suggest some debt to Italian writers. The vernacular arithmetics in France of the fourteenth and fifteenth centuries all show such traces of Italian influence, and it is unlikely that Chuquet's work is an exception. Behind the many Italian works of the period lay the genius of Leonardo of Pisa whose *Liber Abaci* appeared first in 1202 and again, in a second edition, in 1228. The century and a half prior to Pacioli's *Summa* of 1494 was a rich period for Italian arithmetics. A detailed discussion of Chuquet's sources is provided in Chapter 9. This refers also to another unresolved question of historical interest, namely the extent of Chuquet's familiarity with the works of Nicole Oresme.

The *Triparty* is, as Chuquet himself says, in three parts. These parts are broken down into chapters with appropriate headings though Chuquet does not himself supply any table of contents. Marre's table, to accompany publication of the *Triparty* in 1880, was based mainly but not entirely on Chuquet's own headings, and we present it (with some minor variations) in Appendix I.

The first part of the *Triparty* is an arithmetic, fairly conventional for its day, but including names for large numbers and the "rule of intermediate terms". This rule is, somewhat surprisingly, the one item in the whole work which Chuquet claims to be his own original invention. It follows discussion of the "rule of three", the "rules of one and two positions", and the "rule of apposition and remotion", thus concluding the first part, which occupies folios 2r. to 45r. The second part of the Triparty occupies folios 45v. to 81v.. It deals with roots and composite numbers. Here there is simplification of nomenclature and some original notation. Everything is treated as a number, be it integer, fraction, root or a combination of roots. There is no trace of the classical distinction between number and magnitude. Chuquet does not introduce the term "irrational", although he is clearly satisfied that many roots cannot be expressed as fractions exactly. He uses his "rule if intermediate terms" to provide approximations to whatever degree of accuracy is required. After a blank folio 82, the third part occupies folios 83r. to 147r. This is by far the most original part of the work, and in it Chuquet introduces his principles and methods of algebra. Here we find his exponential notation and rules for manipulating exponents, including a highly innovative use of negative numbers. We find also the analogy between arithmetic and geometric progressions, and an extensive discussion of equations which occasionally admits negative solutions. It concludes with the remark that he leaves to "those who would wish to proceed more deeply" the extension of his general principles to higher-degree equations.

After the *Triparty*, folios 148r. to 210r. consist of 166 problems illustrating the applications of the principles expounded in the course of the *Triparty*. Many of these problems have a long history going back at least as far as the *Greek Anthology* reputedly collected by Metrodorus some one thousand years earlier. In Chuquet they are solved by various methods, including the "rule of three" and the "rules of one and two positions", but the significant part of the section on problems is where Chuquet applies his "rule of first terms", that is, his algebra. Chuquet's material on geometry occupies folios 211r. to 262v.. It deals with 'practical' Geometry in the terminology of its day. After a fairly traditional treatment of the measurement of length, area, and volume, the material then becomes to a great extent a text on the application of algebra to geometry. There are extensive borrowings from Eutocius on the graphic extraction of cube roots, but the ethos of the work is the opposite of that of classical geometrical algebra. After an "investigation into the quadrature of the circle", Chuquet concludes his essay into geometry with a discussion of methods for obtaining the capacity of curved vessels. The final section of the manuscript (folios 263r. to 324r.) is a commercial arithmetic. Unfortunately here Chuquet leaves aside his new algebraic methods and restricts himself to a succession of problems solved by the "rule of three". The interest of this final section is therefore not so much mathematical as social and commercial. In it we can find useful evidence about the financial life of fifteenth-century Lyon, about weights and measures, exchange and interest rates, assaying, taxation, and business arrangements between partners in commercial enterprises. It was probably written for the instruction of merchants or their young apprentices, whose mathematical knowledge was limited to basic arithmetic and for whom, therefore, the algebraic content of the *Triparty* would have little or no appeal.

The manuscript begins (folio 2r.) with an ascription to God and a brief indication of the content of the *Triparty*:

This book, to the honour of the glorious and sacred Trinity, is divided into three parts, of which the first deals with numbers in as much as one can number, add, subtract, multiply and divide them, and also of their proportions, progressions and other properties. The second part deals with roots of numbers, and the third is the book of first terms or of the rule of first terms. The first part contains several chapters as appears in what follows, of which the first is

<u>Numeration</u>.

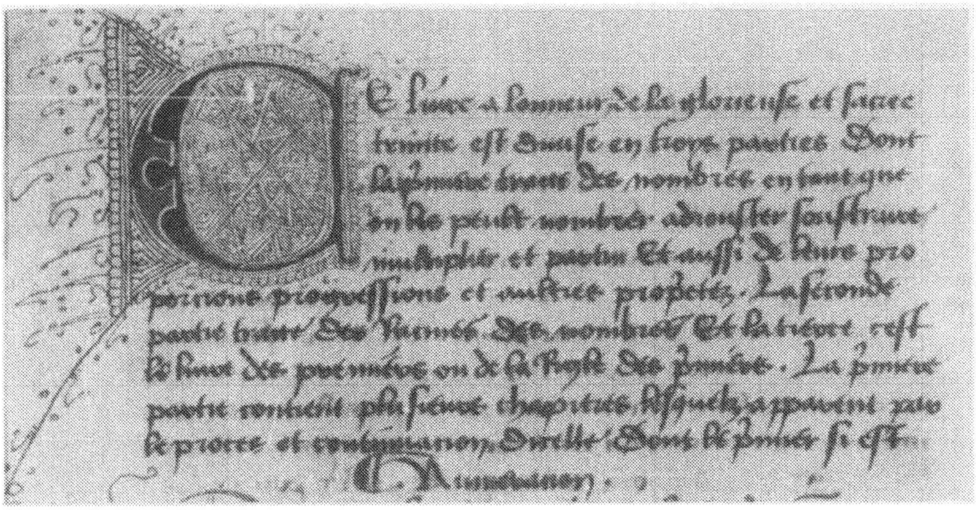
Le livre a lonneur de la glorieuse et sacree
trinite est divise en troys parties Dont
la premiere traicte des nombres en tant que
on les peult nombrer adiouster soustraire
multiplier et partir Et aussi de leurs pro
porcions progressions et aultres proprietez. La seconde
partie traicte des racines des nombres Et la tierce cest
le livre des premiers ou de la rigle des premiers. La premiere
partie contient plusieurs chappitres lesquelz apperent par
le proces et continuacion dicelle Dont le premier si est
Numeracion.

Chapter 3

THE TRIPARTY - FIRST PART

The first part of Chuquet's *Triparty* is concerned with arithmetic and covers, for the most part, similar ground to that in many of the early printed arithmetics. It has four chapters, the first two of which deal respectively with whole numbers and fractions, the latter being called "broken numbers". The third chapter, which is rather more in the scholastic tradition of Boethius and the mediaeval universities, deals with progressions, the classification of numbers, and proportions. The fourth comprises a collection of rules or methods for the arithmetical solution of simple algebraic problems: the rule of three, the rules of single and double false position, a rule for solving certain indeterminate problems, and the rule of intermediate numbers. It is the last of these rules which Chuquet specifically claims for himself as an original contribution in the *Triparty*.

Chuquet begins his arithmetic by introducing the Hindu-Arabic numerals.

Numeration

To number is to represent symbolically by common figures or to express with audible words the number conceived by the understanding. In order to know how to number and make use of this science, it is necessary to know which are the ten figures in this art by means of which one can write and express in figures every number - which are these: 0, 9, 8, 7, 6, 5, 4, 3, 2, 1. Of these, the first from the right is worth or signifies one. The second, having moved to the left, is worth two, the third three, the next four, and so on continuing up to the tenth which by itself has no value or signifies nothing. But, occupying a place, it gives value to those which follow it, and for this it is called cipher or null or figure of no value.

It was quite common for the numerals to be presented in the order which we see here. In the *Liber Abaci*, for example, we find the numerals 9, 8, 7, ..., 1 appearing at the beginning of the work, though in this case Leonardo of Pisa refers to nine rather than ten figures, treating the zero as somewhat a thing apart. Chuquet terms it "chiffre ou nulle", the former being derived from the Arab word *sifr* via the transformations *cifra*, *cyfre*, *ciffre*. (*Zero* is derived from the same Arab source via Latin *zephyrum* and Italian *zefiro*.) Later, *chiffre* was to come into common usage for denoting any of the numerals, but Chuquet restricts its use to denoting zero. This was the normal usage in the fifteenth and sixteenth centuries, though in the mid-sixteenth century we find Jan Trenchant stating that *chiffre* is used to signify "all the figures in the art of arithmetic". It is interesting to note that Chuquet makes no reference to the distinctive properties of zero other than commenting on its ability to "give value", that is, give extra value to a numeral which follows it (on reading from right to left). This was one of the properties of zero which continued for several centuries to cause it to be regarded as being almost magical. It "signified nothing" and, indeed, when added to a number did not increase its value. When placed to the right of a number, however, it increased its value tenfold. Chuquet can thus be regarded as being fairly sophisticated for his day in the way in which, for the most part, he treats zero in the same way as the other numerals.

And note that in this art the figures which are in the right hand place are said to be and may understandably be called first. And the next on moving to the left are called second, and the next following third, and the others fourth and so on continuing without end. A further thing is to know that any one of these figures being first, that is first in order, it is worth once its value, and being second it is worth ten times its value. And if it is third, it represents one hundred times its value, and if fourth one thousand times; if fifth ten thousand times, if sixth one hundred thousand times, and so on always increasing by tenfold proportion. And in order to number a large number more easily, one can divide up the figures six at a time, always beginning at the right. And above the first figure of each group of six, except the first, one can put a small mark. And one should know that all the figures from the first mark to the second, as many as there may be, are all millions, and from the second to the third are millions of millions, and from the third to the fourth are millions of millions of millions, and so on for the other marks, uttering the word 'million' as many times as there are marks. Or one can put 1 in place of the first mark and 2 in place of the second, and 3 in place of the thrid, and 4 in place of the fourth, which will have the same significance as the marks. Or, if you like, the first mark can signify million, the second mark billion, the third mark trillion, the fourth quadrillion, the fifth quillion, the sixth sixlion, the seventh septillion, the eighth ottillion, the ninth nonillion, and so on with the others as far as one wants to proceed. Further, one should know that a million is worth a thousand thousand units, and a billion is worth a thousand thousand millions, and a trillion is worth a thousand thousand billions, and a quadrillion is worth a thousand thousand trillions, and so on with the others. And an example of this follows - a number divided up and punctuated as stated above, all of which number comes to 745324 trillions, 804300 billions, 700023 millions, 654321.

Example: 745324804300700023654321.

This extract begins with a straightforward explanation of place-value, necessary because of the introduction of the Hindu-Arabic numerals. Although these were by now comparatively well-known in Europe, many people were still largely wedded to the Roman system which was basically an additive system in which place-value had no meaning. Chuquet's classification of large numbers was not entirely new. Nine years earlier Jehan Adam had expressed a number of twenty digits in terms of *trimillions*, *bymillions* and *millions*. The word *million* is of Italian origin, from *millione* meaning 'great thousand' in the same way that padrone (patron) originally meant 'great father'. Adam's *trimillion* and *bymillion* and Chuquet's slightly contracted forms *trillion* and *billion*, as well as the latter's *quadrillion*, *quillion*, etc., are all formed naturally in Latin or Italian from *millione* which was in common use in Italy from the fifteenth century. It seems possible therefore, as Chuquet makes no claim to have invented these words, that he found them in a Latin or Italian source at present unknown. The error in the example is not corrected in the manuscript.

Chuquet now turns to the first of the four basic arithmetical operations.

Addition

To add is to combine two or more numbers into one which by itself will be equal to the numbers added. To understand this matter it is necessary to know that in addition there are two sorts of number, that is to say, a simple number and a compound number. A simple number is one which can be written or set down by one of the aforesaid ten figures such as 6, 7, or 0, etc.. A compound number is one which can be expressed by two or more figures such as 10, 12, or 100, 164, etc.. To add, it is first necessary to set down the numbers which one wishes to add one below the other and in such a manner that the first figures are aligned with each other, and the seconds are aligned with the seconds, and each figure is aligned with its like. And then to put together firsts with firsts, seconds with seconds, and so on with the rest. Always in adding the seconds, the thirds, the fourths and the rest one considers the numbers as if they were firsts.

Here Chuquet is just saying that each individual place is added in the same way as the first place. He then deals with the distinction between a simple and a compound sum.

And if by the addition of several figures, be they firsts, seconds or others, we get a simple number, one should put it below and aligned with the figures added. If we get a compound number, one should set down the first figure of that number and keep the other or others to add them with the figures or numbers next following if there are any, and, if not, put it down as if it were the last in order or the last figure of the total number.

The following example is given in the margin by way of demonstrating "all the style and manner of adding":

$$\begin{array}{r} 70830 \\ 60730 \\ \underline{30520} \\ 162080 \end{array}$$

It is then discussed in detail.

And whoever puts together all the first figures of these three numbers, which are 0, 0, 0, he finds 0, which is set down below the firsts. Likewise, whoever adds the seconds, which are 3, 3, 2, he finds 8, which is put below aligned with the seconds. Likewise, whoever adds the thirds, which are 8, 7, 5, he has 20 of which 0 is put below and 2 is kept to add with the fourths, which are 0, 0, 0, with 2 coming to 2, which is set down below. Likewise, the fifths added together come to 16, of which 6 is put below and 1 after because there is nothing more to add. And thus all three numbers added come to a sum of 162080, as it appears in the example.

Three further examples are then given without comment.

In his discussion of subtraction, which now follows, Chuquet is careful to avoid negative numbers. This is achieved by the initial definition.

Subtraction

To subtract is to remove or take away a smaller number from another larger number in order to know by how much the smaller is exceeded by the greater. And in order to know which of two or more numbers is the greater it is necessary to determine if there are more figures in one than in the other, for where there are more there is the greater. And if it is so that there are as many figures in one as in the other, one must then see if one of the last figures is of greater value than the other, for then this number is larger. And if the last figures are equal, one must compare the penultimates or the second penultimates as is necessary, continuing up to the firsts.

Chuquet is counting his numerals from right to left, hence a "last number" is the coefficient of the highest power of ten. Correspondingly, a "first number" is the figure in the units place.

And it is necessary for one to know that in subtraction only two numbers are required, that is the number which one wishes to subtract and the number from which one wishes to subtract it, which two numbers must be written and set down the one under the other and particularly the smaller below the greater, and each figure aligned with its like, and then the first figures subtracted from the first figures, the seconds from the seconds, and each from its like in the manner which follows.

If from a figure, which is first or second or other such, one removes another smaller figure, the remainder should be set down below and aligned with this figure. If one removes its equal, one should put 0 below. And if it is necessary to remove a greater, then one should borrow 10 and add it to the smaller figure which ought to be greater in number. Then one should make the subtraction setting down the remainder below it.

In 'paying back' what has been borrowed, Chuquet chooses to add 1 to the next place of the subtrahend rather than to subtract 1 from the corresponding place of the minuend. This was the usual practice of his day - presumably, having "borrowed", it seemed logical to pay back by the operation of addition, as Chuquet suggests:

And then one should hold 1 on the understanding that one has borrowed and that one should add it with the next figure following of the smaller number, supposing a figure there. Otherwise one should remove or subtract it on its own in doing as aforesaid.

The example 406579 - 38075 is given in the margin and discussed. It involves two "borrowings", the first being the subtraction of 8 from 6.

... *for this one can only say 8 taken from 16, remainder 8 under the fourth figures and 1, which one takes with 3 making 4 taken from 0. One must then take away from 10, there remains 6 under the fifth figures, and 1 which one takes away from 4, there remains 3 under the sixth figures.*

Four further subtractions are written out without comment. In all these, as with additions, Chuquet places his result below the numbers to be subtracted. This had gradually become the more common practice in Italy. However, earlier algorithmic arithmetics were strongly influenced by abacus methods. On the abacus, the numbers upon which an operation was being performed were replaced in the course of the calculation by the final result. This was inevitable because it was generally impracticable to represent more than two numbers at a time. Instructions for written calculations in early works reflected this practice. Leonardo of Pisa introduced the writing of results at the top of a calculation, and this practice continued for several centuries for addition and subtraction, though eventually becoming less and less popular.

In his discussion of multiplication Chuquet stresses the need to know the basic multiplication table, which he gives in the reduced triangular form together with a simple explanation.

Multiplication

To multiply is to increase a number by itself as many times as the multiplying number amounts to. To know how to do this is to note that in multiplication only two numbers are needed, that is the multiplying number and the number to multiply. And they should be set down one below the other and appropriately the larger should be above, and each figure aligned with its like, and from the completed multiplication there results another number containing the number multiplied exactly as many times as there are units in the multiplying number, or containing the multiplying number as many times as there are units in the number multiplied. Further, it is necessary to know by heart the multiplication of each one of the ten figures by itself and also by each of the others, which thing is called the little primer of algorism, which is as follows:

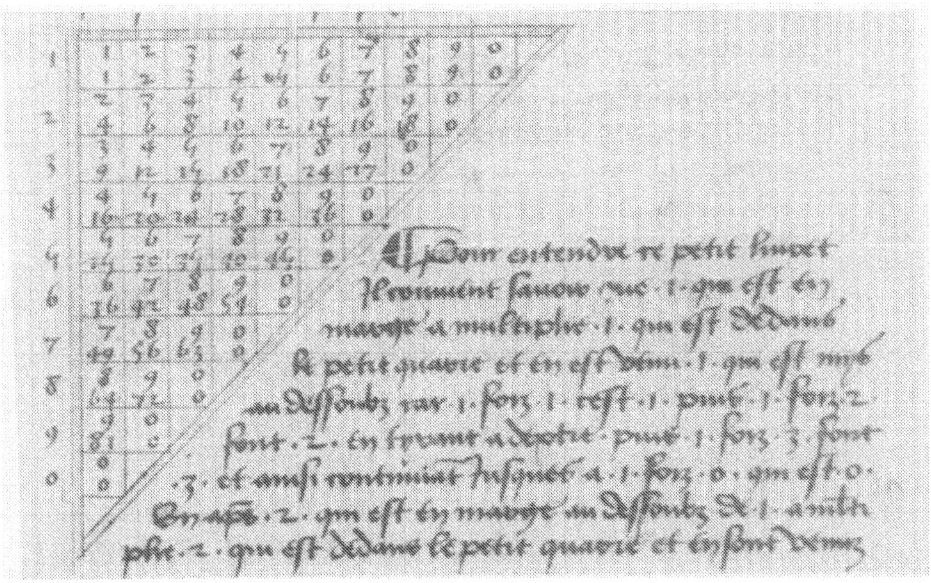

The preceding paragraph may be compared with the corresponding passage in de la Roche. The two are identical with two exceptions: de la Roche replaces "algorisme" with "argorisme", and Chuquet's final phrase "which is as follows" with "which is written in the present page of this folio". The triangular tables of multiplication differ only in that de la Roche's printer has included Chuquet's marginal multipliers as the first column within the framework of the diagram. Perhaps even more striking are the four examples of subtraction at the top of the same folio in de la Roche's work - these involve precisely the same numbers as appear in Chuquet's final examples of subtraction. Chuquet then explains the multiplication table.

To understand this little primer it is necessary to know that 1 which is in the margin has multiplied 1 which is in the small square, and it comes to 1, which is put below, for 1 times 1 is 1; then 1 times 2 makes 2, moving to the right; then 1 times 3 makes 3, and so on continuing up to 1 times 0 which is 0. Afterwards, 2, which is in the margin below 1, has multiplied 2 which is inside the small square and it comes to 4, which is put below, for 2 times 2 makes 4; then 2 times 3 makes 6; then 2 times 4 makes 8, and so on continuing up to 2 times 0 which makes 0. And one must understand the rest thus.

Further, one should know that in multiplying one must observe the rules and statutes put in the chapter on addition as and by the form and manner that hereafter appears. That is to say, that if by the multiplication of one figure by another, be they firsts, seconds or thirds, there is obtained a simple number, one must set it down below the multiplying figure or otherwise in its place as can appear by the examples presented below. And if there is obtained a composite number, one must set down the first figure of this number and keep the second to add it to the product of the next figure following if there is one. If not set it down and place it all alone in its order. To multiply a number of several figures by itself or by another number of one or more figures, one must begin with the first figures of both the multiplier and the number to multiply, and one must multiply by the first figure of the multiplier the first figure of the number to multiply, and subsequently all the others, and one must set down each time what is obtained by the multiplication in accordance with what is stated earlier.

And if a multiplying number has two or more figures, one must do with the others as with the first in always putting the product of the first figure of the number to multiply in line with and below the multiplying figure, and the others following it on continuing to the left. And if a multiplying number has 0, one can put below it another 0 for all its multiplication, and one must do thus if there are several other 0s.

(This last sentence is incomplete in Marre's 1880 transcription.) For a "more ample explanation" of the multiplying process, the following example is written out in the margin:

```
   6043
    502
  -----
  12086
 302150
-------
3033586
```

A detailed discussion then follows.

And so, as one can see, 2 which is the first figure of the multiplier has multiplied 3, which is also the first of the number to multiply, and there is obtained 6, which is put below and aligned with them. Then it has multiplied 4 and there is obtained 8, which is put after 6. Then it has multiplied 6 and there is obtained 12, which is set down after 0. Further, for 0 of the multiplier, 0 is put below it and 8 for all its multiplication. Also 5 times 3 makes 15 of which 5 is set down after 0 and below 5 of the multiplying number, and 1 ten which one keeps. Then 5 times 4 makes 20, and 1 which one has kept makes 21, of which 1 is put after 5 and 2 which one holds. Then 5 times 0 makes 0, and 2 which one holds makes 2, which is put down after 1; then 5 times 6 makes 30, which is set down after the 2. Now to collect and add these multiplications there is first set down 6 for all the first numbers, then 8 after for the seconds, and then 5 for the thirds, and then 3 for the fourths, and again 3, then 0, and 3 is set down at the end. All of which multiplication comes to 3033586. And one is able thus to do all other multiplications, etc.; notwithstanding, for better comprehending the style of multiplying, several other examples are put here below.

Four further examples are then given without comment. In all of these Chuquet strictly follows his main text in placing just one 0 in the next line of multiplication when there is a 0 in the multiplier. He continues with some very simple examples of multiplication by powers of ten and multiples of powers of ten, explaining in each case the number of additional 0s required. In one case there is an error: 80 × 300 is said to require three 0s but the product is given as 2400. These examples are headed "Other brief rules", and under the same heading Chuquet turns to the lattice method.

Another style of multiplication: this is done in a square or quadrangular figure, in which manner one does not keep the tens as one did above and one can begin on the right or on the left, as one can see from the two quadrangles below. The explanation of the larger of these is such as to say that 769504 is multiplied by 83421, of which 8 of the multiplying number has first multiplied 7 of the number to multiply and it comes to 56 which is set down below 7 and aligned with 8. Then it has multiplied 6, amounting to 48 which is set down below 6; then it has multiplied 9 and then 5 and subsequently the others. After that, 3 has multiplied 7 and has come to 21 which is set down below 7 and aligned with 3, and subsequently all the other figures of the number to multiply. And thus one should understand the other figures of the multiplying number. Afterwards, in order to collect and add these multiplications one should begin in the lower square and on the right side where there is a 4, which is set down underneath for the first figure of the total number. Afterwards, one must take 0 and 8, and it amounts to 8 which is set down after 4. Then one must take 5, 0, 6 and it amounts to 11 of which 1 is set down after 8 and 1 which one holds, which it is necessary to add with 9, 0, 0, 0, 1, 2 and it amounts to 13 of which 3 is put after 1. And one should continue thus right up to the upper square on the left side and one will find that the multiplication amounts to 64192793184, as it appears in the larger of these two quadrangular figures.

Chuquet's method of setting out a division is somewhat unusual and differs from those in most of the books on algorism associated with the Hindu-Arabic numerals. After giving some simple examples, he naturally introduces the concept of fraction although strictly this part of his *Arithmetic* treats with whole numbers. Fractions appear first when the dividend is less than the divisor, and it is here that the method of setting out the calculation comes into its own since the numerator and the denominator appear in the correct relative positions though separated by the quotient 0.

Division

To divide is to split up or set out a number in several equal parts, and it should always begin on the left side and finish on the right. In division only two numbers are needed, that is to say, the separator or divisor and the number to be divided. And it can be convenient to put the divisor below the number to be divided, each figure aligned with its like, with two equidistant lines between the number to be divided and the divisor at enough distance from each other for the number resulting from the division to be placed between them. This number is called the part or the quotient because it demonstrates how many times the divisor is contained in the number to be divided. And one should know that the rules of subtraction should be kept here, as one can see later.

This last comment is interesting. Chuquet earlier has associated multiplication with addition. Here he associates division with subtraction. This could suggest that he perceived the division process as effectively repeated subtraction of the divisor from the dividend, though this is nowhere explicitly stated.

One should also understand that the number to divide is either less than the divisor, or equal or greater. If the number that one wishes to divide is less than the divisor then one should put 0 between the two lines, then the number to divide forms the numerator and the divisor makes the denominator, for this fraction is the part or the quotient.

The following examples of this are then given.

3	53	156	5307	96
0	0	0	0	0
4	71	192	9046	1347

He then considers the situation where the dividend and divisor are equal. Here he requires the dividend to be struck out, and of course has the quotient 1 appearing between the two lines. Four examples are given in the margin:

~~7~~	~~12~~	~~875~~	~~509468~~
1	1	1	1
7	12	875	509468

There then remains the situation where the dividend is greater than the divisor -

... *which can arise when the figures of the number to divide are of greater value than those of the divisor, or where there are more figures in the number to divide than in the other. Now when the figures of the number to divide are of greater value than the divisor, and as the two numbers are set down in the aforesaid manner, one must see how many times the divisor is contained in the number to be divided, and one can never find less than 1 or more than 9.*

This seems strange on first reading, but Chuquet is here dealing explicitly with the case where dividend and divisor have the same number of digits. The case where the dividend has more digits is discussed separately afterwards.

Example: Who would divide 9 by 4. You set down the two numbers just as it is shown; one should see how many times 4 is contained completely in 9. It is only 2, which one should set down between the two lines. Then say 2 times 4 makes 8, taken from 9 leaves 1 above 9, and the 9 must be struck out with a line, and thus we get for quotient $2\frac{1}{4}$ *as it appears in the examples given below:*

		2	
1	1	233	32
9	27	873	87023
2	2	3	5
4	13	284	17405

And if there are more numerals in the number to be divided than in the divisor as in dividing 360 by 36, you put down the numbers in the manner aforesaid. One should see how many times 3 is in 3 and also how many times 6 is in 6, and there are 1 times as one should put between the two lines and aligned with the last figure of the number to be divided. Then say one times 3 are three, taken from 3 leaves 0, then one times 6 makes 6, taken from 6 leaves 0, and one must strike out the 3 and the 6 of the number to be divided. Then afterwards one should see how many times 3 is in 0, which is struck out. It is 0 which is put below. And thus we arrive at quotient 10.

The calculation appears in the margin:

360
10
36

Chuquet next discusses 705 ÷ 19.

You set out the numbers in an appropriate way. One should see how many times 1 is in 7 and how many times 9 is in 0, or how many times 19 is in 70. It is contained 3 times, which one should place between the two lines and below 7. Then say 3 times 1 makes 3, taken from 7 leaves 4 above 7. Then 3 times 9 makes 27, taken then from 30 leaves 3 above 0, and 3 which one has borrowed taken from 4 leaves 1 above 4. After this one should see how many times 1 is in 13, or how many times 19 is in 135. It is 7 times, everything considered, and not more, therefore one must put 7 below 0. Then one should say 7 times 1 makes 7 taken from 13 leaves 6, then 7 times 9 makes 63 taken from 65 leaves 2. And thus one arrives at the part $37\frac{2}{19}$ as it appears in the margin.

In the margin we find:

~~10~~
~~43~~2
~~7~~0~~5~~
37
19

Here and elsewhere Chuquet does not strike out the 0. A number which has been struck out is then taken to be 0 - for example, in the division of 360 by 36, 3 is said to go 0 times into ~~0~~. A number of further examples set out in the margin are then discussed. These are 860400 ÷ 300, 182728 ÷ 364, 7500409 ÷ 879, and 6480000 ÷ 324.

Following the pattern with multiplication, Chuquet next discusses the special cases of division by 10 and by powers of 10. Two of the examples chosen involve mixed numbers as quotients. He then introduces the galley method of division, so called because the overall appearance of the calculation was similar to that of a sail. This method originated in India and was transmitted to the Arabs, probably about the ninth century. It came to Europe via Italy during the fourteenth century.

Another way of dividing is that which one can call division by advancing, for the divisor is moved to each figure of the quotient and is put each time aligned with the figures to be divided, striking out the figures of the divisor as one does those of the number to be divided. In proceeding and advancing the divisor each time in turn moving to the right and thus continuing right up to when the first of the divisor is aligned with the first of the number to be divided; in such manner of dividing conveniently one can divide all numbers and all divisors and similarly whenever the divisor is of four or more figures. And by this method is divided ... ... 923260 by 2659, and from it we get the part 347, and $\frac{587}{2659}$.

~~1~~
~~15~~
~~29~~5
~~416~~
~~1252~~8
~~34855~~7
923260
347
~~265999~~
~~2655~~
~~26~~

To follow the sequence of operations leading to the crossings out, it is best to work through the calculation independently, following Chuquet's method.

Chuquet now turns to the various "proofs" of calculations based on remainders, of which the most widely used was the method known as "casting out nines". This had originated with the Hindus and had been transmitted to the West along with other algorithms by the Arabs. Originally, it was used only for checking multiplications and divisions. Additions and subtractions were checked by using the respective inverse operation. This was possible for additions so long as it remained the practice to add only two numbers at a time. When, later, it became common practice to add several numbers together, the remainder checking methods were extended to all four basic arithmetic operations.

<u>The proofs</u>

There are several kinds of proofs such as the proof by 9, by 8, by 7, and so on by other individual figures down to 2, by which one can prove and examine addition, subtraction, multiplication and division. Of these only the proof by 9, because it is easy to do, and the proof by 7, because it is even more certain than that by 9 are treated here. The proof by 9 is nothing else but seeing if the number to be proved can be exactly divided by 9 or not. If exactly, one should put 0 as the proof for that number; if not, one should put the remainder which may be 8 or less down to 1 inclusive. Now to know and recognize easily if a number can be exactly divided by 9 or not, it is necessary to add all the figures of that number, counting them as if they were first figures, or taken by their simple value. And if by the aforesaid addition there comes a number of several figures, one should add them again as above until you reach an individual figure. And if this figure is 9, one should put 9 or 0 as the proof for this number, for if it is divided by 9 the remainder will be 0. If it is 8 or 7 or a smaller figure, one should take this for the proof. Example: if you want to take the proof of the number 8765, you should add 8, 7, 6, 5, which amounts to 26, then again add 2, 6 making 8, and 8 is the proof of this number. And so it should be understood for all other numbers.

The following examples are then given to illustrate the "proof by 9".

13 > 7	217 — 1	41 — 5	128
52	135 > 1	24 — 6	344 — (2.2)
104	082	164 (3	14 — 5
169 — 7		82	24 — 6
		984 (3	

Chuquet then discusses the validity of the proofs, pointing out that the proof by 9 is sufficient to identify an error but not to determine if a calculation is correct.

From the proof by 9 an error can arise in two ways, the one is by remotion [removal] *or addition of one or several 9s or 0s, the other is by mutation of a figure or part of it from one order to another. Example of the first: If someone writes 9539 or 53 for 953, the proof by 9 will always be the same. Also, if someone writes 340 as 3400 or 34, the proof will always be 7, and yet the numbers are different. Example of the second: if 342 or 1143 is written for 243, the proof will always be 0. But when the proof by 9 shows the calculation to be wrong, its demonstration is certain. And therefore when a proof by 9 judges a calculation to be false, then it must necessarily be false; and when it judges that it is true, then it may allow an error. It will rarely fail in the hands of an expert man.*

The proof of 7 is made with the same intention as that by 9. For just as all the 9s are removed from the number to be examined, leaving the proof as remainder, so all the 7s are subtracted and the remainder is the proof by 7. There is a difference, for the proof by 9 can easily be carried out by adding up the figures, but not here. This proof by 7 is wrong less often than that by 9 because 7 has less in common with numbers than 9.

There are other proofs which are done by contraries of one to the other, as addition can be proved by subtraction and also subtraction by addition. Such proofs are treated below at the end of fractions. And thus ends the treatise on whole numbers.

Chuquet now turns to fractions, defining a fraction to be a "part" of unity, thus excluding improper fractions. As we have already seen, he uses the horizontal bar to separate numerator and denominator. The Greeks and Hindus had denoted fractions by writing one number above the other, the horizontal bar occurring first in certain Arab works. The avoidance of improper fractions leads Chuquet to some awkward calculations since improper fractions are more convenient than mixed numbers in multiplication and division. Chuquet's expression "broken numbers" is in accord with *fractio* in the Latin works on algorithm, this being a translation of the Arabic *al-kasr* derived from the stem of the word meaning 'to break'. Leonardo of Pisa uses both *fractio* and *minutum ruptus*. The English word *fraction* occurs in Chaucer (1321), but the expression *broken numbers* can be found in Arithmetics in the English language well into the seventeenth century. Chuquet begins with his formal definition.

A broken number is a part or parts of 1 where there are two numbers, that is to say one above and the other below with a line between the two. As three-quarters, which should be set down thus $\frac{3}{4}$, *of which 3 which is the number above is called numerator, and 4 which is below is called denominator. And know always that the numerator is less than the denominator, for, if it should be equal or greater, it would represent 1 whole or more, and more would not attain to the definition aforesaid. And albeit the whole numbers never have a denominator, it being always understood, and in order to provide more general rules in this treatise on broken numbers, one can give to all whole numbers 1 for denominator, and by this means whole numbers will be denoted by 1 just as the broken numbers are denoted by 2, 3, 4, and by all other numbers, thus:* $\frac{12}{1}$, $\frac{13}{1}$, $\frac{20}{1}$, *etc. for the whole numbers and* $\frac{1}{2}$, $\frac{1}{3}$, $\frac{2}{3}$, $\frac{1}{4}$, $\frac{3}{4}$, *etc. for the broken numbers.*

Chuquet has thus generalized the concept of fraction to include whole numbers, but does not appear to notice that in so doing he has broken his earlier requirement that a numerator should always be less than its denominator. He "enters into the science of broken numbers" with a discussion of reduction. This is ...

... to make alike two or more unequal broken numbers by reducing them to a common denominator, for the diversity and difference of broken numbers comes on account of the denominator or of the denominators. To know how to do this thing, there is a general rule as follows.

Multiply all the denominators one by the other so as to get the denominator common to all, which divide by the individual denominators and multiply each quotient by its numerator. And thus there will be new numerators for the numbers which one reduces, as can be seen below.

Example: Whoever should wish to reduce $\frac{2}{3}$ and $\frac{4}{5}$, it is necessary for the first to multiply 3 by 5 getting 15 for common denominator. Afterwards, you divide 15 by 3 and you will have 5; multiply by 2 and you will find 10, which is $\frac{10}{15}$ for the $\frac{2}{3}$. After you divide 15 by 5 and then multiply by 4 and you will find 12, which is $\frac{12}{15}$ for the $\frac{4}{5}$. And this should be set down as it appears in the margin.

In the margin we find:

$$\begin{array}{ccc} \dfrac{10}{\dfrac{2}{3}} & \qquad & \dfrac{12}{\dfrac{4}{5}} \\ & 15 & \end{array}$$

Chuquet then considers in detail the simultaneous reduction of $\frac{1}{2}$, $\frac{2}{3}$, $\frac{3}{4}$, $\frac{5}{6}$. He obtains the common denominator 144, having followed his rule in this example to "multiply all the denominators one by the other" without regard to their common factors. The reduction is set out in the margin.

$$\begin{array}{cccc} \dfrac{72}{\dfrac{1}{2}} & \dfrac{96}{\dfrac{2}{3}} & \dfrac{108}{\dfrac{3}{4}} & \dfrac{120}{\dfrac{5}{6}} \\ & & 144 & \end{array}$$

In the translations which follow from this point onwards, Chuquet's expression "broken numbers" is rendered as "fractions".

Another example which shows reduction of fractions of fractions. Whoever should wish to reduce $\frac{2}{3}$ of $\frac{1}{4}$ of $\frac{4}{5}$, $\frac{3}{4}$ of $\frac{5}{7}$, and $\frac{1}{2}$ of $\frac{1}{2}$ of $\frac{2}{3}$ of $\frac{1}{3}$. It is necessary first to make one fraction of all the fractions of each part by multiplying numerator by numerator and denominator by denominator. Now for $\frac{2}{3}$ of $\frac{1}{4}$ of $\frac{4}{5}$, multiply 2 times 1 by 4 and it comes to 8 for numerator. Afterwards multiply 3 times 4 by 5 and it comes to 60 for denominator, thus there is $\frac{8}{60}$ which *abbreviates to $\frac{2}{15}$, which is equal to $\frac{2}{3}$ of $\frac{1}{4}$ of $\frac{4}{5}$. And afterwards multiply 3 by 5 and 4 by 7 and you will find $\frac{15}{28}$ for $\frac{3}{4}$ of $\frac{5}{7}$. Next multiply 1 by 1 and then by 2 and again by 1 and there is 2 for numerator; then multiply 2 times 2 by 3 times 3 and there will be 36 for denominator. Thus you will have $\frac{2}{36}$ which abbreviated is $\frac{1}{18}$ for the $\frac{1}{2}$ of $\frac{1}{2}$ of $\frac{2}{3}$ of $\frac{1}{3}$. Take now $\frac{2}{5}$, $\frac{15}{28}$, $\frac{1}{18}$ and reduce according to the rule aforesaid and you will find $\frac{1008}{7560}$ for $\frac{2}{15}$ and $\frac{4050}{7560}$ for $\frac{15}{28}$ and $\frac{420}{7560}$ for $\frac{1}{18}$. And thus fractions of fractions are reduced.*

Chuquet then turns to the advantages of finding the lowest common denominator, and gives an example of his "special rule for reducing any fractions".

Whoever should wish to reduce $\frac{1}{2}$, $\frac{2}{3}$, $\frac{3}{4}$, $\frac{5}{6}$, should multiply 4 by 6 and get 24, which can be exactly divided by 2, 3, 4, 6 which are all the denominators. Or if we multiply 2 by 6 or 3 by 4 there will come 12, which can equally be divided by all the denominators. And therefore one can take 24 for common denominator or, even better, 12, and then do according to the general rule aforesaid, and one will find $\frac{6}{12}$ for $\frac{1}{2}$, $\frac{8}{12}$ for $\frac{2}{3}$, $\frac{9}{12}$ for $\frac{3}{4}$, and $\frac{10}{12}$ for $\frac{5}{6}$. And therefore to find the common denominator one should multiply the larger individual denominators one by the other to see which smaller ones are contained in them. As in the abovesaid reduction, one can take 4 and 6, which are the larger denominators, and leave 2 and 3 for they are contained in these. Or, to be shorter still, one can take the smaller denominators to see whether their multiplication includes the larger. Or take the larger and smaller together, as in the reduction abovesaid, in which are taken 3 and 4, which multiplied one by the other come to 12 for the common denominator as above. And if it happens that one of the individual denominators can be exactly divided by all the others, one can take it as the common denominator and then carry on as said above. Thus, whoever should wish to reduce $\frac{1}{2}$, $\frac{1}{3}$, $\frac{3}{4}$, $\frac{5}{6}$, $\frac{3}{8}$, $\frac{7}{12}$, $\frac{19}{24}$, it is so that 24, which is one of the individual denominators, can be exactly divided by all the others, and so one can take it and make it the common denominator, and then afterwards divide and multiply as the rule of reduction demands. By this means one will find $\frac{12}{24}$ for $\frac{1}{2}$, $\frac{8}{24}$ for $\frac{1}{3}$, $\frac{18}{24}$ for $\frac{3}{4}$, $\frac{20}{24}$ for $\frac{5}{6}$, and $\frac{9}{24}$ for $\frac{3}{8}$, $\frac{14}{24}$ for $\frac{7}{12}$, and $\frac{19}{24}$.

Another special rule for reducing two numbers

Multiply the numerator of one by the denominator of the other, and also the numerator of the other by the denominator of the first. And then still multiply denominator by denominator. Example: Whoever should wish to reduce $\frac{2}{3}$ and $\frac{4}{5}$, it is necessary to multiply 2 ... ... by 5 ... ... and get 10 which one should put above $\frac{2}{3}$. Then one must multiply 4 ... ... by 3 ... ... and get 12 which must be put above $\frac{4}{5}$. Afterwards, multiply 3 by 5 ... ... and get 15 for the common denominator as one can see.

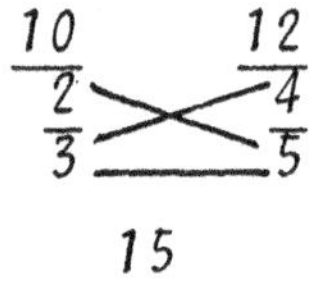

There follow a discussion of cases involving whole numbers and instructions for abbreviating fractions, for which the "general rule" is:

Divide the numerator and denominator exactly by the largest number that you can find and of the quotient of the numerator make the numerator and of that of the denominator make the denominator. Example: Whoever should wish to abbreviate $\frac{54}{81}$, the largest number by which one can exactly divide the numerator and denominator is 27. And thus divide 54 by 27; it gives 2 for numerator. Then divide 81 by 27; it gives 3 for denominator. Now put 2 above 3. We have $\frac{2}{3}$, which value also equals $\frac{54}{81}$.

Chuquet now describes how to find this "largest number".

Divide the denominator by its numerator, and if there remains any number let the divisor be divided by this remaining number. And thus continue until there is nothing left. Know that the last divisor from which 0 remains is the greatest number by which one can abbreviate. And if this last divisor is 1, it is a sign that this broken number cannot be abbreviated. Example: From $\frac{54}{81}$ divide 81 by 54 and there remains 27, then divide 54 by 27 and there remains 0. Now let 27 be taken, for it is the number by which one can abbreviate $\frac{54}{81}$.

The method just described is the standard Euclidean algorithm for finding the greatest common divisor of two integers. It is to be found at the beginning of Book VII of Euclid's *Elements*. Chuquet then suggests, as an alternative method, repeated division of the numerator and denominator by 2, then by 3, and so on up to 10, as often as is possible in each case. Rules are given for recognizing when such divisions are possible.

When the first figure of the numerator and also of the denominator is even, then one can recognise that such numbers can be abbreviated by two ... . And if you add the figures of the numerator as one does it when makeing a proof by 9, it is a sign that this number can be abbreviated by 9 and also by 3 and sometimes by 6. If you find 6, it can be abbreviated by 6 and always by 3. And if you find 3 it is a sign that it can be divided by 3. And thus one should understand for the denominator. And if the first figure of that number is 5 or 0 then one can recognize that the number can be abbreviated by 5. If the first figure is 0, the number can be abbreviated by 10. And several examples of this method are set below. However, not all numbers can be abbreviated in this way, that is, not all the numbers that can be abbreviated by the first rule stated above.

Abbreviate:

by 10	$\frac{3840}{7680}$	*by 9*	$\frac{1890}{4725}$	*by 100*	$\frac{1800}{2700}$
by 8	$\frac{384}{768}$	*by 7*	$\frac{210}{525}$	*and by 9*	$\frac{2}{3}$
by 6	$\frac{48}{96}$	*by 5*	$\frac{30}{75}$		
by 4	$\frac{8}{16}$	*by 3*	$\frac{6}{15}$		
by 2	$\frac{2}{4}$		$\frac{2}{5}$		
	$\frac{1}{2}$.				

Chuquet continues his discussion of abbreviation by reference to special cases, of which $\frac{555}{888}$ reducing to $\frac{5}{8}$ and $\frac{4747}{5959}$ reducing to $\frac{47}{59}$ are quoted as examples. He concludes with the method of checking to see if a fraction can be reduced to a given form.

It is necessary to divide numerator by numerator and denominator by denominator, for if for each division there remains 0 and the two quotients are equal then the given number is the number to which it can be abbreviated. As with $\frac{115}{207}$. I wish to know if it can be abbreviated to $\frac{5}{9}$. To do this I must divide 115 by 5 and 207 by 9, and it comes each time to 23, by which it is seen that this number can be abbreviated by 23 and it will come to $\frac{5}{9}$.

Chuquet now applies the four fundamental arithmetical operations to fractions. For addition and subtraction, he emphasizes the need to find a common denominator. If an improper fraction results from addition, it must be re-expressed as a mixed number. In the addition of mixed numbers, the fractions must be added first. A number of examples are given from which the following is a selection:

Whoever should wish to add $\frac{2}{3}$ with $\frac{3}{4}$, the numbers first reduced are $\frac{8}{12}$ and $\frac{9}{12}$. Now add 8 with 9 giving 17, divide by 12 and you will find $1\frac{5}{12}$ ...

... whoever should add $13\frac{4}{5}$ with $40\frac{2}{3}$, the addition will amount to $54\frac{7}{15}$.

Whoever should add $\frac{2}{3}$ of $\frac{4}{5}$ with $\frac{3}{4}$ of $\frac{5}{7}$ of $\frac{1}{2}$; first, $\frac{2}{3}$ of $\frac{4}{5}$ is $\frac{8}{15}$ and $\frac{3}{4}$ of $\frac{5}{7}$ of $\frac{1}{2}$ is $\frac{15}{56}$. Now reduce $\frac{8}{15}$ and $\frac{15}{56}$ and then add them, and you will find $\frac{673}{840}$ ...

Whoever should wish to subtract a fraction from a whole number, as to subtract $\frac{2}{5}$ from 12, it is necessary to borrow 1, which is worth $\frac{5}{5}$, and to take away $\frac{2}{5}$, and $\frac{3}{5}$ remains; and then 1 which has been borrowed must be taken away from 12, and thus there will remain $11\frac{3}{5}$.

... whoever should wish to take $\frac{1}{4}$ of $\frac{1}{5}$ and subtract it from $\frac{1}{3}$ of $\frac{2}{3}$; first, $\frac{1}{4}$ of $\frac{1}{5}$ is $\frac{1}{20}$, and $\frac{1}{3}$ of $\frac{2}{3}$ is $\frac{2}{9}$. Now reduce $\frac{1}{20}$ and $\frac{2}{9}$ and then subtract $\frac{9}{180}$ from $\frac{40}{180}$ and there will remain $\frac{31}{180}$ ...

The rule for multiplication of fractions is:

Multiply numerator by numerator and denominator by denominator. ... when there is a whole number and fraction together, the whole number must be put into its fractional form, and added with the numerator, and then multiplied ...

A number of examples now follow. In the second of these, we find an expression of the commutative law of multiplication as follows:

Whoever should wish to multiply a fraction by a whole number or a whole number by a fraction, as $\frac{4}{5}$ by 18 or 18 by $\frac{4}{5}$, should multiply $\frac{18}{1}$ by $\frac{4}{5}$, multiplying 18 by 4 and 1 by 5, and this multiplication amounts to $\frac{72}{5}$. Now divide 72 by 5 in order to obtain whole numbers and one will have $14\frac{2}{5}$. Or otherwise, take a fifth of 18 which is $3\frac{3}{5}$ and there remains $14\frac{2}{5}$ as above.

Note the use of denominator 1 in this last example. In the margin of the manuscript, there is inserted the multiplication of a whole number by a mixed number. This is omitted in the 1880 transcription, but it is included in de la Roche's book. Here, 15 is to be multiplied by $16\frac{3}{4}$, which is first re-expressed as $\frac{67}{4}$, the correct result $251\frac{1}{4}$ then being obtained. In a number of his examples, Chuquet gives more than one method of working, as with the following:

Whoever should wish to multiply a whole number and fraction by a whole number and fraction, as $18\frac{2}{3}$ by $13\frac{1}{2}$, it is necessary first to put the whole numbers each into its fractional form and one will have $\frac{56}{3}$ and $\frac{27}{2}$. Now multiply numerator by numerator and denominator by denominator and you will find $\frac{1512}{6}$ which is worth 252. Or otherwise, multiply 18 by 13 to give 234, then 13 times $\frac{2}{3}$ makes $\frac{26}{3}$ which is worth $8\frac{2}{3}$. Afterwards, multiply $18\frac{2}{3}$ by $\frac{1}{2}$, taking $\frac{1}{2}$ of $18\frac{2}{3}$, which is $9\frac{1}{3}$, $8\frac{2}{3}$ with 234; you will have 252 ...

Before turning to general division of fractions, Chuquet inserts a section on halving and dividing numbers into thirds and quarters.

One should know that ... ... one can mediate by multiplying by $\frac{1}{2}$, and put into thirds by multiplying by $\frac{1}{3}$, or by $\frac{2}{3}$ whoever would wish to have $\frac{2}{3}$ of a number. Or put into fourths or fifths, and so on for the other parts, in multiplying by $\frac{1}{4}$ and by $\frac{1}{5}$. One may take $\frac{3}{4}$ and $\frac{4}{5}$ of any number in multiplying by these fractions.

Example: Whoever should wish to mediate or have $\frac{1}{2}$ of $\frac{3}{4}$, should multiply $\frac{3}{4}$ by $\frac{1}{2}$ and we get $\frac{3}{8}$ ... ... And if the numerator of the fraction is even, one can take the $\frac{1}{2}$ without *doing any multiplication. As whoever should wish to mediate $\frac{4}{5}$ takes the $\frac{1}{2}$ of 4 which is 2, and you will have $\frac{2}{5}$ which is the half; and if it is odd, double the denominator ... ...*

The following examples are then given and discussed:

Mediation of 307.
Mediation of $307\frac{5}{8}$.
Taking the third part of $\frac{4}{5}$.
Taking two thirds of $\frac{4}{5}$.
Taking the third part of 308 and two thirds of 308.
Taking the third part of $26\frac{1}{4}$.

To divide by a fraction is to do as follows: reduce the divisor and the number to be divided if they are dissimilar and then divide numerator by numerator.

Chuquet here uses the word "dissimilar" to mean having different denominators.

Example: Whoever should wish to divide a fraction by a fraction, as $\frac{3}{4}$ by $\frac{2}{3}$, should reduce and you will find 9 over $\frac{3}{4}$ and 8 over $\frac{2}{3}$. Afterwards, divide 9 by 8 and you will have 1 and $\frac{1}{8}$. And whoever should wish to divide $\frac{2}{3}$ by $\frac{3}{4}$, divide 8 by 9 and arrive at $\frac{8}{9}$.

Again, whoever should wish to divide a fraction by a whole number, or a whole number by a fraction, as $\frac{3}{4}$ by 13, should reduce and then divide 3 by 52 and you will find $\frac{3}{52}$. And whoever should wish to divide 13 by $\frac{3}{4}$, should divide 52 by 3 and you will have $17\frac{1}{3}$.

The next example is a further marginal addition, and requires 20 to be divided by $5\frac{5}{6}$, and then $5\frac{5}{6}$ by 20. The same method is followed. This marginal addition is again omitted from the 1880 transcription but included by de la Roche. Two further similar examples follow, and the section ends with the division of and by fractions of fractions, of which the example is:

... whoever should wish to divide $\frac{3}{4}$ of $\frac{3}{5}$ of $\frac{1}{2}$ by $\frac{2}{3}$ of $\frac{4}{7}$, first, $\frac{3}{4}$ of $\frac{3}{5}$ of $\frac{1}{2}$ is $\frac{9}{40}$ and $\frac{2}{3}$ of $\frac{4}{7}$ is $\frac{8}{21}$. Now reduce $\frac{9}{40}$ and $\frac{8}{21}$ and then divide 189 by 320 and you will find $\frac{189}{320}$. And whoever should wish to do the contrary should divide 320 by 189 and one will find 1 and $\frac{131}{189}$; and thus for the rest.

Chuquet now explains that one can double, triple and quadruple by dividing by $\frac{1}{2}$, $\frac{1}{3}$ and $\frac{1}{4}$ respectively, and "so for other parts". The examples given are doubling $\frac{3}{8}$, 107 and $2\frac{4}{5}$, and tripling $\frac{4}{5}$, 14 and $14\frac{1}{6}$. On doubling, he comments:

... one can double in another way: if the denominator is even one should mediate it and not change the numerator.

Checks are now given for the various operations with fractions, beginning with abbreviation.

Whoever abbreviates reduced numbers, he returns them to their first state. Example: whoever reduces $\frac{2}{3}$ and $\frac{4}{5}$, he finds $\frac{10}{15}$ and $\frac{12}{15}$. The proof: abbreviate $\frac{10}{15}$ and $\frac{12}{15}$ and you will find $\frac{2}{3}$ and $\frac{4}{5}$ as before.

A corresponding check then appears for abbreviation, and checks are then given for two other pairs of inverse operations: addition/subtraction and multiplication/division. The chapter then concludes with an "epilogue" containing simple arithmetical problems illustrating what has been discussed. These include:

What is the number which, when one will have added $\frac{2}{5}$ to it, the addition all comes to $\frac{5}{6}$? Answer: subtract $\frac{2}{5}$ from $\frac{5}{6}$ and there will remain $\frac{13}{30}$, which is that which one seeks.

What is the number which, when one will have added $7\frac{2}{3}$, the sum comes to $12\frac{1}{4}$? Answer ... ... $4\frac{7}{12}$, which is that which one wants.

What is the number which, when one will have added to it its $\frac{2}{3}$, the addition comes to 20? Answer and rule for such problems: from 2 which is the numerator of $\frac{2}{3}$ make a further numerator, and from 2 and 3 together make a denominator. You will have $\frac{2}{5}$. Now from 20 subtract $\frac{2}{5}$ of it, which is 8, and there will remain 12, which is the required number. Or take $\frac{3}{5}$ of 20 and you will have that which is sought.

What is the number which, when one will have added to it its $\frac{1}{4}$, the addition comes to $13\frac{1}{3}$? Answer by the aforesaid rule: from $13\frac{1}{3}$ subtract its $\frac{1}{5}$ which is $2\frac{2}{3}$ and there remains $10\frac{2}{3}$, which is what one requires. Or, take $\frac{4}{5}$ of $13\frac{1}{3}$ and it will be as above.

The problems include "questions set as subtraction and solved by addition", and also "questions put as multiplication and solved by division", and vice versa.

What is the number which, when one will have taken away its $\frac{2}{5}$, the remainder is 12? Answer and rule for such problems: from the denominator of $\frac{2}{5}$ take away 2, which is the numerator, and 3 remains; thus from $\frac{2}{5}$ we have made $\frac{2}{3}$. Now take $\frac{2}{3}$ of 12, which is 8 and add it with 12; you will have 20, which is the number which one seeks. Or multiply 12 by $\frac{5}{8}$ and it will be done.

What is the number which, when one will have taken away its $\frac{1}{5}$, the remainder is $10\frac{2}{3}$? Answer: from $10\frac{2}{3}$ take the $\frac{1}{4}$ which is $2\frac{2}{3}$, and add them together; you will have $13\frac{1}{3}$. Or, multiply $10\frac{2}{3}$ by $\frac{5}{4}$ and it will be done. All such numbers, both with addition and subtraction, can be investigated by the rule of three which is treated below. As in this question in which it is necessary to find a number of which $\frac{4}{5}$ is $10\frac{2}{3}$. To do this one can say with the rule of three: if 4 gives me 5 how much would give me $10\frac{2}{3}$, etc.?

What is the number which, when it will be multiplied by its $\frac{1}{2}$, the multiplication comes to 72? Answer and rule for such problems: take $\frac{1}{2}$ of 72 which is 36, of which the square root or second root, which is all one, is 6. Now divide 72 by 6 and you will find 12, which is the number that one seeks.

What is the number which, when it will be multiplied by its $\frac{3}{4}$, the multiplication comes to 17? Answer: $\frac{3}{4}$ of 17 is $12\frac{3}{4}$ of which the second root is $R^2 12\frac{3}{4}$. Now divide 17 by $R^2 12\frac{3}{4}$, you will have $R^2 22\frac{2}{3}$, which is the number that one requires.

Next, $\frac{3}{4}$ and $\frac{2}{3}$ of a quarter, what part is this of 1? Answer: add $\frac{3}{4}$ with $\frac{1}{6}$, which is $\frac{2}{3}$ of $\frac{1}{4}$; you will have $\frac{11}{12}$. Divide this by 1 and you will have $\frac{11}{12}$. Now, do as above in saying 3 times 3 makes 9 and 2 makes 11; then 4 times 3 makes 12, which is $\frac{11}{12}$.

The work is illustrated in the margin as follows:

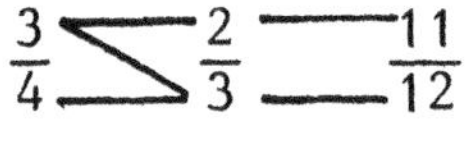

Next, of what number is $5\frac{3}{4}$ the $\frac{3}{7}$? Answer: divide $5\frac{3}{4}$ by $\frac{3}{7}$ and you will have $13\frac{5}{12}$.

What number when it will be divided by $\frac{3}{8}$ yields the part $\frac{2}{3}$? Answer: multiply $\frac{2}{3}$ by $\frac{3}{8}$ and you will find $\frac{1}{4}$, which is the number sought.

What is the number which, when it will be divided by its $\frac{1}{2}$, the quotient is 2? Answer: all numbers when they are divided by their subdouble the quotient is always 2.

What is the number which, when it will be divided by its $\frac{1}{2}$, the quotient is 3 or any number except 2? Answer: no number.

The examples given above are a representative selection of the forty problems included in this "epilogue".

Chuquet next turns to progressions, and gives in words the usual formula

$$s = \frac{(a + l)n}{2}$$

for the sum of an arithmetic progression.

Of progressions of numbers

Progression is a certain arrangement of numbers in which the first is exceeded by the second by as much as the second is exceeded by the third, and subsequently the others if there are any more. And one should know that progressions are made in several and diverse manners. For sometimes they begin at 1 and progress by 1, as 1, 2, 3, 4, etc.. Such is called by the ancients natural progression or continuous progression. Some begin at 1 but progress by a number other than 1, like 1, 3, 5, etc. or 1, 4, 7, etc.. This is called a disjunct or discontinuous progression. Sometimes it happens that neither the first nor the other of the aforesaid progressions begins at 1 but at some other number, and also sometimes it does not progress by 1 but by another number, like 3, 4, 5, 6, 7, etc. or 3, 7, 11, 15, etc.. For all such progressions four rules are made, by which one may easily add all the numbers belonging to the progression. And what is added by one of these rules cannot be added by any of the others. However, here for reason of brevity is here set down a single rule by which all different kinds of progressions are swiftly added. This rule follows.

If the addition of the first with the last is multiplied by half the number of numbers, the multiplication is equal to all the numbers of the progression together.

Example: for 1, 2, 3, 4. Whoever joins 1 with 4 makes 5, which multiplied by half the number of numbers which is 2 comes to 10. And so much does all this progression amount to.

Another example: for 3, 7, 11, 15, 19. Let 3 be added to 19 making 22, which multiplied by $2\frac{1}{2}$, which is half the number of numbers, comes to 55, and so much do the said numbers come to; and thus should one understand the addition of all other numbers in whatever progression they are constituted.

The rest of the material on progressions is written in the margins. This suggests that Chuquet amplified initial explanations after completing the manuscript with substantial marginal material (which Marre incorporated into the main body of the text).

Progressions beginning with the number by which they progress, if they begin at 1 and progress by 1, are of such nature that whoever adds 1 to the number of numbers and multiplies this addition by the subdouble of the number of numbers, this is the total sum of this progression. If the progression begins at 2 and progresses by 2, 1 must be added to the number of numbers and this addition multiplied by the number of numbers; thus one will have the total sum of this progression. If the progression begins at 3 and progresses by 3, 1 should be added to the number of numbers and this addition multiplied by the sesquialter of the number of numbers. If it begins at 4 and progresses by 4, one should add 1 to the number of numbers and multiply this addition by the double of the number of numbers. If it begins at 5 and progresses by 5, add 1 to the number of numbers, and multiply this addition by the double sesquialter of the number of numbers; and in general the addition of 1 to the number of numbers should always be multiplied by half that number of numbers taken as many times as there are units in the number by which the progression begins and continues. As if it progresses by 1, the sum should be multiplied by half the number of numbers taken one time. If it progresses by 2, it should be multiplied by the half taken two times; if by 3, the half taken 3 times, and so for the others.

Further , there is a progression which begins at 1 and progresses by 2 which is of so beautiful a nature that whoever multiplies the number of numbers by itself will find the sum of all the numbers constituted in this progression.

Again, the progressions which begin at 1 and progress by 3 are of such a kind that, whoever multiplies the number of numbers by itself and to the multiplication adds ½ the number of numbers multiplied by that number less 1, will have the total sum of the progression. As 1, 4, 7, 10, 13, 16. The number of numbers is 6, which multiplied by itself comes to 36, then half this number of numbers which is 3 multiplied by 5 which is 1 less than 6 makes 15, added to 36 makes 51 which is the sum of the numbers in the progression. Likewise progressions beginning at 1 and progressing by 4 are such that the number of numbers multiplied by itself, and to the multiplication added that number multiplied by itself, less 1, produces the sum of the numbers constituted in that progression. Thus, if the number of numbers is 5, multiplied by itself makes 25, then whoever multiplies 5 again by 4 which is less than 5 makes 20, which added to 25 makes 45; and that is what all this progression amounts to.

Chuquet turns next to classifications of number into even and odd numbers, and perfect and imperfect numbers. He distinguishes between three kinds of even number and three kinds of odd number. In such classifications he is following Boethius.

Of the division of numbers and of their differences

Each number is even or odd. An even number is that which can be divided into two equal parts without fraction of 1, like 2, 4, 6, 10, 12, etc.. An odd number is that which cannot be divided into two equal parts without fractions of 1, like 3, 5, 7, 9, 11, etc.. Of even numbers there are three kinds, that is, evenly even, evenly odd, and oddly even. An evenly even number is divisible one or more times until it comes to 1, like 2, 4, 8, 16, 32, etc., and these numbers are produced by continued duplation beginning at 1.

Evenly odd numbers are those which, divided into two equal parts, immediately its parts are odd, like 14, 18, 30. Such numbers are produced by duplation of odd numbers.

An oddly even number is divisible several times into equal parts but it cannot go down to 1, like 12, 20, 24, etc.. Such numbers come from duplation of evenly odd numbers once or several times repeated. Odd numbers also have three kinds, that is to say, prime incomposite, second composite, and third composite in itself but compared to another number it is incomposite. A prime incomposite odd number is one which is only made up of several unities like 3, 5, 7, 11, etc., and in general such numbers are those that cannot be exactly divided by any number except by 1. A secondary composite odd number is that which is indeed composed of several units and withal of one or several other numbers, as 9, 15, 21, etc.. 9 is composed of 1 and of 3 also, and 15 is composed of 1 and also of 3 and likewise of 5. Such numbers can be exactly divided by a number other than 1, and come by the multiplication of two odd prime incomposite numbers, as 33 which comes from 3 times 11. An odd number composite in itself but incomposite related to another number, like 9 compared to 25. For although 9 is composite in itself, 25 cannot however be composed of it in any way. And Boethius says all this in his arithmetic.

Another division of numbers

Of numbers, some are perfect and others imperfect. A perfect number is that whose aliquot parts joined together yield precisely that number, like 6, 28, 496, etc., between which there is no intermediate perfect number. The aliquot parts of 6 are 1, 2, 3, which joined together make 6. The aliquot parts of 28 are 1, 2, 4, 7, 14, which assembled make 28. For imperfect numbers, the aliquot parts joined together make more or less than their number of which they are parts, and hence some are called deficient, because their parts added are less, as 16 whose aliquot parts are 1, 2, 4, 8, which make 15. The other imperfect numbers are called abundant because their parts put together make more than themselves, as 12 whose aliquot parts are 1, 2, 3, 4, 6, which together make 16. And one should know that the least aliquot part of any number is 1, and the greatest is half that number. Between these are placed the others.

Of finding perfect numbers

Perfect numbers may be looked for thus. Let there be placed in a continuous sequence several evenly even numbers beginning with 1, like 1, 2, 4, 8, 16, 32, etc.. Now add two or more of these, beginning always with 1 and continuing as far as you wish up to a point where from their addition there comes an odd prime incomposite number, and multiply this by the greatest of the evenly even numbers, that is, the last one added; for the multiplication will produce a perfect number. Example: whoever adds 1 and 2 makes 3, which is prime incomposite, which multiplied by 2 which is the greatest and last of the evenly even numbers added makes 6, which is perfect.

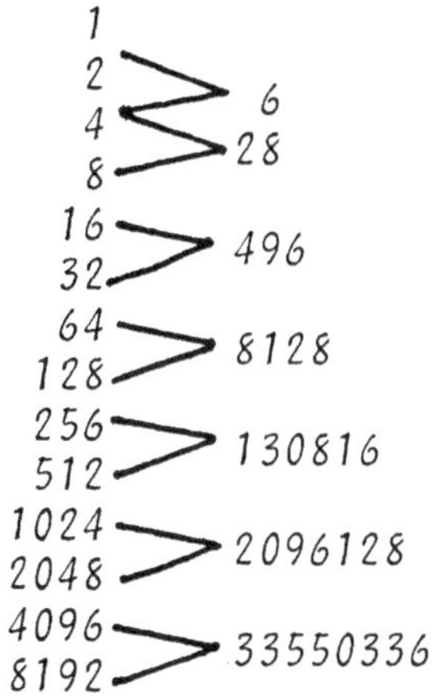

Further, whoever adds 1, 2, 4 makes 7 which is prime incomposite, which multiplied by 4, which is the last evenly even number, added amounts to 28 which is perfect. Thus one may continue to find others. Or alternatively, because the evenly even numbers arranged as above have the property that any one of them taken by itself comes to as much as all its predecessors joined together and 1 more, for this reason take 1 from whatever evenly even number you like, and if the remainder is a prime incomposite number, multiply it by the evenly even number next before it, for the multiplication will be a perfect number. As from 8, which is evenly even, if you take 1 there remains 7; multiply this by 4, which is its precedent and next evenly even number, and you will have 28. From 16, if you take 1, there remains 15 which is not prime incomposite, and for this reason a perfect number *cannot come from it. So now let us take 32, and removing 1 there remains 31 which is a prime incomposite number, which multiplied by 16 amounts to 496 which is perfect. And generally one should know that from 16 with 32 and from 64 with 128, and following that from each pair of evenly even numbers placed next to each other, multiplying one by the other but first subtracting 1 from the greater evenly even number produces a perfect number. And in this way innumerable perfect numbers may be found. But for imperfect numbers there is no rule and little to wonder at.*

This last paragraph is one of the very few places where Chuquet, having correctly stated a rule, then fails to apply it properly. The rule that even perfect numbers can be obtained from the expression

$$(1 + 2 + 2^2 + \ldots + 2^n)2^n = (2^{n+1} - 1)2^n,$$

provided that the sum or difference in brackets is an odd prime, can be found in Euclid. Chuquet, however, forgets the imposed condition in giving his list. Thus we see:

$$130816 = (2^9 - 1)2^8,$$

but

$$(2^9 - 1) = 7 \times 73.$$

Again we see:

$$2096128 = (2^{11} - 1)2^{10},$$

but

$$(2^{11} - 1) = 23 \times 89.$$

His statement that a perfect number can be calculated from each succeeding pair of evenly even numbers is manifestly wrong.

Further, one should know that perfect numbers have only two endings, and when one ends in 6 the next following after or before will end in 8, and then the next in 6, and then in 8 alternately.

This is a further error. It can be easily established from the Euclidean formula that perfect numbers necessarily end in 6 or 8, but not alternately. In the fifteenth century only the first five perfect numbers were known and these do indeed terminate alternately in 6 and 8, as we can see from Chuquet's list (with the two incorrect results removed). The next three, calculated by Jean Prestet in the seventeenth century, break this alternation, for they end in 6, 8 and 8 respectively.

Chuquet then refers to Boethius.

Again, according to what Boethius says, 1 is potentially every number and may be taken to be one of the perfect numbers, not only because it alone is perfect in power and virtue but also in fact by reason of its simplicity, and because it is prime incomposite, and because multiplied by itself it is conserved which is a sign of perfection.

Here Chuquet appears to be including the unit amongst the perfect numbers as defined earlier. This is a clear departure from the Greek tradition which did not regard 1 as a number but as the source of all number. He then discusses the means of determining the factors of perfect numbers.

Manner of finding the aliquot parts of perfect numbers

Mediate the perfect number as many times as you can, until you find an odd prime incomposite number from which this perfect number has been produced. Then again halve this imperfect number, first adding 1 to it, and continue this mediation so many times until you come to 1 inclusively. Know then that all these mediations are the aliquot parts of that perfect number. Example: 496 which is perfect, whose half is 298, whose half is 124, whose half is 62, whose half is 31, which is the prime incomposite number from which this perfect number has been produced. Then whoever adds 1 to that prime incomposite and takes half of this has 16, whose half is 8, whose half is 4, whose half is 2, whose half is 1, which is the smallest and last of the aliquot parts of the said number. Thus we have for its aliquot parts 248, 124, 62, 31, 16, 8, 4, 2, 1 which all together make 496. And thus for all the others.

<u>To know how many aliquot parts each perfect number should have</u>

By what is said above, it appears that the perfect numbers are produced by the multiplication of two numbers, one of which is evenly even and the other odd prime incomposite. And therefore whoever counts that evenly even number and all the others preceding it down to 1 inclusive, and doubles the number of these numbers, and from the double let 1 be taken, the remainder is the number of aliquot parts that a perfect number should have. Example of 496: the evenly even number from which this number is produced is 16 and the others preceding it are 8, 4, 2, 1 which are 5 numbers in all, whose double less 1 is 9. And 496 should have 9 aliquot parts. And thus the others should be understood.

Among the imperfect numbers there are some which are amicable and of marvellous familiarity with each other, for the aliquot parts of one put together yield the other and on the other hand the aliquot parts of the other yield the one. And such are the two numbers 220 and 284 whose aliquot parts are manifest ...

220	284
110	142
55	71
44	4
22	2
20	1
11	220
10	
5	
4	
2	
1	
284	

Chuquet seems to imply that pairs of amicable numbers, other than 220 and 284, are known. A second pair was, however, not known in the West until Fermat calculated the pair 17296 and 18416 in 1636.

Chuquet now turns to the topic of 'proportions'.

Proportion is the disposition which exists between two numbers when one is compared to the other. And it is of two kinds, namely equal proportion and unequal proportion. Equal proportion is when one number is compared to another equal to it, as 1 to 1, 2 to 2, 3 to 3, etc.. Unequal proportion is when two unequal numbers are related to each other, and this proportion again is of two kinds, for one is from the greater to the smaller, and the other from the smaller to the greater. The proportion of the greater to the smaller is when the greater number is compared to the smaller as 6 to 2. And the proportion of the smaller to the greater is, on the contrary, when the smaller is compared to the greater, as 2 to 6. Proportion of the greater to the smaller has five general species, that is, multiplex, superparticular, superpartient, multiplex superparticular, and multiplex superpartient. Multiplex proportion is when one number contains another several times exactly, and this has infinitely many special cases under it, that is, double, like 6 with regard to 3; triple, like 6 in regard to 2; quadruple, like 8 with regard to 2; quintuple, like 10 with regard to 2.

The sextuple is when the greater contains the smaller precisely 6 times, like 12 with regard to 2. And then there are the septuple, octuple, nocuple, decuple, and so for the others. The superparticular is when one number contains the other once and withal an aliquot part of the smaller. An aliquot part is that which has 1 as numerator, or which can be abbreviated as far as 1, like $\frac{1}{2}$, $\frac{1}{3}$, $\frac{1}{4}$. Or like $\frac{3}{6}$, $\frac{3}{9}$, $\frac{4}{12}$, etc.. Or otherwise an aliquot part is that which placed several times exactly yields the whole of the number of which it is an aliquot part. And in this proportion there are infinitely many cases under it. That is, the sesquialter which is when the greater contains the smaller by once and $\frac{1}{2}$ as 3 does 2. The sesquitierce is when the greater contains the smaller by once and $\frac{1}{3}$ as 4 does 3. The sesquiquarte is when the greater contains the smaller by once and $\frac{1}{4}$ as 5 does 4, then there are the sesquiquinte, the sesquisexte, etc.. The superpartiens is when a number contains the other and withal several parts of it which are not aliquot. This has innumerable cases under it, which are, firstly, the superbipartiens third; this is when the greater contains the less once and $\frac{2}{3}$ as 5 does 3. The superbipartiens fifth is when the greater contains the smaller once and $\frac{2}{5}$ as 7 does 5. Then there are the superbipartiens seventh, ninth, 11th, 13th,

15th, 17th, etc.. And there is the supertripartiens fourth, which is when the one contains the other once and $\frac{3}{4}$ as 7 does 4.

And the supertripartiens fifth, which is when the one contains the other once and $\frac{3}{5}$ as 8 does 5. And then there are the supertripartiens sixth, eighth, 10th, 11th, 13th, etc.. Then again there are the superquadripartiens seventh, ninth, 11th, 13th, etc.. And further the superquinpartiens, supersexpartiens, superseptpartiens, etc.. And for each add sixth, seventh, eighth, ninth, etc. or the denomination which should belong to it, and thus you will recognize the infinity of this proportion. The multiplex superparticular is when the greater contains the smaller several times and withal an aliquot part. This also has innumerable cases under it, one of which is the double sesquialter, which is when the one contains the other twice and $\frac{1}{2}$ as 5 does 2. And there is the double sesquitierce, which is when the one contains the other twice and $\frac{1}{3}$, as 7 does 3. Then there is the double sesquiquarte, sesquiquinte, sesquisexte, etc.. There is also the triple sesquialter, which is when the one contains the other thrice and $\frac{1}{2}$, as 7 does 2. Again, there are the triple sesquitierce, sesquiquarte, sesquiquinte, etc.. Yet again, there are the quadruple, quintuple, sextuple, etc., and for each of them add sesquialter, sesquitierce, sesquiquarte, etc., and thus you may understand the multitude of infinities of this proportion. The multiplex superpartiens is when the major contains the minor and with it several non aliquot parts. Under this proportion are contained the double superbipartiens third, the double superbipartiens fifth, the double superbipartiens sixth, etc.. And again there are the triple, quadruple, quintuple, sextuple, septuple, etc.. And to each add superbipartiens, supertripartiens, superquadripartiens, etc., and for denominations you will add to them again third, fourth, fifth, sixth, etc. or such numeration and denomination as may be suitable for them, and you will taste the innumerable species of this proportion. When the greater number is put in front, then the proportions should be given the names given above. And if the smaller is put in front, then to the names given above one should prefix the preposition sub, and that is the proportion of the smaller to the greater. As, 1 to 2 is subdouble proportion, or 2 to 3 subsesquialter as 3 to 2 is sesquialter, and so for the others according to their kind. However, the preposition or postposition of the numbers does not alter their proportion at all, but principally the comparison. For in whatever manner they are placed, if the *smaller is compared to the greater then this proportion should be denominated as sub. The proportion here of the smaller to the greater has just such species, and there are as many of them as those of the greater to the smaller, for the double cannot exist without the subdouble, nor the triple without the subtriple, etc..*

This highly ponderous classification of proportions goes back to Nicomachus' *Introduction to Arithmetic* (2nd century AD), and reflects the way in which the Boethian tradition built up the whole domain of numbers from proportions. From the time of the Pythagoreans, proportions had considerable significance not only within mathematics itself but also in the association of mathematical concepts with the wider world of physical experience, for example with music and architecture. Chuquet has a quite different approach to the domain of number but is apparently unable to avoid introducing the traditional terminology of proportions as codified in the work of Nicomachus. He goes on to present the various proportions schematically. A part of this schema only is shown below.

Of unequal proportions, one may inquire and reply in two ways, that is to say, in general and in particular. In general, as multiplex, or superparticular, or superpartiens, or multiplex superparticular, or multiplex superpartiens. In particular, as double, triple, quadruple, etc. Or sesquialter, sesquitierce, sesquiquarte, etc. Or superbipartiens third, superbipartiens fifth, superbipartiens seventh etc. Supertripartiens fourth, supertripartiens fifth, supertripartiens seventh, etc. and thus for the other species of proportion superpartiens. Or double sesquialter, double sesquitierce, double sesquiquarte, etc. Triple sesquialter, triple sesquitierce etc. and so for other species of multiplex superparticular proportion. Or double superbipartiens third, double superbipartiens fifth, double supertripartiens seventh, double supertripartiens fourth, double supertripartiens fifth, etc. And in the same way one must understand triple superpartiens, quadruple, quintuple, etc. and all the species of multiplex superpartiens and likewise of those proportions for which the preposition sub is to be added and placed in front. Example: whoever should ask what proportion there is between 8 and 2, in general it is multiplex proportion, and in particular it is quadruple proportion. And from 8 to 3 in general is multiplex superpartiens proportion, and in particular it is double superbipartiens

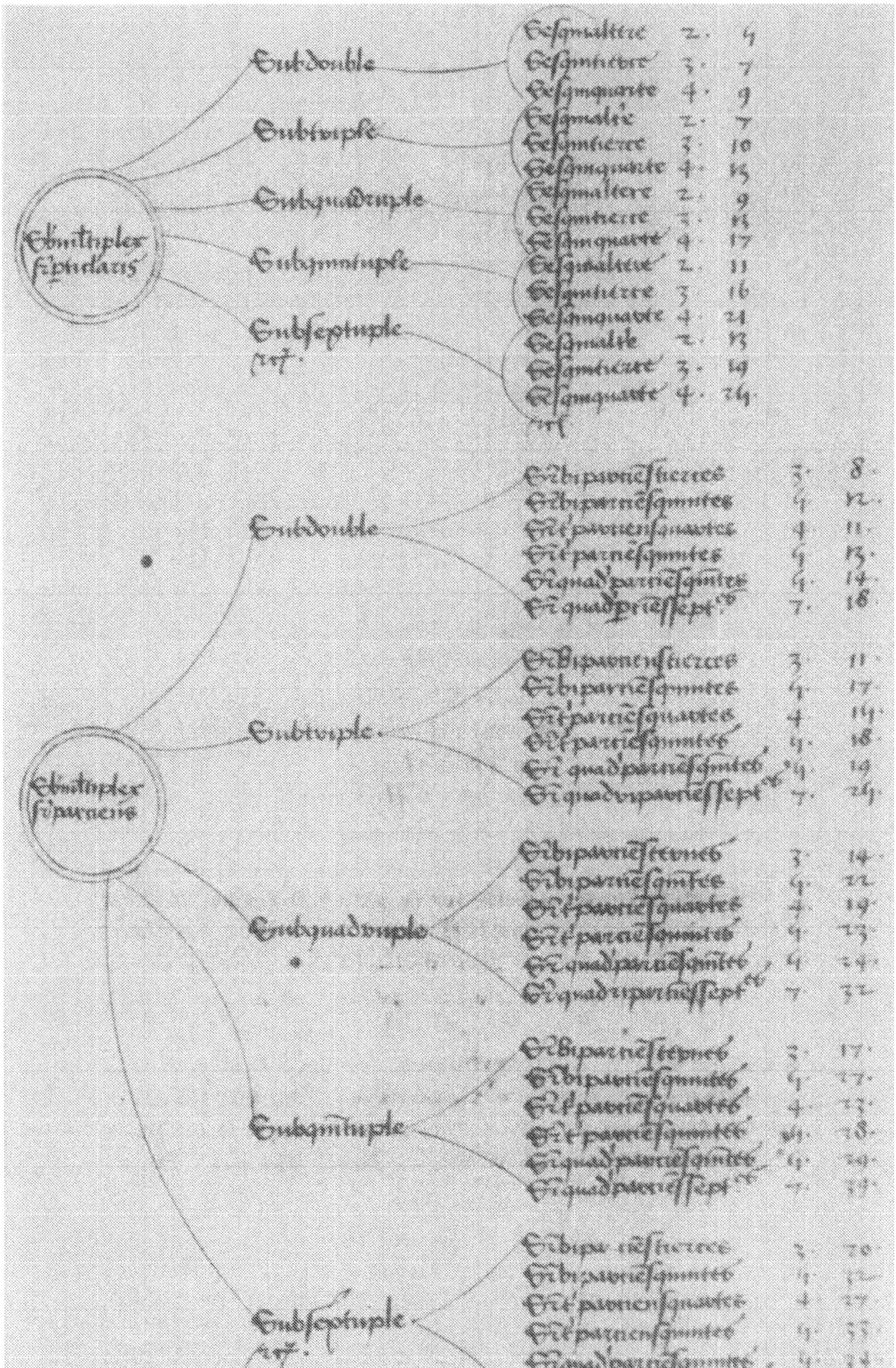

Subdouble
Subtriple
Subquadruple
Subseptuple
Subdouble
Subtriple
Subquadruple
Subseptuple

third. And from 3 to 8 in general is submultiplex superpartiens, in particular subdouble superpartiens third. And so for the others.

How double numbers may be added

Numbers constituted in a continuous sequence in double proportion, beginning with 1 or with any other number, can easily be added in this way: from double the last, let the first be taken away, for what remains is the total sum of those. Example: from 1, 2, 4, 8, 16 let 16 be doubled and from the double take 1 which is the first, and you will have 31 for the total sum of these. Another example: 3, 6, 12, 24. From double 24 which is the last take 3 which is the first. There remains 45 which is the total sum of these.

How one may swiftly add numbers constituted in continuous triple proportion

From the triple of the last take away the first, and take half the residue, for that is what the tripled numbers amount to.

Example: 1, 3, 9, 27. The triple of the last which is 27 is 81. From this take away the first which is 1. 80 remains, half of which is 40, which is what all the said numbers amount to when they are added together.

Another example: 5, 15, 45. The triple of 45 is 135. Take away 5. 130 remain, of which half is 65. And so much do these three numbers amount to, taken together.

Chuquet now gives in words the calculation of the sum of a geometrical progression:

$$\sum_{r=0}^{n-1} ar^n = a(r^n - 1)/(r - 1).$$

General rule for adding easily numbers constituted in a continuous sequence in all multiplex proportions

Let the last number be multiplied by the denominator of the proportion, from which multiplication let the first be removed, be it 1 or another number, whatever it is. And let the residue be divided by 1 less than the denominator of the same. For the quotient will be equal to the proportional numbers which we claim to be adding taken together.

Example: 3, 15, 75, 375, which are constituted in fifth proportion, whose denominator is 5. Now let 375 be multiplied by 5, giving 1875, from which 3 which is the first number must be taken away, and 1872 remains. Divide this by 4 which is 1 less than 5 and from this there comes 468. And so much do the said numbers amount to when they are taken together.

How the general rule given above may serve to add proportional numbers constituted in a sequence in all other kinds of proportion as it does for multiplex

Example: for the sesquialter proportion 8, 12, 18, 27. Now whoever multiplies 27 by $1\frac{1}{2}$ which is the denominator of the proportion reaches $40\frac{1}{2}$ from which 8 must be taken. There remain $32\frac{1}{2}$ which should be divided by $\frac{1}{2}$, which is 1 less than the denominator of the proportion, and one will find 65 for the total sum of the said numbers and thus for the other species of the superparticular.

Example: for the superpartiens 9, 15, 25, which are constituted in superbipartiens third proportion. Whoever multiplies 25 by $1\frac{2}{3}$ which is the denominator of that proportion reaches $41\frac{2}{3}$. Subtract 9 from it, leaving $32\frac{2}{3}$, which is 1 less than the denominator. The division comes to 49 for the total sum.

Example: for the multiplex superparticular 9, 21, 49 which are in double sesquitierce proportion whose denominator is $2\frac{1}{3}$. Which multiplied by 49 comes to $114\frac{1}{3}$. 9 taken away, there remain $105\frac{1}{3}$ which must be divided by $1\frac{1}{3}$ which is 1 less than the denominator, and the division comes to 79 for the total sum of the said numbers.

Example: for the multiple superpartiens. 25, 65, 169 which are constituted in double supertripartiens fifth of which $2\frac{3}{5}$ is the denominator. Which multiplied by 169 comes to $439\frac{2}{5}$, from which 25 must be taken, leaving $414\frac{2}{5}$. When this is divided by $1\frac{3}{5}$, the division comes to 259 for the total sum. And so it must be understood for the other proportions of each species.

Of multiplication and the property of proportional numbers

All proportional numbers constituted in a sequence in whatever proportion, always beginning with 1 and counting that which comes immediately after 1 for the first, and the one after for the second, and consequently for the others. Such numbers thus arranged have the property that whoever multiplies one of them by itself, it comes to the proportional number situated at the place double that of the number multiplied. As, whoever multiplies the 2nd by itself, there comes from it the 4th. And whoever multiplies the 3rd by itself, there comes from it the 6th, and so for the others. And whoever multiplies one of them by one of the others, and adds the two orders in which the two numbers multiplied are situated, will find the place where the number which comes from the multiplication should be situated, that is to say, he finds the how-many'th number that multiplication should produce. As, whoever should multiply the 2nd by the 3rd, there would come from it the 5th number, and whoever should multiply the third by the 5th, there would come from it the 8th. Example in double proportion: whoever multiplies 4, which is the 2nd double, by itself, finds 16, which is the 4th double. And whoever multiplies 8, which is the 3rd double, by itself, finds the 6th which is 64. Also whoever multiplies 2, which is the first double, by 4, which is the 2nd double, there comes from it 8 which is the 3rd, for whoever adds 1 with 2, their orders, has 3. Likewise whoever multiplies 8 which is the third by 32 which is the fifth finds 256 which is the 8th, just as whoever adds their orders, that is 3 with 5, has 8, as appears in the margin. And so for all other multiplex beginning with 1 it should be understood.

Another example in superbipartiens proportion: Whoever multiplies $4\frac{17}{27}$ which is the 3rd superbipartiens by itself finds the 6th superbipartiens which is $21\frac{316}{729}$. Or whoever multiplies the 2nd superbipartiens which is $2\frac{7}{9}$ by the 3rd which is $4\frac{17}{27}$ will find $12\frac{209}{243}$ which is the 5th. For 2 and 3 joined together make 5. And so for other superpartiens. And likewise the other species should be understood.

The two series corresponding to these examples are given in the margin as follows:

1		1	
2 —	1	$1\frac{2}{3}$ —	1
4 —	2	$2\frac{7}{9}$ —	2
8 —	3	$4\frac{17}{27}$ —	3
16 —	4	$7\frac{58}{81}$ —	4
32 —	5	$12\frac{209}{243}$ —	5
64 —	6	$21\frac{316}{729}$ —	6
128 —	7		
256 —	8		

This discussion of proportional numbers represents a significant step forward from Boethian arithmetic in that Chuquet adopts a more general approach applicable to any two terms of a potentially infinite geometrical progression. He also admits non-integer proportions. This can be seen as taking a step towards the later development of logarithms.

To know in which proportion two numbers are constituted

Let the greater be divided by the smaller, for the quotient will be the denominator of the proportion. Inasmuch as if the quotient is 1 precisely they are in equal proportion. If it is 2, in double proportion. If 3, in triple proportion, and so for the other multiplex. If it comes to $1\frac{1}{2}$, they are in sesquialter proportion. If $2\frac{1}{2}$, in double sesquialter proportion. If $3\frac{1}{2}$, in triple sesquialter proportion, and so for the others. If it comes to $1\frac{2}{3}$ then such numbers are arranged in the proportion superbipartiens third. If $2\frac{2}{3}$, they are in the proportion double superbipartiens third. If $3\frac{2}{3}$, in the proportion triple superbipartiens third. And so for the other proportions, superparticular, superpartiens, multiplex superparticular and multiplex superpartiens.

How several proportional numbers, as many as desired, may be found in whatever proportion is desired

Take for the first whatever number you want, and multiply this by the denominator of the proportion in which you want to constitute them, and thus you will have the first proportional. Which if it is multiplied by the said denominator will come to the second; and if the second is multiplied by the said denominator it will come to the third. And thus continue until you have as many proportional numbers as you want to have.

Example in double proportion: Whoever should wish to find several numbers in that proportion of which 3 is the beginning, should multiply 3 by 2 which is the denominator of double proportion, and will have 6, which multiplied by 2 makes 12, which multiplied by 2 makes 24, etc. And so may one do for the other multiplex proportions.

Example in superparticular proportion: I want to find, over 1, several numbers in sesquialter proportion, whose denominator is $1\frac{1}{2}$*. Now, let 1 be multiplied by* $1\frac{1}{2}$ *which comes to* $1\frac{1}{2}$ *which is the second after 1, then multiply* $1\frac{1}{2}$ *by* $1\frac{1}{2}$ *which comes to* $2\frac{1}{4}$ *which is the third. Multiply this by* $1\frac{1}{2}$ *which makes* $3\frac{3}{8}$ *for the fourth. And so for other superparticulars.*

Three further examples are given: in superpartiens, in multiplex-superparticular, and in multiplex superpartiens. The section then concludes with the indication that proportions involving fractions can be expressed in integers alone by means of multiplication by the "greatest denominator" of the fractions. The example given is 1, $1\frac{1}{2}$, $2\frac{1}{4}$, $3\frac{3}{8}$ (in sesquialter proportion) which thus becomes 8, 12, 18, 27.

In the final section of this first part of the *Triparty* Chuquet introduces the rule of three, the rules of one and two false positions, the rule of apposition and remotion, and the rule of intermediate numbers.

Of the rule of three and of its nature and conditions

So that we may ever grow and delve further into the science of numbers and have more ample experience of their proportions, several and diverse rules have been found amongst which the rule of three is greatly to be commended. And then, secondly, there is the rule of one position. After there is that of two positions, and further that of apposition and remotion. And yet again I have added the rule of intermediate numbers. These rules will be compendiously treated, at God's pleasure, so as to enter more quickly the other parts of this book where these rules are perfected, and into further great and instructive matters.

The rule of three is of ancient origin, and has had various names, the most common alternative being the "golden rule". We can trace both the method and the name via Arab works to India. The method can also be found in Egypt (in, for example, the *Rhind papyrus*) and in Chinese works such as the *Nine Chapters*. It is thus of considerable antiquity. It played a considerable part in commercial calculations, and is included in most of the arithmetics of the Middle Ages. Chuquet introduces this rule by explaining its name.

The rule of three is thus called because it always requires three numbers, of which the first two are always constituted in a certain proportion, and in such proportion as they are established, this rule serves to find for the third number its fourth, proportioned to it in the same way as the second is to the first. Note that the four numbers need not necessarily be such that the three are proportional or form a proportion, but whatever the relationship there is between the first and second, the same should be between the third and fourth. And the first and third are always like, and of one condition, and the second and fourth between themselves are like, and different and contrary to the other two. And whoever multiplies the first by the fourth, and the second by the third, the two multiplications are equal. Also, whoever divides the one like by the other and the one unlike by the other, the two quotients are equal. The manner of this rule is: multiply the third number by the second and then divide by the first. Or multiply that which one wants to know by its contrary and then divide by its like.

Or divide the first by the second and let the third be divided by the quotient. Or divide the second by the first and let the quotient be multiplied by the third, and thus one will have the fourth number which one seeks.

In the former of these two alternatives, there is an error in Marre's transcription which reads: "... and let the quotient be divided by the third". This would, of course, make nonsense. There follows an example.

If 8 are worth 12, what would 14 be worth? Or if 8 requires 12 for its proportional, what would 14 require? These three numbers may conveniently be put in this form: if 8/12/14. Multiply 14 by 12 and then divide by 8. You will find 21, and so much are 14 worth; and this is the usual way. Or otherwise divide 8 by 12 and you will have $\frac{2}{3}$, by which divide 14, and you will have 21 as above. Or divide 12 by 8 and you will have $1\frac{1}{2}$, by which multiply 14 and you will again have 21. And thus may one do with all other numbers.

In as much as the numbers of this rule may be found in three different kinds, for sometimes they are whole numbers, sometimes fractions, and sometimes whole numbers and fractions together ...

At this point in the manuscript there is a marginal note (almost certainly in Chuquet's hand) which reads as follows:

If they are whole numbers, it is necessary to do only as is said above. If they are fractions or whole numbers and fractions together, the method and manner of doing admits such variation and difficulty according to the kind of numbers.

This note is omitted in Marre's transcription. In the corresponding passage in de la Roche, it is incorporated into the main text. Chuquet continues the sentence:

... and although always for all different kinds of number one should multiply and divide always as said above, for the variety of numbers the method and manner of doing admits variation and difficulty. For making which thing easy and invariable there is here set down one such manner of doing. The three numbers put in the aforesaid order, for whole numbers without fractions are given 1 below them with a line between the two as a denominator. Whole numbers and fractions together are to be reduced and joined with their fraction. Fractions alone should be left as they are. This done, multiply the denominator of the first number by the numerator of the second, and again by that of the third, and you will have the number which should be divided. Then

afterwards multiply the numerator of the first by the denominator of the second and again by that of the third, and you will have the divisor. Now divide the number to be divided by the divisor, and it will be done, as may appear in several examples which follow.

If 16 is worth 23, what would 12 be worth? Answer: give the numbers denominators and set them out in the above order ... Multiply 1 which is the denominator of 16 by 23, giving 23, and again by 12, making 276 for the number to divide. Then multiply 16 by 1 which is the denominator of 23, and again by 1 the denominator of 12, and you will have 16 for the divisor. Then divide 276 by 16 and you will have $17\frac{4}{16}$*, which abbreviated is* $\frac{1}{4}$*. And 12 would be worth* $17\frac{1}{4}$*.*

The numbers of this example are set out in the margin as follows:

$$\frac{16}{1} \,/\, \frac{23}{1} \,/\, \frac{12}{1} \,.$$

Three further examples follow. In the second and third of these the resolution of mixed numbers into improper fractions is indicated:

$$\frac{4}{5} \,/\, \frac{2}{3} \,/\, \frac{6}{7} \,;$$

$$\frac{25}{12\frac{1}{2}} \,/\, \frac{63}{15\frac{3}{4}} \,/\, \frac{68}{13\frac{3}{5}} \,;$$

$$\frac{62}{20\frac{2}{3}} \,/\, \frac{17}{1} \,/\, \frac{5}{8} \,.$$

These lead to the solutions $\frac{5}{7}$; $17\frac{17}{125}$; $\frac{255}{496}$.

Chuquet then indicates how the results may be checked:

And whoever wants to check such calculations can multiply the first number by the fourth and the second by the third, to see if the two multiplications are equal. Or see if the third number and the fourth are in such proportion as the first and second are. Or turn round the last three numbers back to front, putting the fourth for the first and the third for the second and the second for the third. Then multiply and divide according to that rule, and you will find the first. Example: If 8 is worth 12, what would 14 be worth? Answer: multiply and divide as this rule requires and you will find 21. Check: if 21 is worth 14 what would 12 be worth? Multiply and divide, and you will find 8 which is the first number. Thus may all the others be checked.

Again to have fuller understanding and experience of this rule of three there are set below some questions which are solved by this rule, one of which is this: Which is the number that if someone will multiply it by 5 and divide this multiplication by 7, the quotient will be 13? To do this, it is necessary to find a number which divided by 7 would give the quotient 13, and so let 13 be multiplied by 7, and it comes to 91 which is that number. Then one must find a number which multiplied by 5 comes to 91, and to do this let 91 be divided by 5, and one will find $18\frac{1}{5}$ which is the number one asked for at first. And since 13 has been multiplied, and the multiplication divided by 5 ...

... This problem is none other than to say: if 5 is worth 7, what would 13 be worth; or if 5 is worth 13, what would 7 be worth? It is all one in the rule of three: provided that the divisor is the first, the two others may be set down at pleasure without altering the fourth. The divisor is always known by being similar to the number one wishes to know; it is that for which one wishes to know; it is that for which one wishes to provide its fourth number proportional to it, as stated above.

A number of examples now follow where the initial statement has first to be resolved into the formal structure of the rule of three as Chuquet has given it. These are:

What is the number which divided by 5 and the quotient multiplied by 7 the multiplication amounts to 42? Answer: If 7 is worth 5 what would 42 be worth? ... you will find 30.

If 3 times 4 will lead to 9, what will 4 times 5 lead to? ... If 12 is worth 9, what is 20 worth?

If 7 is the $\frac{1}{2}$ of 12, what is $\frac{1}{3}$ of 9? Answer: ... if 12 gives me $4\frac{2}{3}$ for its third, what will 9 give me? ... you will find $3\frac{1}{2}$ ... Or, alternatively: ... if 12 requires 14 for its whole, what will 9 require? ... you will find $10\frac{1}{2}$ ... take the third and it will be $3\frac{1}{2}$ as before.

If $\frac{2}{3}$ will be $\frac{3}{4}$ of $\frac{4}{5}$, what will be the $\frac{5}{6}$ of $\frac{6}{7}$? Answer: ... if $\frac{4}{5}$ requires $\frac{8}{9}$ for its whole, what will $\frac{6}{7}$ require? ... you will have $\frac{20}{21}$ for its whole, of which take the $\frac{5}{6}$, and you will have $\frac{50}{63}$ for $\frac{5}{6}$ of $\frac{6}{7}$.

If 7 were the $\frac{1}{2}$ of 12 one asks what part $3\frac{1}{2}$ would be of 9. Answer and method: ... if $\frac{7}{12}$ gives me $\frac{1}{2}$, what will $\frac{7}{18}$ give me? Multiply and divide and you will find $\frac{1}{3}$, and thus $3\frac{1}{2}$ would be the third part of 9.

If $\frac{2}{3}$ were the $\frac{3}{4}$ of $\frac{4}{5}$, one asks what part would $\frac{50}{63}$ of $\frac{6}{7}$ be. Answer: ... If $\frac{5}{6}$ requires $\frac{3}{4}$, what will $\frac{25}{27}$ require? ... you will have $\frac{5}{6}$. Thus $\frac{50}{63}$ would be the $\frac{5}{6}$ of $\frac{6}{7}$.

There is a further example in the margin of the manuscript which is not included by Marre. It is similar in form to examples in the main text. Chuquet continues with further applications of the rule of three.

How by the rule of three every number may be divided into several unequal parts constituted into whatever proportion one wishes

The manner of dividing and putting any number into several equal parts is manifest by what has been said of whole numbers and also of fractions. But to put them into unequal parts there is a rule: Multiply the number to divide by each of the proportional numbers, and each time divide by all added together. All this is none other than the rule of three, as may appear in several examples which follow, the first of which is this:

I want to divide 100 into two parts in such proportion as are 7 and 9. And to do this it is necessary to join 7 with 9, and they are 16, for common divisor. Then afterwards multiply 100 by 7, amounting to 700, divide them by 16, and you will have $43\frac{3}{4}$ for the part corresponding to 7. Then let 100 be multiplied by 9 and then divided by 16, and *you will have $56\frac{1}{4}$ for the part corresponding to 9, and it is done.*

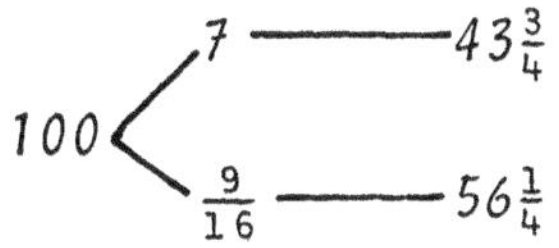

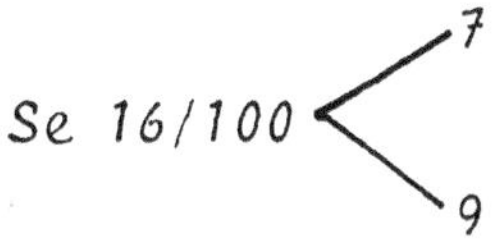

This method is none other than saying by the rule of three: If 16 is worth 100, what would 7 be worth, and again, if 16 is worth 100 what would 9 be worth, etc.? Or otherwise take $\frac{7}{16}$ and $\frac{9}{16}$ of 100 and you will have the proportional parts as demanded.

A number of examples now follow, each dividing 100 into either two or three proportional parts. These include the following:

I wish to divide 100 into three parts of such proportion as are $\frac{1}{2}$, $\frac{2}{3}$, $\frac{1}{4}$ ... reduce the numbers and then take the numerators ... 6, 8, 3, ... which numerators added together are 17 for common divisor. Now multiply 100 by 6, by 8, and by 3, and each time divide by 17 and you will have $35\frac{5}{17}$, $47\frac{1}{17}$, $17\frac{1}{17}$, ...

Further, from 100 I wish to make two parts of which $\frac{3}{4}$ of one is equal to $\frac{2}{3}$ of the other. Or, alternatively, I wish to find two numbers in such proportions as are $\frac{2}{3}$ and $\frac{3}{4}$, which add together to make 100. ... I reduce $\frac{3}{4}$ and $\frac{2}{3}$ in multiplying the numerator of one by the denominator of the other. And I have put the multiplication below the denominator multiplied.

Thus I get 8 and 9, which are of the nature abovesaid. Now I reduce $\frac{2}{3}$ first with $\frac{3}{4}$ ... and I get 8 and 9 as before, which added together make 17 for common divisor.

Then afterwards I multiply 100 by 8 and by 9 and I divide each multiplication by 17, and they come to $47\frac{1}{17}$ and $52\frac{16}{17}$ ...

Further, from 100 I wish to make three parts such that $\frac{2}{3}$ of the first, $\frac{3}{4}$ of the second, and $\frac{4}{5}$ of the third are equal. I put down $\frac{2}{3}$, $\frac{3}{4}$, $\frac{4}{5}$ in this order. Then I multiply 3, which is the denominator of $\frac{2}{3}$ by the numerators of the others ... and it comes to 36 ...

$\frac{2}{3}$	$\frac{3}{4}$	$\frac{4}{5}$
36	32	30
18	16	15

... who would take the half of each one will have 18, 16, 15. ... I ... multiply 100 by 18, 16, 15 and I divide by them all together, which is 49, and there comes $36\frac{36}{49}$, $32\frac{33}{49}$ and $30\frac{30}{49}$.

Further, from 100 I wish to make two parts in such proportion as are $\frac{1}{2}$ plus 6 and $\frac{1}{3}$ plus 4. To do this I add 6 with the $\frac{1}{2}$ of 100, which is 50, and there is 56..., etc. Now I divide 100 into two parts of such proportions as are 56 and $37\frac{1}{3}$, or as are 168 and 112 ... I find 60 for the first and 40 for the other. Similarly, whoever would divide 100 into two proportional parts as are $\frac{1}{2}$ and $\frac{1}{3}$ without adding to them the 6 and the 4, it will come as above. Thus the plus 6 and the plus 4 do nothing for this calculation, and the reason is because 6 and 4 are of such proportion as are $\frac{1}{2}$ and $\frac{1}{3}$ ...

Further, from 100 I wish to make three parts in such proportion as are $\frac{1}{2}$ less 12, and $\frac{1}{4}$ plus 10, and $\frac{1}{6}$ less 8. ... I find $46\frac{130}{245}$, $42\frac{210}{245}$, and $10\frac{150}{245}$.

Further, from 100 I wish to find two parts such that the first when it is divided by 3 and the other by 7 the two quotients are equal. ... I find 30 and 70, which are the parts required.

Chuquet now introduces the rules of one and two false positions, variously known as the rules of single and double position or the rules of false. Chuquet calls them "the rules of one and of two positions". The principle of the *regula falsi*, like that of the rule of three, is of ancient origin. There are examples in Egyptian mathematics in the *Rhind papyrus* and the *Berlin papyrus*. Other examples occur in early Hindu mathematics where it was extended to the double rule in which two successive assumptions are made about the value of the unknown. The two rules passed to the West via Arab works along with other arithmetical methods associated with the Hindu-Arabic numerals. The method is also found in Diophantus' *Arithmetica*. Chuquet's discussion is typical of similar works of the same century.

Of the rule of one position

This rule is so called because the calculations and reasonings which are made by it are found and made by the position of one number taken at will. This rule has two principal parts. The first looks for unknown numbers by means of a known number, taken at pleasure, and in the same way containing the parts proposed in the calculation. The second part likewise investigates diverse proportions of unknown numbers by means of a known number, in terms of yet other numbers artificially found below and sometimes above the position ...

A number of examples now follow:

I want to find a number such that when one will have added its equal, and again its $\frac{1}{2}$, its $\frac{1}{3}$ and its $\frac{1}{4}$, all together come to 17. To do this I take at my pleasure 12. Add to this 12 which is its equal, and again 6, 4, 3 which are the half, the third and the quarter of 12. All come to 37, and I only wanted 17, for which I say by the rule of three: if 37 come to me from 12, from how many will 17 come to me? Then I multiply and divide as the rule of three requires, and find $5\frac{19}{37}$ which is the number I wanted to find, for which if one adds to it $5\frac{19}{37}$ which is its equal number, with $2\frac{28}{37}$, $1\frac{31}{37}$, and $1\frac{14}{37}$ which are its $\frac{1}{2}$, its $\frac{1}{3}$ and its $\frac{1}{4}$, one finds 17.

Further, I wish to find a number such that its $\frac{4}{5}$ will be taken and then $\frac{2}{3}$ of the remainder, the result will be 10. To do this I put 15, from which in taking its $\frac{4}{5}$, which is 12, 3 remains. And further, of 3 in subtracting its $\frac{2}{3}$ there remains 1. I would want 10. By which I say by the rule of *three: if 1 comes to me from 15, what will give me 10? Then I multiply and divide and find 150 ...*

Then I wish to find two numbers [in triple proportion] such that added together and the sub-double sesquialter subtracted from the sum, the remainder is 8. And for this] put 1 and 3 which added are 4, of which the sub-double sesquialter is $1\frac{3}{5}$ which, subtracted from 4, leaves $2\frac{2}{5}$. And I want 8, for which I resort to the rule of three, saying: if $2\frac{2}{5}$ comes to me from 1, from what will I get 8? And by this manner I find $3\frac{1}{3}$ for the subtriple and consequently 10 for the triple, which are the two numbers that I wish to have, whose addition comes to $13\frac{1}{3}$, of which the subdouble sesquialter is $5\frac{1}{3}$, which subtracted from $13\frac{1}{3}$ leaves 8, as I want.

In this last example the words "in triple proportion" are added in the margin of the manuscript.

Further, I wish to find a number such that when it is multiplied by 3, and the result by 5, the final multiplication comes to 18. ... I put 2 ... I find $1\frac{1}{5}$ ...

Further, I wish to find a number such that when divided by 6, and the result divided further by 4, the final quotient is 5. ... I put 48 ... I find 120 ...

Further I wish to find a number such that when multiplied by 5 and again by 7, and then the multiplication divided by 12, the quotient is 8. ... I put 1 ... I find $2\frac{26}{35}$.

Chuquet now provides a "second part" to his discussion of the rule of false position. In it he presents problems where the application of a second rule, namely the rule of three, completes the working. These problems are typical of those found in mediaeval arithmetics where certain stereotyped forms of simultaneous equation are solved by arithmetical methods and hence without recourse to any form of algebraic notation, however primitive.

Of the second part of one position

The positions of this second part, although they are done at will, nonetheless the numbers which should be found above or below the position cannot easily be found without a rule, and the numbers found above the position are of such nature that when one subtracts from the proposed part or parts, always the number of the position remains.

Example: I want to find three numbers of such nature that the first with $\frac{1}{2}$ of the other two comes to 30, and the second with $\frac{1}{3}$ of the other two comes to 30, and the third number with $\frac{1}{4}$ of the other two also comes to 30. To do this I take at will 12. Then above 12 I have to find three other numbers, such that when one takes the $\frac{1}{2}$ away from the first, there remains 12. And when one takes the $\frac{1}{3}$ away from the second, there remains 12. And from the third, if one takes away the $\frac{1}{4}$, there is 12 left. Or otherwise, it is necessary to find three numbers of which the $\frac{1}{2}$ of the one is 12, the $\frac{2}{3}$ of the other is 12, and the $\frac{3}{4}$ of the other is 12. And to do this there is such a rule. Rule: Divide 12 by $\frac{1}{2}$, by $\frac{2}{3}$, and by $\frac{3}{4}$ and you will find 24, 18, 16. Which all three added together come to 58, which should be divided by 1 less than the number of numbers you want to find. Now in fact I want to find three numbers, therefore I have to divide 58 by 2, and 29 comes to me. From this I have to subtract the position which is 12, and also the three numbers found over 12 which are 24, 18, 16 and I am left with for the first, 17, and the other three which remain are 5, 11, 13. These three numbers are of such nature that 5 joined to the $\frac{1}{2}$ of the other two makes 17. And 11 added to the $\frac{1}{3}$ of the other two makes 17. And 13 with the $\frac{1}{4}$ of the other two similarly makes 17. And I wish to have 30, for which I have recourse to the rule of three, saying: if 17 comes to me from 5, from how much will 30 come to me? Then I multiply and divide and find $8\frac{14}{17}$ for the first number. Then afterwards I say: if 17 gives me 11, what will give me 30? And I find $19\frac{7}{17}$ for the second. Then further if 17 gives me 13, what will give me 30? And I find $22\frac{16}{17}$ for the third number...

There are two further examples. The first seeks four numbers such that the 1st plus $\frac{2}{3}$ the others, the 2nd plus $\frac{3}{4}$ the others, the 3rd plus $\frac{4}{5}$ the others, and the 4th plus $\frac{5}{6}$ the others, are all equal to 40. 60 is suggested, and from this 180, 120 and 0 are obtained. The rule of three then gives 24, 16, 8 and 0. The final example is similar, but five numbers are sought. In this case the numbers eventually found are 30, 20, 10, 0 and −10. These two examples are particularly significant in that in the former a zero solution is accepted, and in the latter both a zero and a negative number are accepted. Chuquet comments as follows:

To understand and verify the things said above, one should know that whoever adds 0 or subtracts 0 from some number, the addition or subtraction neither augments nor diminishes. And whoever adds a minus to another number, or subtracts it from one, the addition diminishes itself and the subtraction increases, as whoever adds minus 4 with 10, it gives 6. And whoever subtracts from 10 minus 4, there remains 14. And when one says minus 4 it is as if a person had nothing and still owed 4. And when one says 0 it is simply nothing.

Here Chuquet appears to accept both zero and negatives simply as 'numbers'. In this he was generally ahead of his time, though we do find the occasional negative result of a calculation interpreted as a debt in both Arab and Hindu works, and also in the *Liber abaci*.

There now follows a considerable number of worked examples grouped together with the repeated heading "Other findings of numbers". The modern algebraic equivalent is given for those not set out in detail. Lower-case letters v to z denote the unknowns, and S the sum of the unknowns. Some of the problems are indeterminate, a point of which Chuquet is sometimes aware.

I wish to find four numbers such that all together without the first come to 120, and without the second come to 90, without the third 80, and without the fourth make 75. And to find these, I add 120, 90, 80 and 75, coming to 365 which I divide by 3, which is 1 less than the 4 numbers, and I arrive at $121\frac{2}{3}$, *from which I subtract 120, 90, 80, 75 and there remains to me* $1\frac{2}{3}$ *for the first,* $31\frac{2}{3}$ *for the second,* $41\frac{2}{3}$ *for the third, and* $46\frac{2}{3}$ *for the fourth. These are the numbers which I wished to find.*

$$\left.\begin{array}{l} S - v = 120 \\ S - w = 90 \\ S - x = 80 \\ S - y = 75 \\ S - z = 72 \end{array}\right\} \quad \text{Solution:} \left\{\begin{array}{l} v = 10\tfrac{3}{4} \\ w = 19\tfrac{1}{4} \\ x = 29\tfrac{1}{4} \\ y = 34\tfrac{1}{4} \\ z = 37\tfrac{1}{4} \end{array}\right.$$

$$\left.\begin{array}{l} S - v = 120 \\ S - w = 180 \\ S - x = 240 \\ S - y = 300 \\ S - z = 360 \end{array}\right\} \quad \text{Solution:} \left\{\begin{array}{l} v = 180 \\ w = 120 \\ x = 60 \\ y = 0 \\ z = -60 \end{array}\right.$$

I wish to find three numbers of such nature that the first and the second with the $\frac{1}{2}$ of the third make 20, and the second and the third with the $\frac{1}{3}$ of the first come to 20. And similarly the thrid and the first with the $\frac{1}{4}$ of the second come to 20. And for this I put 12, which I divide by $\frac{1}{2}$, by $\frac{2}{3}$, and by $\frac{3}{4}$ ... ... and I find 24, 18, 16, and further, the $\frac{1}{2}$ of the third added with the other two is the first number 18, the $\frac{1}{3}$ of the first added with the other two is the second number 16, and for this also the $\frac{1}{4}$ of the second added with the other two is the third number 24. Thus I have 18, 16, 24 which added together make 58, of which I take away my position which is 12, and 46 remains to me. Now I have three numbers, that is to say 18, 16, 24, which are of such condition that the two firsts joined with $\frac{1}{2}$ of the third comes to 46, and the second and the third with the $\frac{1}{3}$ of the first makes 46. And I wish only 20. So I have recourse to the rule of three in saying: if 46 gives 18 as the first, 16 as the second, and 24 as the third, what will 20 give? And in this way I find $7\frac{19}{23}$, $6\frac{22}{23}$ and $10\frac{10}{23}$, which are the numbers that I wished to find.

Three indeterminate problems now follow for which Chuquet provides only particular solutions.

$$\left.\begin{aligned} w + \tfrac{1}{2}x &= 20 \\ x + \tfrac{1}{3}y &= 20 \\ y + \tfrac{1}{4}z &= 20 \end{aligned}\right\} \qquad \text{Solution:}\left\{\begin{aligned} w = z &= 12\tfrac{4}{5} \\ x &= 14\tfrac{2}{5} \\ y &= 16\tfrac{4}{5} \end{aligned}\right.$$

$$\left.\begin{aligned} w + \tfrac{1}{2}x &= 30 \\ x + \tfrac{2}{3}y &= 30 \\ y + \tfrac{3}{4}z &= 30 \end{aligned}\right\} \qquad \text{Solution:}\left\{\begin{aligned} w &= 20\tfrac{20}{29} \\ x &= 18\tfrac{18}{29} \\ y &= 17\tfrac{2}{29} \\ z &= 17\tfrac{7}{29} \end{aligned}\right.$$

Chuquet then comments:

Of such problems as the two abovesaid, some are circular. And this is when the last coressponds to the first, and the parts proposed to be added are in a diminishing progression such as $\frac{1}{2}$, $\frac{1}{3}$, $\frac{1}{4}$ etc. Although such problems may be done by the method stated immediately above, yet here is presented a special rule to do this, which is as follows. Rule: make your position that of the multiplication of the denominators of the fractions, to which add 1. That which comes of this will be the common number. Next, to find the other particular numbers.

Find the first below the position and then subtract it from the common number, for what remains is the part which the first demands of the second, by which one can know the second, and thus continue to the finding of the others. Example: I wish to find three numbers such that the first with $\frac{1}{3}$ of the second comes to 20, and the second with $\frac{1}{4}$ of the third number comes to 20, and the third number joined with $\frac{1}{5}$ of the first, the addition is 20. And to do this I multiply the three denominators one by the other, which are 3, 4, 5, and 60 comes to me which is my position. To which I add 1, making 61 which is the common number. Next, to find the first number below the position, I multiply 60 by $\frac{3}{4}$ as in the preceding problems and there comes to me 45 for the first number. This I subtract from 61, and I have left 16 which is $\frac{1}{3}$ of the second, from which the second number is 48. In the same way I take 48 from 61, and 13 remains, which is $\frac{1}{4}$ of the third number, and thus the third number is 52. Further I subtract 52 from 61, and I am left with 9, which is the $\frac{1}{5}$ which the third asks of the first, by which the first is 5 times 9 which makes 45, as stated above. Now I have found three numbers, that is, 45, 48, 52, of which the first added with $\frac{1}{3}$ of the second, and the second joined with $\frac{1}{4}$ of the third, and the third added with $\frac{1}{5}$ of the first, each addition comes to 61, and I wished it to be 20. Wherefore I go back to the rule of three, saying: if 61 gives me 45 for the first number, 48 for the second, and 52 for the third, what will 20 give me? And in this way I find $14\frac{46}{61}$, $15\frac{45}{61}$, $17\frac{3}{61}$, which are the three numbers I desired to have.

$$\left.\begin{aligned} x + 40 &= 2(y + z) \\ y + 40 &= 3(z + x) \\ z + 40 &= 4(x + y) \end{aligned}\right\} \quad \text{Solution:} \left\{\begin{aligned} x &= 3\tfrac{61}{73} \\ y &= 9\tfrac{23}{73} \\ z &= 12\tfrac{44}{73} \end{aligned}\right.$$

$$\left.\begin{aligned} w + 50 &= 3(x + y + z) \\ x + 50 &= 4(y + z + w) \\ y + 50 &= 5(z + w + x) \\ z + 50 &= 6(w + x + y) \end{aligned}\right\} \quad \text{Solution:} \left\{\begin{aligned} w &= \tfrac{200}{941} \\ x &= 3\tfrac{527}{941} \\ y &= 5\tfrac{745}{941} \\ z &= 7\tfrac{363}{941} \end{aligned}\right.$$

$$\left.\begin{aligned} w + 26 &= 2(x + y + z) \\ x + 26 &= 3(y + z + w) \\ y + 26 &= 4(z + w + x) \\ z + 26 &= 5(w + x + y) \end{aligned}\right\} \quad \text{Solution:} \left\{\begin{aligned} w &= \tfrac{78}{123} \\ x &= 2\tfrac{66}{123} \\ y &= 4\tfrac{54}{123} \\ z &= 5\tfrac{87}{123} \end{aligned}\right.$$

$$\left.\begin{aligned} w + z &= 2(x + y) = 26\tfrac{2}{3} \\ x + z &= 3(y + w) = 30 \\ y + z &= 4(w + x) = 32 \end{aligned}\right\} \quad \text{Solution:} \left\{\begin{aligned} w &= 2\tfrac{1}{3} \\ x &= 5\tfrac{2}{3} \\ y &= 7\tfrac{2}{3} \\ z &= 24\tfrac{1}{3} \end{aligned}\right.$$

Further, I want to find three numbers and one other in addition for the fourth which are of such condition that the first with the fourth comes to 17, the second with the fourth comes to 18, and the third with the fourth comes to 19. And to find them I can set a position at will, provided it be *less than the least of the three numbers given above. Well, let us put 12, which position I subtract from 17, from 18 and from 19, and I am left with 5, 6, 7, which are the three numbers, and 12 is the fourth. And if anyone should put a position other than 12, he would find other numbers of the same condition, wherefore such calculations may have innumerable responses.*

I wish to find three numbers and one other in addition for the fourth which are in such proportion that the two first joined with the fourth are the quadruple of the third, and the second and the third with the fourth are the double of the first, and the third and the first with the fourth are the triple of the second. And to find them I consider the submiltiplex of the quadruple, double and triple proportion which are $\frac{1}{4}$, $\frac{1}{2}$, $\frac{1}{3}$. From these I make $\frac{4}{5}$, $\frac{2}{3}$, $\frac{3}{4}$ by multiplication of the denominators by the numerators and by addition of the numerators with the denominators as has been said several times before. Then I multiply these three last denominators the one by the other, that is 5, 3, 4, making 60, of which I make my position. And since from $\frac{4}{5}$ up to the whole is $\frac{1}{5}$, I take the fifth of 60 which is 12 for the third number. Then from $\frac{2}{3}$ to 1 whole is $\frac{1}{3}$, which I multiply by 60 and 20 comes to me. Again from $\frac{3}{4}$ to $\frac{4}{4}$ which is 1 whole there is $\frac{1}{4}$, which I multiply by 16, and 15 comes for the second number. Which three numbers I add together and 47 comes to me, which I take away from 60 and 13 remains to me for the fourth number, which is the end of this problem.

$$\left.\begin{aligned} x + y + 18 &= 4z \\ y + z + 18 &= 2x \\ z + x + 18 &= 3y \end{aligned}\right\} \quad \text{Solution:} \begin{cases} x = 27\frac{9}{13} \\ y = 20\frac{10}{13} \\ z = 16\frac{8}{13} \end{cases}$$

$$\left.\begin{aligned} w + x + y + 8 &= 2z \\ x + y + z + 8 &= 3w \\ y + z + w + 8 &= 4x \\ z + w + x + 8 &= 5y \end{aligned}\right\} \quad \text{Solution:} \begin{cases} w = 40 \\ x = 32 \\ y = 26\frac{2}{3} \\ z = 53\frac{1}{3} \end{cases}$$

$$\left.\begin{aligned} w + 16 &= 2x \\ x + 16 &= 3y \\ y + 16 &= 4z \end{aligned}\right\} \quad \text{Solution:} \left\{\begin{aligned} w &= 24 \\ x &= 20 \\ y &= 12 \\ z &= 7 \end{aligned}\right.$$

Chuquet comments that this problem is of a kind which is "not of necessary conclusion".

$$\left.\begin{aligned} x + 18 &= 2y \\ y + 18 &= 3z \\ z + 18 &= 4x \end{aligned}\right\} \quad \text{Solution:} \left\{\begin{aligned} x &= 7\tfrac{1}{23} \\ y &= 12\tfrac{12}{23} \\ z &= 10\tfrac{4}{23} \end{aligned}\right.$$

The last of these problems follows.

Further, I wish to find four numbers of such nature that the first added with 23 is the double sesquialter of the second, and the second with 23 is the triple sesquitierce of the third, and the third with 23 is the quadruple sesquiquarte of the fourth. To do this I take 9 for the first as in the problem immediately above which I add with 23 and it comes to 32, which I divide by $2\frac{1}{2}$ which is the denominator of the double sesquialter proportion, and I find $12\frac{4}{5}$ for the second number. I put this with 23 and it comes to $35\frac{4}{5}$, which I divide by $3\frac{1}{3}$ which is the denominator of the triple sesqui-tierce proportion and it comes to $10\frac{37}{50}$ for the third number. This with 23 makes $33\frac{37}{50}$ which I divide by $4\frac{1}{4}$, which is the denominator of the quadruple sesquiquarte proportion and it comes to the part $7\frac{798}{850}$ for the fourth number. And whoever would take 20 for the common number and at will 10 for the first, and then do as in the second preceding problem, one will find 12 for the second, $9\frac{3}{5}$ for the third, $6\frac{82}{85}$ for the fourth. And so in such problems one can put two positions or two numbers at one's pleasure. That is to say, the common number and the first particular number, by which such problems do not lead to a necessary conclusion save that the common number and the first particular number are made specific.

Chuquet now turns to the rule of two (false) positions. The explanation in the verbal form in which it is presented is somewhat tortuous, and hence illustrates well the need for an effective algebraic notation.

What cannot be found by the rule of one position, and likewise by the first part of that rule, is accomplished as far as possible by the rule of two positions. And beyond this one should know that whatever position is brought into being, deducted from and manipulated as the process of the calculation may require, the result will be precisely the number that was sought, or more or less than that. If the result is precisely the number asked, the position is true, and it is not necessary to follow it further. If it comes to more or less, one should put beside that position plus or minus, like 4 plus or minus 7, etc.. And then, starting anew, one should take another position at will and likewise containing the opposed parts different from the first position and do with this as above. And thus one will have two positions and two pluses or two minuses.

Or one plus and one minus. And this done one should multiply the plus or minus of one of the positions by the other position, and equally the other position should be multiplied by the plus or minus of the one, and thus one will have four pluses or four minuses, or two pluses and two minuses, which four numbers should be treated according to this rule: Plus and plus, minus and minus, let us subtract. Plus and minus, let us add.

That is to say, the first plus or minus of the one and the other position should be subtracted the one from the other to have the divisor, that is to say, plus from plus and minus from minus. And if there is a plus with the one and a minus with the other, then the plus should be added with the minus to be the divisor. And from the two numbers one should do similarly to have the number to divide. That is to say, plus should be subtracted from plus and minus, and plus should be added with minus. Then the number to divide is divided by the divisor. For the quotient is that which one demands, since it is the number which can be attained by this rule, as it may appear in several examples which follow hereafter ...

What Chuquet has written amounts to the following. Suppose we have to solve for x an expression $p(x) = a$. Make two guesses x_1, x_2 and calculate $p(x_1)$, $p(x_2)$, arriving at $p(x_1) = a + e_1$ and $p(x_2) = a + e_2$. The correct solution x is then given by the expression $(x_1e_2 - x_2e_1)/(e_2 - e_1)$. Thus in Chuquet's first example, which follows, x_1 is taken to be 3 and x_2 to be 4. It is required to solve

$$[x + (2x + 4) + 3(2x + 4)] - 7 = 30.$$

Substitution of x_1 for x on the left-hand side yields 36, i.e. 30 + 6, and substitution of x_2 yields 45, i.e. 30 + 15. The correct value of x is therefore (3.15 - 4.6)/(15 - 6), which is 7/3.

I wish to find a number such that when it is added with its double and 4 over, and again added with the triple of that double plus 4, and 7 is removed from all the addition, the remainder is 30. To do this, I take 3, which doubled and added with 4 makes 10. Then whoever triples 10 makes 30. Now, whoever adds 30, 10, 3 makes 43, from which one must take 7. There remains 36, and I want only 30, from which this is by 3 plus 6. Afterwards for the second position I take 4, which doubled and added to 4 makes 12. This tripled makes 36. Now whoever adds 36, 12 with 4 has 52, from which I take 7, leaving 45, and I want only 30, from which this is by 4 plus 15.

Now, from plus 15 take plus 6 and I am left with 9 for divisor. Afterwards I multiply: 4 by 6 makes 24, and 3 by 15 makes 45, from which I remove 24 and I am left with 21 for the number to divide. Now I divide 21 by 9, and there comes to me $2\frac{1}{3}$ which is the number to divide. Now I divide 21 by 9, and there comes to me $2\frac{1}{3}$ which is the number that I require.

24
By 3 plus 6
By 4 plus 15
45

The next example leads to both e_1 and e_2 being negative.

Next, of 15 I want to make two portions, of which the one multiplied by 9, the other by 13, the two multiplications together make 160. And to do this I suppose that one of them be 12, which multiplied by 9 comes to 108. And thus the other will be three, which multiplied by 13 makes 39. Then I add 39 with 108 and find 147, and I want 160. By which I have by 12 minus 13 for the first position. Afterwards for the second I take 10 for one of the parts of 15, and therefore the other will be 5. Now, whoever multiplies

10 by 9, they make 90, and 5 by 13 makes 65, which added to 90 makes 155. And I want 160. Wherefore I have by 10 minus 5. Now I take 5 from 13, and am left with 8 for divisor, then I multiply 10 by 13 making 130 and 12 by 5 making 60 which I take from 130, and I am left with 70, which I divide by 8, and obtain $8\frac{3}{4}$ for one of the parts of 15, and in consequence $6\frac{1}{4}$ for the other.

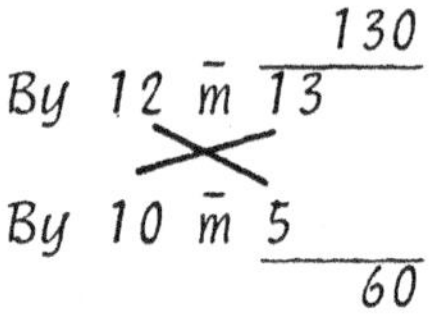

A third example, similar to the second example above, now follows. It illustrates what happens when the differences associated with the two positions are of opposite sign. The marginal calculation, comparable with the two problems above is:

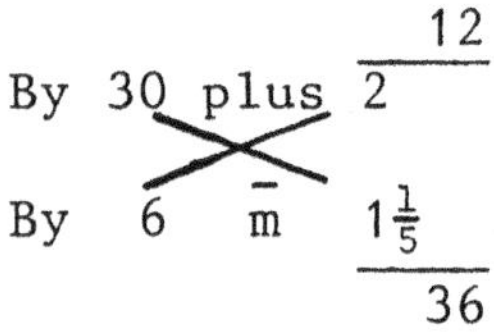

Chuquet concludes by commenting that these three examples make "the nature and property of this rule of two positions sufficiently clear". He then introduces "apposition and remotion".

<u>Of apposition and remotion</u>

This is not properly a rule in the sense of a rule for relief of labour, but a manner of consideration, which is this: that one should take away from one of the parts and put into the other, and continue this for several repetitions, until one reaches one's goal.

Example: I want to make of 12 three parts of which one multiplied by 2, the second by 1 and the third by $\frac{1}{2}$, all these three multiplications make 12. After I have investigated this by taking here and putting there, as expedient, I have found 2 for the first part, 6 for the second, and 4 for the third.

Others have been obliged to find a rule and another method of doing it than the above, which is this: As, for example, I want to find three numbers which added together make 12, and which, when the first is multiplied by 8, the second by 5 and the third by 3, the three multiplications added together make 60. And to do this 12 should be multiplied by the smallest of the three multiplying numbers which is 3, giving 36, which taken from 60 leaves 24. Then let this 3 be subtracted from the other two multipliers which are 8 and 5, and there remain 5 and 2. Now let 24 be divided by one of these remainders, that is, by 5, and 4 comes for the first of the three numbers, and there are 4 left which should be divided by the other remainder which is 2, and 2 comes for the second number. Thus the first number which is 4, and the second which is 2, joined together make 6; which subtract from 12, and 6 remains for the third number.

Again I want to find three numbers which all together make 12. Of these the first multiplied by 4, the second by 3 and the third by 2, the three multiplications together make 36. And to do this following the rule set above, I multiply 12 by 2 which is the least of the three multipliers, and it comes to 24, which subtracted from 36 leaves 12. Now from 4 and from 3 I take 2, and am left with 2 and 1. And by this remainder 2 I divide 12 which remains from 36, and I get 6 and remainder 0. And so I only take 5 for the first number, and there remains 2 from the division which I divide by 1, and I get 2 for the second, and in consequence 5 will be the third number. Thus I have 5, 2 and 5 for the three numbers I was looking for.

Two further examples have been added in the margin, namely:

$$\left.\begin{aligned} x + y + z &= 12 \\ 6x + y + z &= 35 \end{aligned}\right\} \quad \text{Solution:} \begin{cases} x = 3 \\ y = 4 \\ z = 5 \end{cases}$$

$$\left.\begin{aligned} x + y + z &= 20 \\ 6x + 2y + \tfrac{1}{2}z &= 20 \end{aligned}\right\} \quad \text{Solution:} \begin{cases} x = 1 \\ y = 3 \\ z = 16 \end{cases}$$

No comment is made on these. Chuquet is clearly not attracted to apposition and remotion for he adds:

This rule cannot be extended for the discovery of four or more numbers. Also one should know that all such calculations have several answers, as many as you want, as appears in the application of the general examples in the third part of this book. Wherefore apposition and remotion is a science which has little to recommend it.

The first part of the *Triparty* now concludes with Chuquet's particular invention.

The rule of intermediate numbers

This rule serves to find as many numbers intermediate between two neighbouring numbers as one desires. By its means it is possible to find many more numbers and to do more calculations than are found by the rule of three or by one position or by two positions. And to understand and know how to practise this rule, one should know that $\frac{1}{2}$ is the first and beginning among the fractions, and from it arise and issue forth two natural progressions, of which one proceeds by increasing, as $\frac{1}{2}$, $\frac{2}{3}$, $\frac{3}{4}$, $\frac{4}{5}$, etc. and the other proceeds by diminishing, as $\frac{1}{2}$, $\frac{1}{3}$, $\frac{1}{4}$, $\frac{1}{5}$, etc. These things being understood, the rule follows.

Numerator is added to numerator, and denominator to denominator. This means that when one wishes to find the first intermediates between two neighbouring whole numbers, one should add $\frac{1}{2}$ to the smaller whole number, and thus one will have an intermediate number, greater than the smaller extreme and smaller than the greater extreme. As between 3 and 4 the intermediate number, and the first one, is $3\frac{1}{2}$. And whoever wants to find several intermediates between 3 and $3\frac{1}{2}$ should add to 3 $\frac{1}{3}$ or $\frac{1}{4}$ or $\frac{1}{5}$, etc., as $3\frac{1}{3}$, $3\frac{1}{4}$, $3\frac{1}{5}$, etc. And the more one proceeds by this progression, the more one approaches the smaller extreme which is 3. And whoever wants to have several intermediates between $3\frac{1}{2}$ and 4 should add to 3 $\frac{2}{3}$ or $\frac{3}{4}$ or $\frac{4}{5}$ etc. and thus the more one proceeds the more one approaches the greater extreme which is 4. And thus between two neighbouring whole numbers, innumerable intermediates can be found, some declining to the smaller extreme and others tending to the greater. And again between two neighbouring intermediate numbers innumerable intermediates *may be found, tending in the same way to whichever extreme is desired, by adding numerator to numerator and denominator to denominator as the rule says. As, whoever would want to find an intermediate between* $3\frac{1}{2}$ *and* $3\frac{1}{3}$ *should add 1 to 1, which are the two numerators, and they come to 2 for numerator, and then 2 to 3 which are the two denominators, making 5 for denominator. Thus I have* $3\frac{2}{5}$ *as an intermediate between* $3\frac{1}{2}$ *and* $3\frac{1}{3}$*. For* $\frac{2}{5}$ *is more than* $\frac{1}{3}$ *and less than* $\frac{1}{2}$*. Again, whoever would want to find an intermediate between* $3\frac{1}{2}$ *and* $3\frac{2}{5}$ *should do as the rule says, and one will have* $3\frac{3}{7}$*. And whoever would want to find an intermediate between* $3\frac{1}{3}$ *and* $3\frac{2}{5}$ *should manipulate according to the rule, and one will have* $3\frac{3}{8}$*. And in this way one may continue to search for intermediate numbers, until one has found what one was looking for.*

To understand the style and manner how this rule may be applied, I wish to find by its means a number which multiplied by itself, and that number added to the multiplication, all comes to $39\frac{13}{81}$*. And to find it I have set down two neighbouring whole numbers of which one makes more and the other less, like 5 which multiplied by itself comes to 25, which added with 5 makes 30 which is less than* $30\frac{13}{81}$*, and 6, which multiplied by itself and added with 6 makes 42, which is more than* $39\frac{13}{81}$*. Thus it appears that the number which I seek is intermediate between 5 and 6. Now, to find this intermediate , I take* $5\frac{1}{2}$*, which multiplied by itself and added with* $5\frac{1}{2}$ *makes less than* $39\frac{13}{81}$*. And because I have less I proceed by the augmenting progression, and will take* $5\frac{2}{3}$ *which multiplied by itself and added with* $5\frac{2}{3}$ *are less than what I want. Therefore I proceed again to augment, and take* $5\frac{3}{4}$*, which I multiply and add as above and find again less than* $39\frac{13}{81}$*. And so I shall proceed as above, taking* $5\frac{4}{5}$*, which I multiply and add, and I find more than* $39\frac{13}{81}$*. Thus the number that I seek is between* $5\frac{3}{4}$ *and* $5\frac{4}{5}$*. And to find it I add numerator with numerator and denominator with denominator. Thus I have* $5\frac{7}{9}$*, which multiplied by itself and added with* $5\frac{7}{9}$*, all comes to* $39\frac{13}{81}$ *which is what I asked.*

The calculation has the following marginal clarification:

5 —— $\bar{m}$

6 —— plus

$5\frac{1}{2}$ $\bar{m}$

$5\frac{2}{3}$ $\bar{m}$

$5\frac{3}{4}$ $\bar{m}$

$5\frac{4}{5}$ plus

$5\frac{7}{9}$ 0

Chuquet has rightly claimed that this rule of intermediate numbers allows solutions of problems which are not amenable to the rule of three, or to the rules of single or double false position. It is now known that it is possible to obtain all the rational numbers by repeated use of Chuquet's procedure. His example here, however, which is just the equation $x^2 + x = 39\frac{13}{81}$, has an exact solution $x = 5\frac{7}{9}$. The other exact solution, which is negative, is not found. This example does not display the full power of the method, though Chuquet's first sentence does suggest that he was aware of its power. Later applications of the rule, when it is used for approximating irrational solutions, tend to confirm this. Thus, despite its apparent simplicity, the rule represents an understanding of numbers remarkable for its time. It provides a fitting climax to the first part of his manuscript.

Chapter 4

THE TRIPARTY - SECOND PART

The second part of the *Triparty* is concerned with roots. It begins with an introduction in which Chuquet defines various kinds of roots including surds and roots of surds, here called "compound roots". It is in this introduction that we first encounter Chuquet's special notation for roots along with the use of underlining to indicate collections of terms which we would today put in brackets. Thus, in the last line of folio 46*r*. below, there is the expression which would now be written as $\sqrt{14 + \sqrt{180}}$ or as $\sqrt{}(14 + \sqrt{}180)$.

Chuquet's symbol ℞ and other similar variations of the capital letter R were widely used by the early sixteenth century. A similar symbol is found, for example, in Pacioli's *Summa*, the work with which Chuquet's manuscript should especially be compared. Such symbols represented the first letter of the Latin word *radix* (root) used to translate Arabic *jidr*, which had the general meaning 'lowest part' and could hence refer to the root of a plant. However, the various R-symbols did not necessarily signify a square root; in some cases they stood for the first power of unknown quantities. Chuquet confines his symbol to roots, generalising it by the use of indices; thus the index 2 gives it the specific meaning 'square root' as can be seen in the extract from the manuscript reproduced above. Although Nicole Oresme had introduced special notation for fractional powers in his *Algorismus proportionum* written in the mid-fourteenth century, Chuquet confines his indices to integers. For each of printing, Chuquet's symbol will be represented by the Roman capital R in the translation which now follows.

The second part of the book deals with roots and compound numbers

A root of a number is a number which, multiplied by itself one or several times according to the demands and nature of the root, produces precisely the number of which it is the root. Or otherwise the root of a number is such that, written and set down two or more times one under the other or one beside *the other, and then the first multiplied by the second, and what comes by the third if there is a third, and again by the fourth, and again by the others if there are others, the last product shall be equal to the number, or shall produce the number whose root it is. And one should know that there are infinitely many kinds of roots, for some are second roots, others third roots, others fourth roots, others fifth, and thus continuing without end. There are no first roots. And if anyone should wish to include these for the sake of continuing the order, it is appropriate to say that a first root is understood as any simple number. As, whoever should say the first root of 12, which may be written thus, putting 1 above R in this manner: $R^1 12$, it is 12. And $R^1 9$ is 9, and so for all other numbers.*

A second root is that which, put down in two places one under the other and then multiplied one by the other, produces the number whose second root it is...

Here there is a marginal interpolation:

As 4 and 4 which, multiplied one by the other make 16, thus the second root of 16 is 4.

The text continues:

... which may be noted by putting 2 above the R as whoever should wish to write the second root of 16 may thus put $R^2 16$. Such roots are called square roots by the ancients. A third root is that which set down in three places and then the first multiplied by the second, and what results by the third, the last product is the number whose root it is, as 4 put in three places thus: 4, 4, 4 and then multiplied one by the other amount to 64, which one can write thus: $R^3 64$ is 4. And $R^3 8$ is 2. Such roots are called cubic roots by the ancients.

Chuquet is careful here to introduce purely arithmetical terminology, since geometrical terminology would have limited his generalizations. Thus he replaces the geometrical terms "square" and "cubic" by "second" and "third".

A fourth root is one which, put down in four places and then multiplied one by the other as stated above, makes up the number whose fourth root it is. As, 2, 2, 2, 2 which multiplied one by the other up to the fourth make 16. Thus the fourth root of 16, which may be noted down thus: $R^4 16$, is 2. Such roots are called by some square roots of square roots. A fifth root is one which, put down in five places and then multiplied one by the other up to the fifth, the last multiplication is equal to the number whose root it is, as 3, 3, 3, 3, 3 which multiplied one by the other in the aforesaid manner makes 243. Thus the fifth root of 243, which may thus be noted: $R^5 243$, is 3. And $R^5 32$ is 2. A sixth root should be put thus: R^6, and a seventh root R^7. And so should other roots be understood, by multiplying them six times or seven times or as many times as the nature of the root requires. Let all such roots as those named above be called simple roots.

There are other kinds of roots than the simple roots mentioned above, which may be called compound roots, as in the case of 14 plus $R^2 180$, whose second root is $3\ \bar{p}\ R^2 5$; or 7 plus $R^2 40$, whose second root is $R^2 2$ plus $R^2 5$, or $R^2 5\ \bar{p}\ R^2 2$, it is all one. Of these compound roots, some are second roots as aforesaid, others are third roots. There are also fourth, fifth, and all other orders, as for simple roots. Such roots of compound numbers may be bound together by a line and noted down in this manner, as the second root of $14\ \bar{p}\ R^2 180$ may be put thus: $R^2\,\overline{14\ \bar{p}\ R^2 180}$, which one should understand in this way: the second root of 180 should be added with 14, and then the R^2 must be taken again of all the addition. Or $R^2\,\overline{14\ \bar{m}\ R^2 180}$, which should be regarded as the second root of 180 subtracted or taken away from 14, and again the R^2 should be taken of what remains. Or $R^3\,\overline{20\ \bar{p}\ R^2 60}$, which should be understood thus: the R^2 of 60 should be added to 20 and then of the whole sum one should again take the third root. Or $R^4\,\overline{20\ \bar{m}\ R^2 60}$, which one should understand as the R^2 of 60 subtracted from 20 and of the remainder one should take the fourth root. Also $R^5\,\overline{20\ \bar{m}\ R^2 60}$ which one should understand as the R^2 of 60 taken from 20 and then of the remainder one should extract the fifth root. And thus it is necessary to understand of all other numbers and similarly of sixth, seventh and other roots.

Of such bound roots the nature is that the first root on the left is a second root or third or fourth or fifth or another, according to what is denoted, and the root following it and all the others if there are any more are [second] roots [of R^2 which are said to be R^4]. That is to say, if the first root towards the left is R^2, all the others moving *towards the left are* R^4, *as in* $R^2\underline{20\ \bar{m}\ R^2 60}$, *the* $R20$ *is a second root and the* $R60$ *is of the nature of a fourth root although it is noted down as* R^2.

Chuquet is careful here to say "of the nature of a fourth root". He does not inted to imply that the expression can be interpreted as $\sqrt{20} - \sqrt[4]{60}$. His point is that, if the expression is squared to give $20\ \bar{m}\ R^2 60$, a square root is still involved. (The square brackets in the translation above indicate words which are ruled through in the manuscript.)

Also $R^2\underline{20\ \bar{p}\ R^2 17\ \bar{m}\ R^2 13\ \bar{p}\ R^2 12}$. *The* $R20$ *is of the nature of the* R^2; *all the other roots, that is to say* $R17$, $R13$, $R12$, *are of the nature of the* R^4. *And if the first root on the left is a third root, the others are of the nature of a sixth root for they are* R^2 *of* R^3.

Similarly, if the first root were fourth and the other or the others were denoted with R^2, *they would be eigth roots because they are* R^2 *of* R^4. *And likewise, if the first were fifth, the others after them would be of the nature of* R^{10}, *and thus it must be understood.*

One should know further that such bound roots may be converted and, put in another way than that stated above, as $R^2\underline{30\ \bar{p}\ R^2 120}$ *which may be put without changing its value thus* $R^2\ \underline{R^2 120\ \bar{p}\ 30}$. *In this manner the 30 here is of the nature of* R^2 *and the 120 is of the nature of* R^4. *And one must also understand that the* $R^2 120$ *and 30 should be added together and then one should take the* R^2 *of the sum. Also* $R^3\underline{30\ \bar{p}\ R^2 120}$ *can be transformed thus,* $R^3\ \underline{R^2 120\ p\ 30}$, *which means thus: it is that,* $R^2 120$ *and 30 being added together, the* R^3 *of the sum is what is denoted; the 30 there is of the nature of* R^3 *and the 120 is of the nature of* R^6.

Chuquet then considers $R^4\underline{30\ \bar{p}\ R^2 120}$ in a similar way, and adds:

However, numbers or roots of numbers denoted with the word minus cannot be transformed, as $R^2\underline{30\ \bar{m}\ R^2 18}$ *and its like. Nevertheless, many roots are, as the aforesaid, denoted with this word minus, as* $R^2\ \underline{R^2 15\ \bar{m}\ 2}$ *which one should understand thus: it is that one must take away 2 from the* $R^2 15$ *and* R^2 *is what one must have of the remainder.*

Other examples similarly discussed include $R^2\ \underline{R^2 7\ \bar{p}\ R^2 5}$, $R^2\ \underline{R^2 7\ \bar{m}\ R^2 5}$, $R^3\ \underline{R^2 13\ p\ R^2 7\ m\ R^3 10}$, $R^3\ \underline{R^2 13\ \bar{p}\ R^3 7}$. In the last case, Chuquet comments that the root of 13 is "of the nature of R^6" and the root of 7 is "of the nature of R^9". He adds:

Many other and innumerable orders of compound roots may be found amongst the numbers which are here left aside for those who would wish to delve further.

This second part of this book contains six chapters of which the first concerns reducing two or more unlike roots to a like one.

The second chapter concerns simplifying the roots and extracting them.

The third teaches the adding of them together.

The fourth separates them one from the other.

The fifth multiplies them.

And the sixth divides them.

The first chapter concerns reducing the unlike roots to a like one.

The likeness or unlikeness of roots comes from their denominations alone, as $R^2 12$ and $R^2 17$ are like, for the one and the other are R^2. But $R^2 12$ and $R^3 12$ are unlike, for one is a second root and the other is a third root, and so for other denominations of root.

The manner of reducing two unlike roots to a like is this: multiply the number of one of the roots by itself once or several times according to the nature and denomination of the other root. And then multiply the number of the other according to the denomination of the root of the one. And then again multiply denomination by denomination and you will have the common denomination. Or thus convert the one root into the other and the other into the one and it will be done.

Example: I want to reduce $R^2 6$ and $R^3 7$. To do this it is necessary to multiply 6 and reduce it to a third root, so *that one will have $R^3 216$. Then 7 must be multiplied and reduced to a second root, and one will have $R^2 49$. Afterwards it is necessary to multiply 2 by 3, which are the denominations of the roots, and one will have 6 for the denomination common to the one and to the other. Thus $R^2 6$ is reduced to $R^6 216$ and $R^3 7$ is reduced to $R^6 49$, without varying their value, for $R^2 6$ is worth as much as $R^6 216$, and $R^3 7$ as $R^6 49$; it is all one.*

And if it happens that the greater denomination of one of the roots contains exactly the lesser denomination, then one can divide the greater by the lesser and then multiply the number of the root of least denomination by itself once or several times according to the unities of the quotient of the above division, and then the root of the lesser denomination will be reduced to the likeness of the root of greater denomination. Example: I want to reduce $R^2 5$, $R^3 10$ and $R^6 7$. For the first, divide 6, which is the greatest denomination by 2 which is the denomination of R5, and we obtain 3. Now multiply 5 three times, or reduce it to a third root, and you will have $R^6 125$ for and instead of $R^2 5$. Afterwards, divide 6 by 3 which is the denomination of R10 and we get 2. Now multiply 10 and reduce it to a second root and you will have $R^6 100$ in place of $R^3 10$. And thus the roots of the lesser denomination are reduced to the likeness of the greatest denomination, that is to say, to a sixth root.

Sometimes it happens that a root has several denominations, as second root of second root, or second root of third root, or second root of fourth root, or third root of fifth root, and so for the others. To avoid confusion and to make them intelligible, it is desirable to reduce them to one denomination in this manner.

Multiply the denominations of the root you want to reduce by the other or by the others, and you will have the single denomination of the root.

Example: I want to reduce $R^2\ R^3 13$ to one denomination. Multiply 2 by 3 which are the denominations, making 6 for the denomination of the root of 13, which should be set down thus: $R^6 13$, which is worth as much as $R^2\ R^3 13$.

Also whoever should wish to reduce to one denomination the third root of the second root of the fifth root of 12, which one can write as $R^3\ R^2\ R^5 12$, multiply 3 by 2 and again *by 5, giving 30 for the denomination of the root which one can denote as* $R^{30}12$. *And thus must similar ones be understood.*

Bound roots of several denominations: Some cannot be reduced to one denomination without changing their value, such as $R^2\ \underline{R^2 7\ \bar{m}\ 2}$ *or* $R^3\ \underline{R^2 15\ \bar{p}\ 4}$ *and the like. There are others whose denomination can be reduced, such as* $R^2\ \underline{R^2 13\ \bar{p}\ R^2 5}$ *which is* $R^4\underline{13\ \bar{p}\ R^2 5}$, *and such as* $R^3\ \underline{R^2 13\ \bar{m}\ R^2 5}$ *which is* $R^6\underline{13\ \bar{m}\ R^2 5}$. *For a fuller exposition of these bound roots such as* $R^2\ \underline{R^2 13\ \bar{p}\ R^5 5}$ *and their like, one should know, as said earlier, that they can be understood in two ways. The first is that* $R^2 5$ *is added to 13, and of the addition the* R^2 *of the* R^2 *is what is meant, and in this sense the number can be put thus:* $R^4\underline{13\ \bar{p}\ R^2 5}$.

The other way of understanding such roots is that $R^2 13$ *and* $R^2 5$ *are added together and the* R^2 *of all this addition is what is wanted. Such roots should thus be left as* $R^2\ \underline{R^2 13\ \bar{p}\ R^2 5}$. *They may be called bound roots of the second intention, as the other similar roots which are understood differently are of the first intention. However, bound roots set in this book from here on are all understood and taken in the first intention.*

In principle, Chuquet's notation is clear and unambiguous, and hence need not lead to problems of interpretation. However, we see that he has two possible interpretations of, for example, $R^2\ \underline{R^2 13\ \bar{p}\ R^2 5}$. For him, this can mean either $\sqrt[4]{13 + \sqrt{5}}$ or $\sqrt{13 + \sqrt{5}}$, which he describes as being of the "first intention" and "second intention" respectively. Subsequently, he interprets them logically in the latter manner in direct contravention of the final sentence above. Clearly in this sentence the word "first" is a slip of the pen for "second". He then turns to further cases of reducing roots.

By the method for reducing simple roots, compound roots may be reduced. As, for example, I want to reduce $R^2\underline{5\ \bar{p}\ R^2 3}$, *against* $R^3\underline{4\ \bar{p}\ R^2 6}$. *To do this, multiply* $R^2\underline{5\ \bar{p}\ R^2 3}$ *threefold so that it may be reduced to a third root; you will have a third root of a second root which is* $R^6\underline{170\ \bar{p}\ R^2 7500\ \bar{p}\ R^2 2352}$. *Then reduce* $R^3\underline{4\ \bar{p}\ R^2 6}$ *to a second root by multiplying it by itself and you will have a second root of a third root which is* $R^6\underline{22\ \bar{p}\ R^2 384}$.

I wish further to reduce $R^2\,\underline{R^2 3\ \bar{m}\ 1}$ *against* $R^3\,\underline{R^2 5\ \bar{m}\ 1}$. *To do this it is necessary to multiply and reduce* $R^2\,\underline{R^2 3\ \bar{m}\ 1}$ *to a third root; thus one will have a third root of R second, which is equipollent to* $R^6\,\underline{300\ \bar{m}\ 10}$. *Then after, one must reduce* $R^3\,\underline{R^2 5\ \bar{m}\ 2}$ *to a second root by multiplying it by itself, and thus one will have a second root of a third root, which is equipollent to* $R^6\,\underline{9\ \bar{m}\ R^2 80}$, *which is the reversal of* $R^3\ \bar{m}\ \underline{R^2 80\ \bar{p}\ 9}$.

This final clause is incomprehensible unless it is assumed that there is an error in the manuscript, which should presumably read "$R^6\,\underline{\bar{m}\ R^2 8\ \bar{p}\ 9}$".

Further, I wish to reduce $3\ \bar{p}\ R^2 5$ *against* $R^2\,\underline{2\ \bar{p}\ R^2 7}$, *and to do this I multiply* $3\ \bar{p}\ R^2 5$ *by itself; it comes to 14 plus* $R^2 180$ *of which the* R^2 *is* $R^2\,\underline{14\ \bar{p}\ R^2 180}$ *which is a bound root like the other.*

The second chapter deals with how roots may be extracted and simplified

Simplification of roots is nothing other than extraction of the same until they are reduced to numbers or as near as can be done. As fourth roots, which are sometimes simplified to a number and sixth roots sometimes are simplified to third roots, or to second roots, and often to a number. And thus it may be understood for other roots, as appears more plainly hereafter.

How second roots may be extracted or simplified

One should know that among second roots, which are otherwise called square roots, some may be extracted and others not. Roots whose numbers terminate in 2, in 3, in 7 or in 8 can never be simplified. And of the roots that can be simplified their numbers are called squares, like 4 whose root is 2, and 9 whose root is 3, and 16 whose root is 4, and 25 whose root is 5, and so for the others. The roots of numbers contained between two neighbouring true squares can also never be abbreviated, like the roots of numbers being between 4 and 9, which are 5, 6, 7, 8; and between 9 and 16, which are 10, 11, 12, 13, 14, 15, and so for the others, but they must be denoted and put in this manner: $R^2 10$, $R^2 11$, $R^2 12$, $R^2 13$, *etc..*

One should also understand that whoever multiplies a

square number by a square number, the number produced by the multiplication is always square. Also, whoever multiplies two numbers, whatever they may be, in quadruple proportion, the number of the multiplication is square. And likewise whoever divides them, one by the other, the quotient is always 4 or $\frac{1}{4}$ which are square numbers. And whoever reflects well on square numbers, finds that the root of one multiplied by the root of the other produces a mean proportional, as, from 4 and from 9, whose roots are 2 and 3, which multiplied by each other make 6 which is called the mean proportional, for the relationship between 4 and 6 is the same as between 6 and 9. And thus must the like be understood.

To extract, therefore, the square root of a number is nothing but to search for a number which multiplied by itself produces precisely the number whose root it is, if the proposed number is a true square.

Chuquet has here used the expression "mean proportional", a Greek concept associated with the calculation of roots, though the arithmetic methods involved go back to Babylonian times. The mean proportional x between any two lengths a and b satisfies $a : x = x : b$, thus $x = \sqrt{ab}$ and if $a = 1$, $x = \sqrt{b}$.

Chuquet turns next to a rule for extracting square roots, dealing first with exact roots for which he uses the familiar method, and then with inexact or "imperfect" roots for which he utilises his rule of intermediate terms to obtain approximations.

Rule for extracting square roots or second roots

It is first necessary to divide the figures of the number whose root one wishes to seek two by two, beginning at the right part and finishing at the left, and to put two lines below this number, equidistant and sufficiently far from each other. And under the first figure of the last order, be it single or one of a pair, one should put within the two lines a figure signifying that, multiplied by itself, it comes to as much as the value of the figures or figure of this last order, or the nearest that can be done, and this product should be taken from the same order, each figure from its like according to their differences. Afterwards one should double the root already found of the last order, and put this double below the two lines, in such a way that the first figure of this double is below and aligned with the second figure of the penultimate order, and the tens, if 10s there are, is after it, below the root already found, that is to say, aligned with the first figure of the last order.

And then observe and reflect how many times this double is contained in the figures which are in its place, including the remainder of the last order if there is one, and the quotient number should be placed in front of the doubled root and aligned with the first figure of the penultimate order, considering also whether that figure of the quotient number is exactly contained in the number which is above it and in its place with the help of the precedents; and that figure put as quotient and as root of this penultimate order should be put between the two lines and aligned with the first figure of that order. And then this quotient should multiply each figure of the doubled number with the figure placed in front of that double, and each product should be taken from the figures which are above and in the place of the multiplied figure. And that done, one should work with the next penultimate order, with the remainder of the penultimate if there is one, doubling the root of two orders aforesaid while moving forward in the aforesaid form, multiplying the double of the root as well as the quotient and subtracting as described above. And if it should happen that the double with the figure placed in front of it is not contained once in the number which is in the same place and above it, one should put 0 for the root of that order. And in this manner one should proceed up to the first order inclusive. And know that if there remains 0, the number from which the root is extracted is square. If there remains another number, this is a sign that it is not square. However the remainder should be less than the last divisor. For the extraction of roots is done, as aforesaid, by dividing the proposed number by several and diverse divisors and as many times as there are orders into the number above upon which one works, or by as many divisors as there are figures in the root...

The following example is given in the margin:

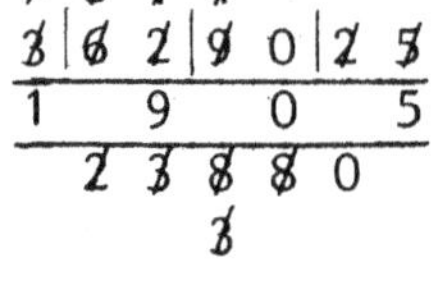

Another example: for more adequately demonstrating the method of the extraction of second roots, there is put here the manner of extracting the root of 94/21/80/73/55, divided two by two.

A step-by-step explanation of this calculation is given, various stages of the calculation itself appearing in the margin as follows:

(i)

$$\begin{array}{l} 1\ 3 \\ \cancel{9}\ \cancel{4}\,|\,2\ 1\,|\,8\ 0\,|\,7\ 3\,|\,5\ 5 \\ \hline \ \ 9 \end{array}$$

(ii)

$$\begin{array}{l} \ \ \ \ \cancel{6}\ 1 \\ \cancel{1}\ \cancel{3}\ \cancel{6}\ 2 \\ \cancel{9}\ \cancel{4}\,|\,\cancel{2}\ \cancel{1}\,|\,8\ 0\,|\,7\ 3\,|\,5\ 5 \\ \hline \ \ 9\ \ \ 7 \\ \hline \ \ 1\ 8 \end{array}$$

To arrive at stage (ii) from stage (i), the 9 between the lines is doubled and 18 written below the lower line as shown. A largest number n has then to be determined which, when multiplied by $180 + n$, is not greater than 1321. Taking n as 7, this number is then placed between the lines as shown and multiplication of 187 by 7, commencing from the left, is begun. The product $7 \times 1 = 7$ is subtracted from 13, and the remainder 6 is written above the 3 and the 13 is struck out. The product $7 \times 8 = 56$ is then subtracted from 62 and the remainder 6 is written above the 2 and the 62 is struck out. Finally, the product $7 \times 7 = 49$ is subtracted from 61, and the remainder 12 is written with the 2 above the 1 above the 7 and the 1 above the 6 in the next place to the left. This now corresponds to the stage in the method taught in schools in the present century:

$$\begin{array}{r l} & \ \ 9\ \ \ 7 \\ 9\,) & 9\ 4\ 2\ 1\ 8\ 0\ 7\ 3\ 5\ 5 \\ & 8\ 1 \\ 1\ 8\ 7\,) & 1\ 3\ 2\ 1 \\ & 1\ 3\ 0\ 9 \\ & \ \ \ \ 1\ 2 \end{array}$$

(The alignment of numbers in Marre's transcription is often wrong, and can be particularly confusing in 'galley' calculations such as that being presented here). The remaining stages are:

(iii)

```
   6 1
 1 3 6 2
 9 4|2 1|8 0|7 3|5 5
   9   7   0
   1 8 9 4
     1
```

(iv)

```
        1
     6 1 6 1
   1 3 6 2 4 6 3 7
   9 4|2 1|8 0|7 3|5 5
     9   7   0   6
     1 8 9 4 4 0
       1 1 9
```

(v)

```
             1
             2
           1 6 9   1
     6 1 6 1 1     2
   1 3 6 2 4 6 3 7 3
   9 4|2 1|8 0|7 3|5 5
     9   7   0   6   5
     1 8 9 4 4 0 1 2
       1 1 9 9 4
           1
```

Thus the last divisor is 194125. The root obtained is 97065 and there is a remainder 193130 at the end of the calculation showing that the original number is not a perfect square.

The style and manner of extracting second roots of fractions is that one should extract the root of the numerator, and put that on one side; and the root of the denominator and set it below the root of the numerator which was set aside, and one will have the root of that number. As for example, whoever should wish to have the second root of $\frac{4}{9}$, must take the root of 4 which is 2, then the root of 9 which is 3, which should be set below 2, and one will have $\frac{2}{3}$ which is the second root of $\frac{4}{9}$.

Also, whoever should wish to extract the second root of a whole number and fraction should put the whole number as a fraction, and join it with the numerator of the fraction, and from all this extract the root and set it to one side. And also that of the denominator. Afterwards one should divide the root of the numerator by that of the denominator, and it will be done. Example: I wish to extract the root of $12\frac{1}{4}$. It is necessary to put the 12 into fourths by multiplying it by 4, and it is 48, added with 1 is $\frac{49}{4}$. Now, from 49 extract the second root, which is 7, and from 4 which is 2. And then divide 7 by 2 and you will have $3\frac{1}{2}$ for the second root of $12\frac{1}{4}$.

The Extraction of imperfect roots

As is aforesaid, all numbers are not true squares, inasmuch as one cannot have a precise second root for them. For their roots multiplied by themselves always come to more or less than the numbers whose roots they are. And therefore they are called imperfect roots, whose extraction is only labour without profit. Nevertheless for the completeness of this book is set down a manner of seeking them as near to perfection as it is possible. And to enter into the practice it is first necessary to know that there are two forms of progressions to be used in this case, that is to say, progression by augmentation, as $\frac{1}{2}$, $\frac{2}{3}$, $\frac{3}{4}$, $\frac{4}{5}$, etc., and progression by diminution as $\frac{1}{2}$, $\frac{1}{3}$, $\frac{1}{4}$, $\frac{1}{5}$, etc.. Now, to extract all imperfect roots one may do as follows. As, for example, whoever should wish to extract the imperfect second root of 6, it is necessary to work first in the form and manner aforesaid for square numbers, dividing the figures of the proposed number two by two, as many as there are, and to work neither more nor less than is said above. Then the root of 6 is 2, for 2 times 2 makes 4, and there remains yet 2. Then, as 2 for root does not suffice to approach 6 sufficiently, and also whoever should take 3 for root would take too much, and therefore the R^2 of 6 is a certain number intermediate between 2 and 3. To find this one should use the rule of intermediate terms placed at the end of the first part of this book, and take for the first intermediate $2\frac{1}{2}$ which, multiplied by itself, comes to $6\frac{1}{4}$ which is $\frac{1}{4}$ more than 6. And so let us take less, proceeding by the progression by diminution, and let us try if $2\frac{1}{3}$ multiplied by itself comes to more than 6. Now it is the case that it comes to 6 minus $\frac{5}{9}$. Now, as we have found two roots of which the one makes greater and the other less, it is necessary for us to find an intermediate term between $2\frac{1}{2}$ and $2\frac{1}{3}$ by adding numerator with numerator and denominator with denominator, and there comes $2\frac{2}{5}$. Now test the root by multiplying $2\frac{2}{5}$ by itself, and we find 6 less $\frac{6}{25}$. It is necessary then to find another intermediate term between $2\frac{1}{2}$ and $2\frac{2}{5}$ by adding as above, and one will have $2\frac{3}{7}$ which, multiplied by itself, comes to 6 minus $\frac{5}{49}$. And by this manner one may proceed, adding the less with the greater or the greater with the less, until one approaches very close to 6, a little more or a little less, and until it is sufficient. And one should know that the more one should continue in this way, the nearer to 6 one would approach. But one would never attain it precisely. And from all this follows *the practice, in which the good and sufficient root of 6 is found to be $2\frac{89}{198}$, which root multiplied by itself produces 6 plus $\frac{1}{39204}$.*

The various stages are listed below the explanation, those between $2\frac{3}{7}$ and $2\frac{89}{198}$ being:

$2\frac{4}{9}$	$\bar{m}$	$\frac{2}{81}$
$2\frac{5}{11}$	$\bar{m}$	$\frac{3}{121}$
$2\frac{9}{20}$	*plus*	$\frac{1}{400}$
$2\frac{13}{29}$	$\bar{m}$	$\frac{5}{841}$
$2\frac{22}{49}$	$\bar{m}$	$\frac{6}{2401}$
$2\frac{31}{69}$	$\bar{m}$	$\frac{5}{4761}$
$2\frac{40}{89}$	$\bar{m}$	$\frac{2}{7921}$
$2\frac{49}{109}$	*plus*	$\frac{3}{11881}$

Chuquet then adds one further approximate root for "whoever should wish to seek further". This is several stages beyond the previous calculation.

$2\frac{881}{1960}$ *plus* $\frac{1}{3841600}$.

Again after this are set down the imperfect roots of several numbers which have been found by the rule of intermediate numbers as aforesaid, the first of which is the root of 2 which is $1\frac{169}{408}$, which, multiplied by itself comes to 2 plus $\frac{1}{166464}$.

R^2 *of 3 is* $1\frac{571}{780}$	*plus*	$\frac{1}{608400}$
R^2 *of 5 is* $2\frac{161}{682}$	*plus*	$\frac{1}{465124}$
R^2 *of 7 is* $2\frac{7853}{12192}$	*plus*	$\frac{1}{148644864}$
R^2 *of 8 is* $2\frac{985}{1189}$	*plus*	$\frac{1}{1413721}$
R^2 *of 10 is* $3\frac{1405}{8648}$	*plus*	$\frac{1}{74960964}$

or: $R^2 10$ is $3\frac{228}{1405}$ minus $\frac{1}{7974025}$

R^2 of 11 is $3\frac{379}{1197}$ plus $\frac{1}{1432809}$

R^2 of 12 is $3\frac{181}{390}$ plus $\frac{1}{152100}$

R^2 of 13 is $3\frac{109}{180}$ plus $\frac{1}{32400}$

R^2 of 14 is $3\frac{2667}{3596}$ plus $\frac{1}{12931216}$

Of fractions, some are square on the part of the numerator only like $\frac{16}{19}$. Some on the part of the denominator, like $\frac{17}{25}$. And there are others which are not square either in the one or in the other, like $\frac{5}{7}$. And there are others which are square in the one part and in the other, like $\frac{4}{9}$, whose R^2 is $\frac{2}{3}$, as has been said above.

The manner of extracting the imperfect roots of those which are not square either on the one side or on the other is that one should extract the imperfect root of the numerator and that of the denominator in the form and manner aforesaid in the extraction of the root of 6. Then reduce the one against the other if they are unlike, and from the root of the numerator make numerator, and from what of the denominator make denominator, and it will be done.

To avoid the trouble and tedium that one can have in searching for roots of such numbers as are not square either in the one or in the other, one can make one or the other square, whichever one wishes, in this way: whoever should wish to make $\frac{5}{7}$ square on the part of the numerator must multiply 5 by itself, making 25 for the numerator, which is a square number, then 5 times 7 making 35 for the denominator. And whoever should wish to make it square on the part of the denominator would have to say 7 times 7 make 49, which is a square number, for the denominator, then 7 times 5 make 35 for the numerator, and thus it is $\frac{35}{49}$. Now it is easier to work on $\frac{25}{35}$ or on $\frac{35}{49}$ than on $\frac{5}{7}$, because for the first two one imperfect root must be found, and with $\frac{5}{7}$ two must be sought. And in this manner one may make square all other fractions.

The roots of fractions which are squares on the part of the numerator or of the denominator are found thus: that is, the root of the square part is taken and set aside, and the imperfect root of the part which is not square is found as *given above for that of 6, then one must be reduced against the other and it should be done as above.*

The roots of whole numbers and fractions are found thus: that is to say, one may put the whole numbers into their fractions and add the numerator, then it is necessary to seek the perfect root of the numerator and also that of the denominator according to the aforesaid rules. After, one must divide the root of the numerator by that of the denominator and one will have that which one seeks.

And whoever should wish that one could seek the imperfect roots of whole numbers and fractions in this manner, that is to say, that one should extract the root of the whole number in the way according to the rule of perfect roots. Then with the root of the whole number one can add $\frac{1}{2}$ *and after see if this root multiplied by itself amounts to more or less than the number proposed of which one seeks the root; and if more, one should in place of* $\frac{1}{2}$ *add* $\frac{1}{3}$*, and proceed by progression by diminution. If less, one should add to it* $\frac{2}{3}$*, and proceed by progression by augmentation, and do neither more nor less than at the investigation of the imperfect root of 6 given before.*

Also, to seek the imperfect root of any fraction, one should begin at $\frac{1}{2}$*, then multiply* $\frac{1}{2}$ *by itself and see if the multiplication comes near enough to the number whose root one seeks or if it comes much more or less. If more, one should seek by the progression by diminution; if less by the progression by augmentation, and continue thus until one has found two neighbouring fractions of which the one makes more and the other less, and then add the greater to the less and the less with the greater, and continue as far as it suffices. And whoever in this way, would seek the square root of* $\frac{2}{3}$ *will find* $\frac{89}{109}$*, which, multiplied by itself, comes to* $\frac{2}{3}$ *and* $\frac{1}{35643}$ *more. And whoever would seek further will find* $\frac{881}{1079}$ *which, multiplied by itself, comes to* $\frac{2}{3}$ *and* $\frac{1}{3492723}$ *more. Also, whoever in like manner would seek the second root of* $\frac{3}{4}$ *will find* $\frac{2521}{2911}$ *for the root, which, multiplied by itself, comes to* $\frac{3}{4}$ *and* $\frac{1}{33895684}$.

This use of the rule of intermediate terms for approximating roots, though it appears also in the work of de la Roche, seems to have had little subsequent influence. The method is only very occasionally used by later mathematicians, as for example in a Latin work by the French mathematician Butéon, published in 1559, where successive approximations are found for the square root of 13. Chuquet's claim for its invention, however, has not as yet been successfully challenged.

How cube or third roots may be extracted and simplified

The third roots which may be simplified cannot be recognized by their terminations, that is, by the first figure of the number towards the right-hand part, for of all terminations there are some which can be simplified. The numbers contained between two neighbouring cubes, as, between 1 and 8, or between 8 and 27, or between 27 and 64, are not true cubes, and therefore their third or cube roots cannot be simplified. However, to have any knowledge of cubic numbers one should know that whoever divides some unknown number by any known cube, if the quotient is cubic then the number divided is cubic. Or if it is multiplied by such and the product is a cubic number, likewise the number multiplied is cubic. Also between two cubes, whether neighbouring or not, there are two mean proportionals, that is to say, the greater mean and the lesser mean. The greater mean comes from the multiplication of the root of the lesser cube by the square of the greater. The lesser mean comes from the multiplication of the root of the greater cube by the square of the lesser, as appears... for 8 and 27, whose lesser mean proportional is 12, and the greater is 18.

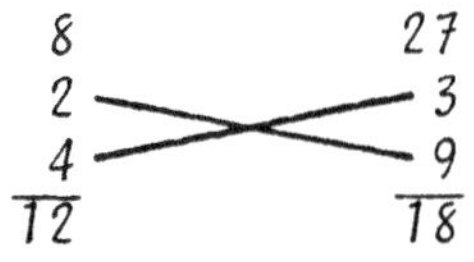

By analogy with the single mean proportional, the concept was extended by the Greeks to cubic problems. Thus, between any two lengths a and b there are two mean proportionals x and y satisfying $a : x = x : y = y : b$, giving $x = \sqrt[3]{a^2b}$, and if $a = 1$, $x = \sqrt[3]{b}$. Chuquet then comments that 1 is an exact square, cube, fourth, etc., and continues:

To extract the third or cube root of a number is to seek a number which, when multiplied triply, that is to say, multiplied by itself and then this product multiplied by that number, this second product is to be equal to the proposed number of which one has extracted the root. The manner of doing this is as follows.

It is necessary to divide the figures of the proposed number three by three, beginning at the right hand and moving to the left, as one does two by two in the extraction of second roots. And then it is necessary to begin to work from the left as follows.

When the number has been divided as aforesaid, one should take away the greatest third power or cube contained in the first triad, whether the triad be complete or not, and then write above this the remainder if there is one. And then set down below between the two lines the extracted root of that order. Afterwards for the second triad moving to the left one should seek a figure which, multiplied in the manner which follows, approaches as near as may be done to the figures of the second order with remainder of the first, if there is a remainder.

Here, Chuquet has written "left" for "right". In continuing this passge, Chuquet uses the term "simple" to indicate numbers having only a single digit. This was a common designation for single-digit numbers in arithmetics of Chuquet's time, but he does not refer to it in the introductory sections of the manuscript.

The manner of searching for the figure or root of the second order is that this figure, whatever it is, should be put in front of the root of first order in such a way that the first root be tens and the other be simple, and by this number one should multiply the triple of the root, and then secondly one should once more multiply the product already made by the figure set in front of the root. To which multiplication, thirdly, one should add the third power or cube of that figure in such a way that the simple figure of the cube occupies the first place, and the first figure of the other number be tens. And this sum which was last found should be as close as possible to the figures of the second order with the remainder of the first if there is one. And one should take away from its like, writing above what remains, if anything remains, and put the figure below the second order in front of the root of the first and aligned with the first figure of the triad. Thus one will have the root of the two orders.

In the same way, for the third order and for all the other triads if there are any more, one should likewise seek a figure whose sum produced from that figure as stated above, and in the same manner, and with that figure put in front of the roots which have been found, and by this number one should multiply the triple of all the figures found and taken as roots; and then the number which comes from the multiplication must secondly be multiplied by that figure put in front; and then thirdly add the cube of that figure as was done in the second triad, and then subtract all this sum of numbers from above, that is to say, from this third triad together with the remainder of the others if there is a remainder, always taking each away from its like as the art of subtraction requires. And continue thus, until for each triad one has a figure as root. And if it happens in the working that I cannot hold place in front of the root of the other triad or triads, one should put 0 for the root of the triad which one is seeking. And then work for the others if there are more. And one should know that after the extraction of roots of all the triads, if the remainder is 0, the number is a true third power or true cube. If there is any number left, it is not such.

Example: whoever should wish to extract the third root of 4/913/087; after the figures have been divided as aforesaid, it is necessary to take away the greatest third contained in the first triad, which is 1, and there remains 3 above 4, and to put 1 between the two lines for the root of the first triad. Then for the second triad 7 is put in front of 1, which is the root already found, and it comes to 17. Now for the first multiplication, multiply 17 by 3 which is the triple of the root, making 51. Then for the second multiplication multiply 51 by 7 which is the figure placed in front of 1, and it comes to 357. To this product must be added the third or cube of 7, which is 343, in such a way that 3, which is the first and simple figure, occupies the first place by itself, and the 4 with the 7 the second, and then the 3 with the 5; it all comes to 3193 which must be subtracted from 3913 which are the figures of the second triad with the remainder of the first, and thus there remains 0 and 7 which may be put between the two lines for the root of the second triad. In the same way, for the third triad, if any figure of value were put in front of the 17 it would come to more, without doing the multiplications, than are the figures of the third triad, which are only 87. Whence 0 must be put for the root of this triad, *and thus one will have for the root of the greatest cubic contained in the proposed number 170...*

```
4|9 1 3|0 8 7
1     7     0
-------------

   17
    3           170
  ---
   51
    7
  ---
  357
  343
 ----
 3913
```

For clarification and easing of the aforesaid practice, after one has found the root of the first triad, and sometimes also the root of the second triad, by the method aforesaid, one should know to find easily the roots of the other triads afterwards, if there are any more following on the right. It is necessary to put 0 in front of the root already found as, if you have 5 or 56 for root, by this means you would have 50 or 560, which you should multiply by the triple of the root. Now, let us take it that the root was 56, the triple would be 168, which one should multiply by 560. The multiplication comes to 850080, which should be put below the figures of the third order so that 0 is below the penultimate figure of that third triad, and 8 is after it, moving to the left, and the others accordingly. And then see how many times 8, which is the last figure of 850080, is contained in the figure or figures which are above and aligned with 8, considering the same for the other figures as one does on dividing. Thus one will see what figure should be put in the place of the 0 which was placed in front of 56. This figure should be multiplied by the triple of the root which is 168, and what comes from this must be added to 850080, and all this sum must again be multiplied by this figure. And again to this last multiplication there must be added the cube of this figure in the usual way...

In the manuscript the words "*figure fault a*" appear lightly ruled out between "last" and "multiplication" in this sentence. In Marre's transcription the same sentence appears twice, though on the first occasion without the final phrase "in the usual way".

... And then take away each figure from its like. And thus may the roots of the other triads be investigated, if there are any. However, when it comes to the last triad of the proposed number, if this number is a true cube, one may consider the first figure of that number on the right hand, for if it is 1 then 1 should be taken from the root of this triad; if 2, one should take 8, for the cube of 8, which is 512, ends in 2. If 3, one should take 7, because the cube of 7, which is 343, ends in 3. If 4, one should take 4 for the reason aforesaid. If 5, one should take 5. If 6, one should take 6. If 8, 2. If 9, 9. And if 0, 0. So in this manner third roots are easily found.

The manner of extracting the third roots of fractions is that it is necessary first to extract the root of the numerator and put it aside, then to extract that of the denominator and put it under that of the denominator set aside, and it will be done.

Here there is another slip of the pen: Chuquet writes "denominator" for "numerator".

Example: whoever should wish to find the third root of $\frac{8}{27}$, *it is necessary to take the root of 8, which is 2, and that of 27, which is 3, and one will have* $\frac{2}{3}$ *for the root of* $\frac{8}{27}$.

Also, whoever should wish to extract the third root of some whole number and fraction, such as $190\frac{7}{64}$, *it is necessary first to put the whole number into its fractional form and add it to the numerator, and there will be 12167 for numerator, of which one must take the third root, which is 23. Then one must take the root of 64, which is 4, for the denominator. Then divide 23 by 4; you will have* $5\frac{3}{4}$ *for the third root of* $190\frac{7}{64}$.

Imperfect cube roots, that is to say of numbers which are not true cubes, can be found in the form and manner in which one seeks imperfect square roots. However, it is only wasted time and labour without use or other necessity. For such roots, then, because they can neither be simplified nor extracted should be left as they are and denoted, as has been said, that is to say, as $R^3 9$, $R^3 10$ *or* $R^3 12$. *And so for all other numbers which are not true thirds or cubes.*

How fourth roots may be extracted or simplified

One should know that among the numbers some are true fourths, for they have true fourth roots, and others are not. The numbers whose first figure on the right hand is 2 or 3 or 4 or 7 or 8 or 9 are never true fourths, but those which terminate in 1, in 5, in 6 or in 0 are often found to be true fourths. Also, to know for certain whether a number is a true fourth, let the proposed number be divided by some known fourth number, and then let the quotient be considered, for if it is a fourth then the number divided will be a true fourth. The quotient is always of the nature of the number divided and of the divisor. And this is a general statement for all orders of number, just as with seconds, thirds, fourth, fifths or others.

Afterwards, it is necessary to divide up the proposed number of which one wishes to extract the root four by four, just as for third roots one divided three by three and for second roots two by two, putting two lines under this number, the same distance from each other, to collect the root there. And then afterwards for the first order or quartet on the left, whether or not it be complete, it is necessary to find a figure of value which multiplied fourfold, that is to say once by itself and what comes further multiplied by itself, this product may be taken from this quartet in such a way that there is nothing left or as little as possible. This is to be put between the two lines below that quartet. Then for the first quartet on the right, if the proposed number ends in 1, one may take 1 for the root of that quartet or 3 or 7 or 9. If it ends in 5, one should take 5 as root. If it ends in 6, one should take 2 or 4 or 6 or 8. And if the first quartet is 0000, one should take 0 for root.

For the middle quartets, if there are three or more quartets in the proposed number, one should put down for each quartet a figure of value such that with the root of the first order on the right one may exactly divide by that divisor.

And again this quotient may be exactly divided by this same divisor, and from this third division the last quotient should be equal to the divisor. Thus this divisor is the true fourth root of the proposed number. And if the quotient is less then one should diminish the middle figure or figures of the divisor, and if it is greater one should increase them. Then divide the number proposed as above. And if by the diminution or augmentation of the middle figures one cannot reach the state where the quotient and divisor of the third division are equal, one should vary the root of the first quartet on the right, putting there 1 or 5 or 6 or 0. And continue this until it is enough.

Example: I wish to extract the fourth root of 30/4980/0625. The number divided four by four figures in the manner aforesaid, one should take the fourth root of 30, which is 2. This, multiplied to the fourth, amounts to 16, which taken from 30 leaves 14 above 30; thus we have 2 for the root of this order. Afterwards, for the first quartet on the right, which is 0625, as it ends in 5 we take 5 for the root of this order. And for the middle quartet, which is 4980, if we take 2 we will have 225 for the root of the proposed number. Now it is necessary to see if this root is true, by dividing 3049800625 by 225 and it leaves 100. And because there remains some number it is a clear sign that 225 is not the root of this number, for the root is always contained exactly in its number. And if we divide this number by 245, there will remain 200, which signifies that 245 is also not its root. If we take 235 for divisor, we shall find in this case 12977875, which one divides further by 235 and one will find then for quotient 55225, which it is necessary to divide again by 235 and one will find 235. And since the quotient and divisor of this third division are equal, it is shown that 235 is the true root of the proposed number. And thus should be understood all others.

Another style of doing this: extract the second root, and then from the second root extract the second root again, and it will be done. Example: from 3049800625, whose R^2 is 55225. Then whoever extracts the square root of 55225 finds 235 for the fourth root of the proposed number.

There are some fourth roots which cannot be simplified to a number, but can indeed be simplified to a second root. Like R^4 1369 which simplified by extraction of second roots comes to $R^2 37$, which cannot be further simplified. And $R^2 784$, which simplified by extracting the second root of 784 comes to $R^2 28$ which is equal to $R^4 784$. And so for the like.

In the manuscript Chuquet has "728" in place of the last "784".

There are many fourth roots which cannot be simplified completely or in part, like $R^4 10$, $R^4 11$, $R^4 17$, and infinitely many others.

How fifth, sixth, seventh and other roots may be simplified

To extract or simplify all manner of roots, it is good and useful to have before your eyes the following table which may be called the little manual of roots. In this little manual one can see in what figures the numbers may end, which have precise second, third, fourth or fifth roots, or others up to 10 and further if desired. And by this one is saved great labour. In the first order of this little manual from the left are put the figures of value, that is to say, 1, 2, 3, etc. up to 10 reading downwards. And aligned with each of these 10 figures, moving to the right, are put those numbers having exact roots. And in the upper part of this table one can find the nature and denomination of their roots, that is to say, if they are second, third, fourth, or others, as follows.

	Second root	Third root	Fourth root	Fifth root	Sixth root	Seventh root
1	1	1	1	1	1	1
2	4	8	16	32	64	128
3	9	27	81	243	729	2187
4	16	64	256	1024	4096	16384
5	25	125	625	3125	15625	78125
6	36	216	1296	7776	46656	279936
7	49	343	2401	16807	117649	823543
8	64	512	4096	32768	262144	2097152
9	81	729	6561	59049	531441	4782969
10	100	1000	10000	100000	1000000	10000000

	Eighth root	Ninth root	tenth root
1	1	1	1
2	256	512	1024
3	6561	19683	59049
4	65536	262144	1048576
5	390625	1953125	9765625
6	1679616	10077696	60466176
7	5764801	40353607	282475249
8	16777216	134217728	1073741824
9	43046721	387420489	3486784401
10	100000000	1000000000	10000000000

By this table one may know that the tenth root of 1024 is 2, the ninth root of 512 is 2; the eighth root of 256 is 2; ... ... the third root of 8 is 2, the second root of 4 is 2. And so may the rest of the table be understood.

By this table, one may also understand how fifth numbers having a true perfect root may terminate in all manner of figures. Thus by their first figures one cannot discern fifths from others. However, if they are 100000 or below, one may know them from the little manual or table given above. If they are above 100000, one may choose them by dividing them by some known fifth number, as stated above for fourth roots, and similarly for sixths by dividing them by any known sixth number. And also the sevenths by dividing them by some seventh put in the little manual aforesaid, or by another greater number if need be. And likewise the eighth and other numbers may be chosen and known by this same intention. For if the quotient be a fifth, sixth or other number, the number divided will be fifth, sixth or other, because the quotient is always of the nature of the divisor and the number divided. And whoever wants to know them by conjecture by their first figure contemplates the little manual given above, for there he may see all the figures in which all orders of numbers may terminate.

After one has true knowledge that the proposed number is a true fifth, or true sixth, seventh or other, if one wishes to extract the fifth root one should divide up that number five by five in the manner aforesaid. If the sixth root, six by six. If the seventh root, one should divide the figures of the proposed number seven by seven, and so on for the others. And then from the first order towards the left one should extract the fifth or other root, aided by the little manual given above, which will teach what figure one should take. After that, for the first order towards the right and for the others in between one should handle it in the same way as has been said for fourth roots, considering the ending of the proposed number. And then one should divide that number by this root, and then again the quotient, up to four divisions for fifth roots, or up to five divisions for sixth roots, and so continuing for the others. If the last quotient is equal to the divisor, know that you have found the fifth or other root of the proposed number. If it is unequal, you must then increase or diminish the intermediate figures of the root, and if need be, also vary the first figure on the right; and then divide and continue to that which sufficies.

Fifth roots are immediately simplified to numbers, like the fifth root of 59049, which is 9, and the fifth root of 3125, which is 5, and so for the others.. And likewise seventh, eleventh, thirteenth, seventeenth and nineteenth roots, and in general all roots whose denominations have no relationship with 2, with 3, or with any other number. All such roots are simplified, when they can be simplified, immediately to a number. But those whose denominations have relationship with some number other than 1, such roots are sometimes simplified to a number, and at times cannot be simplified as far as a number, but can indeed be simplified to a root of lesser denomination. Like a sixth root, which sometimes is simplified as far as a number, other times as far as a third root or as far as a second root; and on eighth root which can be simplified to a fourth root or to a second root and on occasion to a number. Ninth roots can on occasion be simplified to a third root or to a number. And twelfth roots which can often simplify to a sixth root or a fourth root, or third or second, and at times to a number. And thus it must be understood for other roots, according to whether their denominations have participation with several numbers, so they may be simplified in many ways.

Example: whoever should wish to extract the sixth root of 531441 may first extract the second root of that number according to the rule of second roots, and will find $R^3 729$. Then whoever takes the third root of 729 has 9, which is the sixth root of the proposed number. And whoever should wish may first extract the third root of the proposed number according to the rule of the third roots, and will find $R^2 81$; then whoever takes the second root of 81 which is 9, he has the sixth root of the proposed number as above. And by this means one can simplify all sixth roots which may be simplified as far as a number. There are some sixth roots which can be simplified only as far as a second root, like $R^6 2197$, which by extraction of the third root comes to $R^2 13$. And likewise $R^6 841$, which, simplified by extraction of the second root comes to $R^3 29$. And thus should the like be denoted.

There are some eighth roots which are simplified to a fourth root such as $R^8 169$, which, simplified by extraction of the second root, comes to $R^4 13$. There are others which can simplify to a second root, such as $R^8 614656$, which, simplified by extraction of the second root comes to $R^4 784$, which, further simplified, comes to $R^2 28$. And there are some which can be simplified to a number, such as $R^8 43046721$, which, simplified by extraction of the second root, comes to $R^4 6561$, which, simplified again by extraction of the second root, comes to $R^2 81$, which, simplified again by extraction of the second root comes to 9, which is the eighth root of 43046721. There are others which cannot be simplified by any contriving, such as $R^8 17$, $R^8 18$, $R^8 19$, and infinitely many others.

Similar paragraphs then follow, containing discussions of ninth, tenth and twelfth roots. These are omitted here.

From what has been said above one may understand which roots are such that without any intermediary they cannot immediately be simplified other than to a number, and which there are that one can simplify to some number through an intermediary. And also those that one can simplify as far as some root of lesser denomination. And how, in every order of root, there are some that cannot be simplified by any contriving. Also, from what has been said above one may understand how fourteenth, fifteenth, sixteenth, eighteenth, twentieth and other roots may be simplified.

The reason why simplification of roots has been treated is for and to the end that one might have more ample knowledge of this, for the extracted and simplified root is more easily understood than the unsimplified, and it is also simpler to deal with, that is, to add or subtract, multiply or divide, with or by another number. And if it happens that some root cannot be simplified exactly as far as a number, then it is necessary to deal with it by some circumlocution, as for 3, which does not have an exact root, which we cannot call anything other than the root of 3 which one may write thus: R3. And if it is a second or third or other root, one may also put the denomination above R in this way: $R^2 3$, $R^3 3$ or $R^4 3$, etc. as is aforesaid.

How compound roots may be simplified

Whoever should wish to extract the second root of 14 plus $R^2 180$, it may be done in this manner. It is necessary to mediate that number, and thus one will have 7 plus $R^2 45$. Afterwards 7 must be multiplied by itself, and one will have 49, from which it is necessary to take 45, leaving 4, whose second root which is 2 should be added to 7 and also subtracted from 7; thus one will have 9 for the addition and 5 for the subtraction. Adter the addition and subtraction, it is necessary to take the second roots, thus one will have 3 and $R^2 5$, And because the proposed number is composed with term plus, for this reason we should say that 3 plus $R^2 5$ is the second root of 14 $\bar{p}$ $R^2 180$. And if it were 14 minus $R^2 180$, the second root would be 3 $\bar{m}$ $R^2 5$. And thus all others should be understood. However, one should know that if the difference between 49 and 45, which is 4, were not a true square, that is, that its root could not be extracted exactly, one cannot extract their root from such compound numbers. As, for example, whoever should wish to extract the second root of this number here, 6 $\bar{p}$ $R^2 7$, it would be necessary in order to do this to take half of 6 which is 3, multiplied by itself makes 9, from which it is necessary to take the quarter of $R^2 7$ which is $1\frac{3}{4}$, and leave $7\frac{1}{4}$. And since one cannot extract the R^2 of $7\frac{1}{4}$, it is a sign that 6 $\bar{p}$ $R^2 7$ is not a true square, and that one cannot abbreviate or extract its root, but must call it, or write it, in this manner: $R^2 \underline{6\ \bar{p}\ R^2 7}$, bound by a line underneath, which is to be understood as the second root of 7 added to 6, and then take again the square root of the whole sum. And where 7 was previously a second root, now it is R^2 of R^2, that is, a fourth root, and 6 is R^2, which was previously a number. And from this come the bound roots.

Another extraction: whoever should wish to extract the second root of 12 plus $R^2 140$, it is necessary to mediate 12 and one gets 6 which, multiplied by itself, comes to 36, from which it is necessary to take away 35, which is the half of $R^2 140$ or the quarter of 140, leaving 1 of which the R^2 is 1, and when this is added to 6 and subtracted from 6, one will have 7 and 5, of which the second roots, which are $R^2 7$ plus $R^2 5$, are the second root of 12 plus $R^2 140$. And thus other similar ones must be understood.

How bound roots may be simplified

This section heading together with a short paragraph introducing the simplification of bound roots appear in the margin of the manuscript, and are omitted from Marre's transcription even though they appear to be in Chuquet's hand. This kind of insertion does perhaps raise the question of whether or not the manuscript is a copy of an earlier version. It seems open to question whether or not this could have been merely a subsequent editorial addition by the author. The main text continues:

Again, of roots composed of two orders as those aforesaid, there are some which may be simplified in another way than the aforesaid, and sometimes as far as a simple root. Like $R^2\underline{15\ \bar{p}\ R^2 16}$, which can be simplified by taking the second root of 16 which is 4, added to 15 makes 19, whose R^2 is $R^2 19$. There are others which can be simplified to a number such as $R^2\underline{13 + R^2 9}$, which can be simplified thus in taking the $R^2 9$, which is 3, which added to 13 comes to 16, of which the R^2 is 4. And to this comes $R^2\underline{13\ \bar{p}\ R^2 9}$ when it is simplified.

Some numbers are composed of four parts just as those immediately preceding are composed of 2, and some are composed of three parts. Those composed of three parts cannot be simplified at all. Of those that are composed of four parts, some may be simplified by extraction of their second roots, like $R^2\underline{12\ \bar{p}\ R^2 48\ \bar{p}\ R^2 80\ \bar{p}\ R^2 60}$. The style of simplifying or extracting the second root of such numbers is this: It is necessary to find three numbers such that added together they make 12, and in such proportion that whoever divides the quarter of 48 which is 12, and the quarter of 80 which is 20 by the first of these numbers will find the other two numbers; which two numbers being found, if they are multiplied one by the other, the product should be equal to the quarter of 60, which is 15. And then take the second roots of these three numbers, and it will be done. Now let us take 4 for the first of the three numbers by which it is necessary to divide 12, the $\frac{1}{4}$ of 48, and there comes 3 for the second number. Then it is necessary again to divide 20, which is the fourth of 80, by 4 and there comes 5 for the third number. Now it is necessary to see if these three numbers have the property aforesaid. First, whoever adds 4, 3, 5 gets 12. Then, whoever multiplies 3 by 5 gets 15, which is equal to a fourth of 60. Thus are found the three numbers whose second roots are 2 and $R^2 3$ and $R^2 5$. And since the first two roots of the proposed number, that is to say, $R^2 48$ and $R^2 80$, are denoted with this term plus, for this reason $R^2 3$ and $R^2 5$ will be plus. And thus the second root of the proposed number, that is, of $R^2 12\ \bar{p}\ R^2 48\ \bar{p}\ R^2 80\ \bar{p}\ R^2 60$, is $2\ \bar{p}\ R^2 3\ \bar{p}\ R^2 5$. And so must the like be understood.

And one should know that if $R^2 48$ and $R^2 80$ had been denoted with the term minus, then the root of the number proposed above would be $2\ \bar{m}\ R^2 3\ \bar{m}\ R^2 5$. And if one had been denoted plus and the other minus, then the one should have been denoted plus and the other minus.

I wish further to extract the R^2 of $R^2 16\ \bar{m}\ R^2 96\ \bar{p}\ R^2 160\ \bar{m}\ R^2 60$. To do this we take 8 from 16, and by 8 it is necessary to divide the fourth of 96 which is 24 and there comes 3 for the second number. Then one must divide 40, which is the fourth of 160, by 8 and there comes 5 for the third number. Now, whoever adds these three numbers, he has 16. And whoever multiplies the second by the third gets 15 which is the fourth of 60. Now it is necessary to take the second roots of these three numbers and one will have $R^2 8\ \bar{m}\ R^2 3$ plus $R^2 5$ for the second root of the proposed number.

Some of the numbers mentioned above are such as cannot be simplified like $15\ \bar{p}\ R^2 10\ \bar{p}\ R^2 13\ \bar{p}\ R^2 6$. As one cannot obtain the root of such numbers, one should set them down and note them thus: $R^2 15\ \bar{p}\ R^2 10\ \bar{p}\ R^2 13\ \bar{p}\ R^2 6$. From this come the bound and compound roots of several parts.

Further, of roots composed of three or several parts, there are some which can be simplified by another way than by the aforesaid, such as $R^2 17\ \bar{p}\ R^2 16\ \bar{p}\ R^2 13\ \bar{m}\ R^2 25$ which one can simplify by taking $R^2 16$ which is 4 and adding this to 17 *and there is 21, and by taking* $R^2 25$ *which is 5, which one should subtract from 21 and 16 will remain. Thus the root of this number, when it is simplified, comes to* $R^2 \underline{16\ \bar{p}\ R^2 13}$. *There are others which can be simplified to a simple root, such as* $R^2 \underline{10\ \bar{p}\ R^2 49\ \bar{p}\ R^2 25}$ *which one can simplify by taking the roots of 49 and of 25, which, added to 10, come to 32. Thus this root when it is simplified comes to* $R^2 22$. *There are others which can be simplified to a number, as it is with* $R^2 \underline{17\ \bar{m}\ R^2 16\ \bar{p}\ R^2 9\ \bar{p}\ R^2 81}$, *which one can simplify thus by extracting* $\bar{m}\ R^2 16$, *which is* $\bar{m}$ *4, which one must take away from 17 leaving 13, then one must extract the root of* $\bar{p}$ *9 and of* $\bar{p}$ *81, which are 3 and 9, which, added with 13, make 25 of which the* R^2 *is 5, and to so much this root comes when it is simplified. There are others roots which can be simplified not by the method of extraction of roots but by adding or subtracting them, like* $R^2 \underline{13\ \bar{p}\ R^2 18\ \bar{p}\ R^2 2}$, *which may be simplified by adding* $R^2 18$ *and* $R^2 2$, *which together make* $R^2 32$, *so that the simplified root comes to* $R^2 \underline{13\ \bar{p}\ R^2 32}$. *And* $R^2 \underline{13\ \bar{m}\ R^2 20\ \bar{p}\ R^2 5}$ *which, simplified by addition of* $\bar{m}\ R^2 20$ *and* $\bar{p}\ R^2 5$, *comes to* $R^2 \underline{13\ \bar{m}\ R^2 5}$. *And* $R^2 \underline{27\ \bar{p}\ R^2 12\ \bar{m}\ R^2 3}$ *which simplifies by adding* $\bar{p}\ R^2 12$ *and* $\bar{m}\ R^2 3$, *which are together plus* $R^2 3$, *then whoever adds* $\bar{p}\ R^2 3$ *with* $R^2 27$ *he has* $R^2 48$. *And to so much this root comes when it is simplified. The manner of adding and subtracting the roots will be manifest in the chapters on addition and subtraction.*

Further, of compound numbers there are those which are compounded otherwise than the above in that the root is put before the number, as $R^2 7\ \bar{p}$ *3, of which the second root should be denoted thus* $R^2\ \underline{R^2 7\ \bar{p}\ 3}$, *which is a bound root and its* R^3 *can be put down thus:* $R^3\ \underline{R^2 7\ \bar{p}\ 3}$. *Its fourth root can also be put* $R^4\ \underline{R^2 7\ \bar{p}\ 3}$, *and thus for the others. Of such roots there are some which can be simplified to a simple root, as* $R^2\ \underline{R^2 81\ \bar{m}\ 3}$ *which, simplified by extraction of the second root of 81, which is 9, and from 9 you take away* $\bar{m}$ *3, there remains* $R^2 6$ *which is equal to* $R^2\ \underline{R^2 81\ \bar{m}\ 3}$.

Chuquet also here considers $R^3\ \underline{R^2 121\ \bar{m}\ 4}$ reducing to $R^3 7$, $R^4\ \underline{R^2 64\ \bar{m}\ 5}$ reducing to $R^4 3$, $R^5\ \underline{R^2 144\ \bar{m}\ 3}$ reducing to $R^5 9$, $R^2\ \underline{R^2 144\ \bar{m}\ 3}$ reducing to 3, $R^3\ \underline{R^2 144\ \bar{m}\ 4}$ reducing to 2, $R^4\ \underline{R^2 625\ \bar{m}\ 9}$ also reducing to 2, and $R^5\ \underline{R^2 625\ \bar{p}\ 7}$ again reducing to 2.

The third chapter. How roots may be added and put together

Before two or more roots can be added together, they must first be reduced to a like form in the case where they are unlike. And then add them together according to the rules which follow, one of which is this.

If the double of the multiplication of one number by another is added to their two squares, the root of what results is equal to the two numbers added together. Example: whoever multiplies 5 by 7, the multiplication comes to 35 whose double is 70, which it is necessary to add to 25 and 49 which are the squares of 5 and 7. They all come to 144, whose second root is 12, which is equal to 5 and 7 taken together.

By this proposition one may add several roots, both simple and compound. As, whoever should wish to add $R^2 2$ to $R^2 18$, one multiplied by the other comes to $R^2 36$, which must be doubled making $R^2 144$ which is 12, which should be added to 2 and 18 which are the squares of $R^2 2$ and $R^2 18$, making altogether $R^2 32$. And so much $R^2 2$ and $R^2 18$ come to when they are added together.

And if the two roots which one wishes to add are equal and like in plus or in minus, then one may double one of them by multiplying it by 2 times 2 which makes 4, and take the root of the product; as, whoever should add $R^2 7$ with $R^2 7$ multiplies 7 by 4, making $R^2 28$. And so much does this addition come to.

And whoever should wish to add three equal and like roots in plus or minus together, then one of those roots should be multiplied by 3 times 3. And if four roots, one must be multiplied by 4 times 4. And if five roots, it must be multiplied by 5 times 5. And so for the others.

If one of the two equal roots one wishes to add is plus and the other is minus, as $\bar{p}$ $R^2 7$ and $\bar{m}$ $R^2 7$, added together they make 0.

For explanation of the aforesaid rule, one should know that when two roots are multiplied the one by the other, if the number produced from the multiplication has no exact *root that one can simplify as far as a number, such roots cannot be added to a simple root. However, according to the aforesaid rule they can always be added to a compound and bound root. As, whoever would add $R^2 3$ to $R^2 5$. It is necessary to multiply one by the other and get $R^2 15$, which one should double, coming to $R^2 60$, whose root is R^2 $R^2 60$ which multiplied by itself comes to $R^2 60$. Also multiply $R^2 3$ and $R^2 5$ by themselves, and you will have 3 and 5 which added to $R^2 60$ makes 8 plus $R^2 60$, whose R^2 is $R^2 \underline{8\ \bar{p}\ R^2 60}$. And so much $R^2 3$ and $R^2 5$ come to, added together.*

Another style and manner of doing this

Whoever divides a number by another and adds 1 to the quotient, and then multiplies this sum by the divisor, will find the number divided and the divisor added together. Example: whoever divides 24 by 6, the quotient is 4. Add 1 to this, making 5, which multiplied by 6 comes to 30. And so much the number divided amounts to with its divisor which is 6.

By this conclusion one may add all different kinds of roots, be they second, third, fourth, or others, be they simple or compound, provided that the root of the quotient may be simplified to a number.

By way of an example, I wish to add $R^2 2$ to $R^2 18$. It is necessary first to divide 18 by 2, and it comes to $R^2 9$, which simplified is 3, to which 1 must be added, making 4. This should be reduced to a R^2, and it will be 16, which should be multiplied by $R^2 2$ which is the divisor, and one will have $R^2 32$. And so much does this addition come to.

Ten examples now follow, some of which are quite complicated. (In the Marre transcription of these there are a number of errors.) They are:

$$\sqrt[3]{6} + \sqrt[3]{48} = \sqrt[3]{162}$$
$$\sqrt[4]{7} + \sqrt[4]{567} = \sqrt[4]{1792}$$
$$\sqrt[5]{8} + \sqrt[5]{8192} = \sqrt[5]{25000}$$
$$\sqrt{(7 + \sqrt{5})} + \sqrt{(175 + \sqrt{3125})} = \sqrt{(252 + \sqrt{6480})}$$
$$\sqrt[3]{(2 + \sqrt{6})} + \sqrt[3]{(54 + \sqrt{4374})} = \sqrt[3]{(128 + \sqrt{24576})}$$
$$\sqrt[4]{(2 + \sqrt{5})} + \sqrt[4]{(164 + \sqrt{32805})} = \sqrt[4]{(512 + \sqrt{327680})}$$
$$\sqrt[5]{(3 + \sqrt{2})} + \sqrt[5]{(96 + \sqrt{2048})} = \sqrt[5]{(729 + \sqrt{13122})}$$
$$\sqrt{(\sqrt{5} + 3)} + \sqrt{(\sqrt{1280} + 48)} = \sqrt{(\sqrt{3124} + 75)}$$
$$\sqrt[3]{(\sqrt{7} - 2)} + \sqrt[3]{(\sqrt{28672} - 128)} = \sqrt[3]{(\sqrt{109375} - 250)}$$
$$\sqrt[4]{(\sqrt{7} - 12)} + \sqrt[4]{(\sqrt{45927} - 162)} = \sqrt[4]{(\sqrt{458752} - 512)}.$$

There are many roots which divided one by the other produce a quotient which cannot be simplified to a number, and in consequence they cannot be added, or at least it is a science which has not yet been discovered. For which reason it is expedient to use circumlocutions and compositions of number. As, whoever should wish to add $R^2 3$ to $R^2 7$, since the quotient of $R^2 7$ divided by $R^2 3$, which is $R^2 2\frac{1}{3}$, cannot be simplified to a number, and thus they cannot be added in the aforesaid manner, it is necessary to add them by this term plus*, and to say that $R^2 3$ and $R^2 7$ added together make $R^2 3$* plus *$R^2 7$. And just as not all roots can be added by means of this term* plus*, likewise not all roots can be subtracted save by means of this appelation* minus*. As, whoever should wish to take away $R^2 3$ from $R^2 7$ would have to say that the remainder is $R^2 7$ minus $R^2 3$. And from this there come and are produced compound numbers with plus and with minus.*

Chuquet thus begins this third chapter by stating the familiar identity $\sqrt{(a^2 + 2ab + b^2)} = a + b$ and applying it to cases where a and b are themselves roots, explaining at the same time how to take a multiplying coefficient within a root. After discussing some special cases, he points out that unless the product of the two roots to be added is a perfect square, it is not possible to simplify the sum of a simple root. He then suggests another method of tackling such additions of which the general statement is:

$$\sqrt[n]{a} \pm \sqrt[n]{b} = \left(\sqrt[n]{\frac{a}{b}} \pm 1\right)\sqrt[n]{b}.$$

In the various examples he allows a and b to be "compound", i.e. to be mixed surds, which again underlines the generality of his conception of number. He begins a further section of the chapter by introducing general rules for adding "all numbers and even compounds ones".

How simple and compound numbers with plus and minus can be added

The style and manner of adding all numbers and even compound ones is this. Place the numbers that one wishes to add thus, with their pluses and minuses in a straight line, coupling them together by the appellation plus*. And then simplify them if they can be simplified , and it will be done, but before one can indeed add such numbers not susceptible of simplification, one must know this following notable fact.*

Plus and plus, minus and minus, add. Plus and minus, subtract.

This to say that plus should be added to plus, and minus to minus. And if it is necessary to add plus to minus, the addition is done by taking the smaller number from the greater. Or otherwise, whoever adds plus with plus, there comes from it plus; and minus with minus produces minus. And whoever adds plus to minus or vice versa, if the plus exceeds the minus, it comes to plus. If it is exceeded by the minus, it comes to minus, as may appear by serveral examples which follow.

Eighteen sums now follow beginning with some trivial additions. The intention is to show that for mixed surds the permissible operation is that of adding together the rational parts of each, thus, for example, $(\surd 15 - 3) + (7 + \surd 5)$ yields $4 + \surd 5 + \surd 15$. Chuquet concludes:

The addition of all other orders of roots can be done by such manner as is aforesaid.

The fourth chapter is how roots may be subtracted one from the other

Before one root of numbers can be taken from another without circumlocution of plus or minus, it is necessary that they be like. And if they are unlike one should reduce them to one denomination, and then proceed according to the following rules, the first of which is thus.

If the double of the product of one number by another is subtracted from their two squares joined together, the root of what remains is that by which the greater of those numbers exceeds the lesser.

Example: whoever multiplies 7 by 5 gets 35, which doubled makes 70, then the square of 7 which is 49 and that of 5 which is 25, joined together make 74. From which take 70, there remains 4 of which the second root is 2. And so much remains from 7 when one has taken away 5. By this proposition one may subtract one simple or compound root from another.

I wish to subtract $R^2 2$ from $R^2 18$. To do this I multiply $R^2 2$ by $R^2 18$, making $R^2 36$. The double of this comes to $R^2 144$, *which is 12. After this I multiply $R^2 2$ and $R^2 18$, each by itself, making 2 and 18 which joined together make 20, from which I take 12, and I am left with $R^2 8$.*

Another rule

Whoever divides one number by another, and from the quotient takes away 1, the remainder multiplied by the divisor produces a number equal to the remainder of the number divided when the divisor has been subtracted from it.

Example: whoever should wish to subtract 4 from 12 divides 12 by 4, and the division comes to 3. When 1 is taken away, 2 remains, which multiplied by 4 makes 8. And so much is left of 12 when one has removed 4. By this manner one may subtract many roots, simple and compound.

I wish to subtract $R^2 3$ from $R^2 48$. To do this I divide $R^2 48$ by $R^2 3$, and from the division I have $R^2 16$ which is 4. From this I take away 1, leaving 3, which multiplied by $R^2 3$, which is the divisor, comes to $R^2 27$. And so much remains when one takes $R^2 3$ from $R^2 48$. And thus one must understand for all other roots.

If the roots were equal in number, and alike in plus or in minus, then the one subtracted from the other leaves 0. And if they were equal in number and dissimilar in plus or in minus, then one should subtract them in this way: as, whoever would take plus $R^2 7$ from minus $R^2 7$, there remains $\bar{m}\ R^2 7\ \bar{m}\ R^2 7$, which simplified is $\bar{m}\ R^2 28$; or, whoever would take $\bar{m}\ R^2 7$ from plus $R^2 7$, there would remain $R^2 7\ \bar{p}\ R^2 7$ which simplified is $R^2 28$.

In the manuscript the words "in plus" in the second sentence of this paragraph are repeated. This is a further small piece of evidence suggesting the Chuquet may be making a copy of an earlier version or, possibly, making a fair copy from notes. Repetition frequently occurs, when copying, if the writer is temporarily distracted from his work.

The style and manner of subtracting one simple or compound number from another simple or compound number is this.

Rule: Put the number from which you wish to subtract first as it is with its plus and its minus. Then after, in moving to the left, place the number you wish to subtract changing its plus to minus and its minus to plus. And then simplify if it can be simplified. But to use this rule it is necessary first to know the notable fact which follows.

Plus and plus, minus and minus, subtract,
Plus and minus, add.

Or otherwise, whoever takes plus or minus from plus, there remains plus unless the greater plus is taken from the smaller plus, when there remains minus. And whoever takes minus or plus from minus, there remains minus, unless the greater minus is taken from the lesser minus, when there remains plus. A greater plus is when a greater number marked with plus must be subtracted from a smaller number also marked with plus, as, if plus 12 were to be removed from plus 9, there would remain m 3. And the greater minus, likewise, as if minus 12 should be taken from $\bar{m}$ 9, there would remain $\bar{p}$ 3.

All these remarks here are nothing but changing the plus to minus and the minus to plus in the number that one wishes to subtract, without altering those of the number from which the subtraction is done.

Example: I wish to subtract 7 $\bar{p}$ $R^2 5$ from 12. To do this, I put down 12, and after 12 I put 7 $\bar{p}$ $R^2 5$, changing what is plus into minus and vice versa. Thus I am left with 12 m 7 $\bar{m}$ $R^2 5$, which should be simplified by adding $\bar{m}$ 7 to plus 12, making $\bar{p}$ 5. Thus there remains 5 $\bar{m}$ $R^2 5$.

Nine further examples are given. They are $12 - (7 - \sqrt{5})$; $12 - (18 + \sqrt{12})$; $12 - (18 - \sqrt{12})$; $(12 + \sqrt{5}) - (12 - \sqrt{5})$; $(\sqrt{39} + 3) - (\sqrt{7} - \sqrt{5})$; $(\sqrt{6} - \sqrt{2}) - \sqrt{3}$; $(\sqrt{44} - 2) - (\sqrt{31} - 3)$; $(\sqrt{17} + 3) - (\sqrt{15} - 3)$; $(\sqrt{17} - \sqrt[3]{5} + \sqrt[4]{7}) - (\sqrt[3]{15} + \sqrt{12} - \sqrt[5]{13})$.

Bound roots are subtracted in the aforesaid manner, changing the pluses into minuses and the minuses into pluses in the number one wishes to subtract, except for the plus and minus signs attached inside those roots, for they are not transformed at all. But the pluses and minuses which are outside are altered in the above said manner.

Whoever from 15 would subtract $R^2\underline{5\ \bar{p}\ R^2 3}$, you put down the numbers as aforesaid; there remains 15 $\bar{m}$ $R^2\underline{5\ \bar{p}\ R^2 3}$.

There are four further examples: $\sqrt{(7 + \sqrt{3})} - \sqrt{(5 - \sqrt{3})}$; $\sqrt{(7 + \sqrt{3})} - [-\sqrt{(7 + \sqrt{3})}]$ for which the result is given in the form $\sqrt{(28 + \sqrt{48})}$; $\sqrt{(7 - \sqrt{3})} - \sqrt{(7 - \sqrt{3})}$; $\sqrt[3]{(13 + \sqrt{10})} - \sqrt[4]{(\sqrt{6} - 2)}$, together with the comment:

And thus one understands of all other orders of number and of roots.

The fifth chapter which is of the multiplication of roots

If the multiplying number and the number to be multiplied are not of one nature, one should reduce them so that they are like. And then multiply one by the other and it will be done.

That is to say that if one of the numbers were a root of a number and the other were a number, then the number should be put as a root, or the root should be reduced to a number by its extraction if it can be done. Or if one is a second root and the other a third root, or other, then the roots should be reduced to one denomination and then multiplied as above. And if one of the parts is a bound root and the other a compound root which is not bound, the unbound root should be multiplied by itself and then this product should be bound, so that it may be like the other. And one should know for explanation of what is said above that when one multiplies a number by a number, the product is a number. And whoever multiplies a root by a root, it comes to a root of the same species, that is to say, third, fourth, second or other.

Example: whoever should wish to multiply $R^2 3$ *by 5 or vice versa. It is necessary first to reduce 5 to a second root, and one will have 25... Now... you will have* $R^2 75$*...*

Whoever should wish to multiply $R^2 3$ *by itself or by* $R^2 3$*, the multiplication comes to 3 or* $R^2 9$*. For all second roots multiplied by themselves produce the number whose roots they are*

Whoever should wish to multiply $R^2 5$ *by* $R^2 7$*, it is necessar to multiply 5 by 7 coming to* $R^2 35$*.*

Whoever should wish to multiply $R^3 5$ *by 3, it is necessary first to multiply and reduce 3 to a third root and one will have* $R^3 27$*, which multiplied by* $R^3 5$ *comes to* $R^3 135$*. Thus whoever multiplies* R^3 *by itself or by* $R^3 5$*, it comes to* $R^3 25$*. And* $R^3 7$ *multiplied by* $R^3 10$ *comes to* $R^3 70$*.*

The following further examples are given: $\sqrt[4]{} \times 5$; $\sqrt[4]{7} \times \sqrt[4]{7}$; $\sqrt[4]{5} \times \sqrt[4]{7}$; $\sqrt[5]{8} \times 2$; $\sqrt[5]{8} \times \sqrt[5]{8}$; $\sqrt[5]{7} \times \sqrt[5]{3}$; $\sqrt[6]{7} \times 2$; $\sqrt[6]{10} \times \sqrt[6]{10}$; $\sqrt[6]{7} \times \sqrt[6]{5}$.

For explanation of the aforesaid matters one should know that any root multiplied by itself or as required by the root, produces the number whose root it is. Thus whoever multiplies $R^2 13$ by itself, the multiplication comes to 13. Also, whoever multiplies $R^3 13$ thrice, that is to say by itself, and then multiplies that which ensues again by it, the multiplication comes to 13. And $R^4 13$ multiplied four times, that is to say first multiplied by itself comes to $R^2 13$. And then $R^2 13$ multiplied by itself comes to 13. And $R^5 13$ multiplied five times comes to 13, and thus with the other roots. One should also know that a fourth root multiplied by itself once comes to R^2 as is said before of $R^4 13$ which, multiplied by itself, comes to $R^2 13$. And $R^6 13$ multiplied by itself comes to $R^3 13$. Or, multiplied thrice comes to $R^2 13$. And $R^8 13$ multiplied by itself comes to $R^4 13$. And, generally, all roots when denominations are even are ruled by this ordinance, that is to say that when the denomination of the root is divided by two, the quotient is the denomination of the root when it multiplied once by itself. Without altering the number of the root, the denomination may be altered, like $R^{10} 8$ which, when it is multiplied by itself once, comes to $R^5 8$, for whoever divides 10 by 2, the division comes to 5 for the denomination of the root. Roots whose denominations are odd are outside this consideration, for the number of the root should be multiplied and the denomination should not be altered at all. As, $R^3 12$ multiplied by itself amounts to $R^3 144$. And $R^5 9$ multiplied by itself comes to $R^5 81$. And so for the others. And if the roots one wishes to multiply be unlike, that is to say the one were a second root and the other a third root, one should reduce them to one denomination and then do as above.

Chuquet, like his contemporaries, does not regard roots as fractional powers, even through such powers can be found in the works of Oresme written around 1360. Chuquet was certainly aware of some of Oresme's writings, but there is no evidence that he had encountered the fractional powers, which were confined to the *De proportionibus proportionum* and the *Algorismus proportionum*.

Multiplication of compound numbers

Further, to multiply compound numbers it is first of all necessary to know the following noteworthy fact, which is this:

Whoever multiplies plus by plus and minus by minus, it comes to plus. And whoever multiplies plus by minus or vice versa, it comes always to minus

Example: whoever should wish to multiply $R^2 5\ \bar{p}\ 3$ by 6, it is first necessary to multiply $R^2 5$ by 6 times 6 which comes to $R^2 180$, then one should multiply plus 3 by 6, making plus 18. Thus the multiplication comes to $R^2 180$ plus 18.

A number of worked examples now follow. These include the following:

Whoever should wish to multiply $R^2 8\ \bar{m}\ R^2 3$ by $R^2 5$, it is necessary first to multiply $R^2 8$ by $R^2 5$ coming to $R^2 40$, then one should multiply $\bar{m}\ R^2 3$ by $R^2 5$ coming to $\bar{m}\ R^2 15$. Thus this multiplication comes to $R^2 40\ \bar{m}\ R^2 15$.

Whoever should wish to multiply $R^2 3\ \bar{p}\ R^2 7$ by $R^2 5\ \bar{p}\ 2$, it is necessary first to multiply $R^2 3\ \bar{p}\ R^2 7$ by $R^2 5$ coming to $R^2 15$ plus $R^2 35$. Afterwards, one should multiply $R^2 3\ \bar{p}\ R^2 7$ by plus 2 coming to $R^2 12\ \bar{p}\ R^2 28$. Thus this multiplication comes in all to $R^2 15\ \bar{p}\ R^2 35\ \bar{p}\ R^2 12\ \bar{p}\ R^2 28$.

Whoever should wish to multiply $R^2 7\ \bar{p}\ R^2 2$ by $R^2 7\ \bar{m}\ R^2 2$, it is necessary first to multiply $R^2 7\ \bar{p}\ R^2 2$ by $R^2 7$ coming to 7 plus $R^2 14$. Afterwards, one should multiply $R^2 7\ \bar{p}\ R^2 2$ by $\bar{m}\ R^2 2$ coming to $\bar{m}\ R^2 14\ \bar{m}\ 2$. Which you simplify by adding $\bar{p}\ R^2 14$ to $\bar{m}\ R^2 14$, coming to 0, then $\bar{m}\ 2$ with $\bar{p}\ 7$, coming to plus 5. And this multiplication comes to 5.

Whoever should wish to multiply $R^2 192\ \bar{m}\ R^2 48\ \bar{p}\ R^2 39$ by $R^2 48\ \bar{m}\ R^2 39$. It is necessary first to multiply $R^2 192\ \bar{m}\ R^2 48\ \bar{m}\ R^2 39$ by plus $R^2 48$ and one will have $R^2 9216\ \bar{m}\ 48\ \bar{p}\ R^2 1872$. And then to multiply by $\bar{m}\ R^2 39$, and one will find $\bar{m}\ R^2 7488\ \bar{p}\ R^2 1872\ \bar{m}\ 39$. Thus this multiplication comes to $R^2 9216\ \bar{m}\ 48\ \bar{p}\ R^2 1872\ \bar{m}\ R^2 7488\ \bar{p}\ R^2 1872\ \bar{m}\ 39$. Which simplified comes to 9. And this is what this multiplication comes to. The manner of simplifying this multiplication is that one should first add $\bar{m}\ 48$ and $\bar{m}\ 39$, coming to $\bar{m}\ 87$. Then add *plus $R^2 1872$ which plus $R^2 1872$ coming to $\bar{p}$ $R^2 7488$ which one must add with $\bar{m}$ $R^2 7488$, and it comes to 0 for those which are equal and the one plus and the other minus. Afterwards, extract the R^2 of 9216 which is $\bar{p}$ 96. Now add plus 96 with $\bar{m}$ 87 and you will find 9. And thus should one simplify the like when they can be simplified. And similarly first as it is said to second roots one must understand of third roots and other orders of roots.*

Other examples – not given here in full – are: $(7 - \sqrt{5}) \times 4$; $(\sqrt{7} + \sqrt{5}) \times 3$; $(\sqrt{7} - \sqrt{2}) \times (\sqrt{5} + \sqrt{3})$; $(\sqrt{7} + \sqrt{2})^2$; $(\sqrt{13 + 7 - \sqrt{6}}) \times (\sqrt{6} - \sqrt{3})$. The last of the examples of unbound compound roots is:

Again, whoever should wish to multiply $R^2 3$ $\bar{p}$ $R^3 4$ by $R^4 2 - R^5 7$, it is necessary first to multiply $R^2 3$ $\bar{p}$ $R^3 4$ by $R^4 2$ coming to $R^4 18$ plus $R^{12} 2048$. After one must further multiply $R^2 3$ $\bar{p}$ $R^3 4$ by $\bar{m}$ $R^5 7$, and it comes to $\bar{m}$ $R^{10} 11907$ $\bar{m}$ $R^{15} 351232$. Thus this multiplication comes to $R^4 18$ plus $R^{12} 2048$ $\bar{m}$ $R^{10} 11907$ $\bar{m}$ $R^{15} 351232$.

Bound roots may be multiplied in the following manner: as whoever should wish to multiply $R^2 5$ $\bar{p}$ $R^2 3$ by 6, it is necessary first to multiply $R^2 5$ by 6 times 6 coming to $R^2 180$. Afterwards, $\bar{p}$ $R^2 3$ must be multiplied by 36 times 36, and one will have $R^2 3888$. Thus this multiplication comes to $R^2 180$ $\bar{p}$ $R^2 3888$.

Twenty-one worked examples now follow amongst which the following are included:

Whoever should wish to multiply $R^2 5$ $\bar{m}$ $R^2 3$ by $R^3 7$. It is necessary first to reduce $R^2 5$ $\bar{m}$ $R^2 3$ to a third root by multiplying it initially by itself. It comes to $R^2 28$ $\bar{m}$ $R^2 300$ which it is necessary further to multiply by $R^2 5$ m $R^2 3$. It comes to $R^2 170$ $\bar{m}$ $R^2 7500$ $\bar{m}$ $R^2 2352$, and since this is a third root of a second root it is therefore $R^6 170$ $\bar{m}$ $R^2 7500$ $\bar{m}$ $R^2 2352$. Afterwards, $R^3 7$ must be reduced to a second root and thus this will be a second root of a third root which is $R^6 49$, which one must then multiply by $R^6 170$ $\bar{m}$ $R^2 7500$ $\bar{m}$ $R^2 2352$. And one will find that this multiplication comes to $R^6 8330$ $\bar{m}$ $R^2 18007500$ $\bar{m}$ R 5647152...

Whoever should wish to multiply $R^3 7$ $\bar{m}$ $R^2 5$ by 4 $\bar{p}$ $R^2 6$. It is necessary first to reduce 4 $\bar{p}$ $R^2 6$ to a third root by multiplying it three times, that is to say by itself, coming to 22 $\bar{p}$ $R^2 384$ which one should further multiply by 4 plus $R^2 6$, coming to 136 $\bar{p}$ $R^2 6144$ $\bar{p}$ $R^2 2904$, of which the third root is R^3 136 $\bar{p}$ $R^2 6144$ plus $R^2 2904$, which simplified and adding $R^2 6144$ with $R^2 2904$ comes to $R^2 17496$; thus it comes to R^3 136 $\bar{p}$ $R^2 17496$, which one should now multiply by $R^3 7$ $\bar{m}$ $R^2 5$ in this manner. Multiply first $R^3 136$ by $R^3 7$ coming to $R^3 952$. Then reduce 7 to a second root which is $R^2 49$, which it is necessary to multiply by plus $R^2 17496$, coming to $R^2 857304$. Afterwards 136 must be reduced to a R^2 which is $R^2 18496$, which it is necessary to multiply by $\bar{m}$ $R^2 5$, coming to $\bar{m}$ $R^2 92480$. Then it is necessary to multiply $\bar{m}$ $R^2 5$ by plus $R^2 17496$, coming to $\bar{m}$ $R^2 87480$. Thus this multiplication comes to $R^2 952$ $\bar{p}$ $R^2 857304$ $\bar{m}$ $R^2 92480$ $\bar{m}$ $R^2 87480$ which one should simplify if it can be simplified.

Whoever should wish to multiply $R^4 7$ $\bar{p}$ $R^2 5$ by $R^3 2$. It is necessary first to reduce $R^3 2$ to a fourth root, which is $R^{12} 16$. Then after it is necessary to reduce $R^4 7$ $\bar{p}$ $R^2 5$ to a third root by multiplying it first by itself, coming to $R^4 54$ $\bar{p}$ $R^2 980$, which one must further multiply by $R^4 7$ $\bar{p}$ $R^2 5$, coming to $R^4 378$ $\bar{p}$ $R^2 48020$ $\bar{p}$ $R^2 14580$ $\bar{p}$ $R^2 4900$, which one should simplify by extraction of the second root of 4900 which is 70 which one must add to 378, and one will have 448.

Then it is necessary to add $R^2 48020$ to $R^2 14580$ and one will find $R^2 115520$. Thus this number simplified comes to $R^4 448$ $\bar{p}$ $R^2 115520$. And since it is a fourth root which is reduced to a third root, it is then R^3 of R^4 which is $R^{12} 448$ $\bar{p}$ $R^2 115520$, which it is necessary to multiply by $R^{12} 16$, and one will find that this multiplication comes to $R^{12} 7168$ $\bar{p}$ $R^2 29573120$.

The other worked examples are: $(\sqrt{5} - \sqrt{3}) \times \sqrt{7}$; $(\sqrt{5} + \sqrt{7})^2$; $(\sqrt{7} + \sqrt{3}) \times (\sqrt{7} - \sqrt{3})$; $(\sqrt{5} + \sqrt{8}) \times (\sqrt{5} - \sqrt{8})$; $(\sqrt{7} + \sqrt{3}) \times (\sqrt{5} + \sqrt{2})$; $(\sqrt{5} + \sqrt{7}) \times (2 + \sqrt{3})$; $(\sqrt{7} + \sqrt{5}) \times (\sqrt{6} - \sqrt{2})$; $(\sqrt[3]{7} + \sqrt{5}) \times 3$; $\sqrt[3]{(7 - \sqrt{5})} \times \sqrt[3]{4}$; $\sqrt[3]{(7 - \sqrt{5})} \times \sqrt[3]{4}$; $\sqrt[3]{(7 + \sqrt{5})} \times \sqrt[3]{(7 - \sqrt{5})}$; $[\sqrt[3]{(7 + \sqrt{5})}]^2$; $\sqrt[4]{(7 + \sqrt{5})} \times 2$; $\sqrt[4]{(7 - \sqrt{3})} \times \sqrt{5}$; $\sqrt[4]{(7 - \sqrt{5})} \times \sqrt[4]{3}$; $\sqrt[4]{(7 - \sqrt{5})} \times (2 + \sqrt{3})$; $[\sqrt[4]{(7 + \sqrt{5})}]^2$; $\sqrt[4]{(7 + \sqrt{5})} \times \sqrt[4]{(7 - \sqrt{5})}$. Chuquet adds:

And thus it is necessary to understand fifth, sixth, and other roots. In every case, whoever multiplies a sixth root by itself it comes to a R^3 without change of number. As $R^6 7$ $\bar{p}$ $R^2 5$ multiplied by itself comes to $R^3 7$ $\bar{p}$ $R^2 5$. And an eighth root multiplied by itself comes to a fourth root without changing the number itself. And a tenth root, which in the same manner comes to a fifth root. And a twelfth root to a sixth root. And likewise for other roots whose denomination is even. It must be understood that the other roots, whose denominations are odd, are excepted from this rule, for when they are multiplied once by themselves the denomination does not vary at all but the number is multiplied.

Other bound roots can be treated in the manner which follows. As whoever should wish to multiply $R^2\ \underline{R^2 5\ \bar{p}\ 7}$ by 3. It is necessary first to reduce 3 to a R^4 so that it will be of the nature of $R^2\ R^2 5$, and it comes to $R^4 81$, which must be multiplied by $R^2 5$, coming to $R^2 405$. Then one must multiply plus 7 by 3, reduced to a second root which is $R^2 9$, which multiplied by 7 comes to $\bar{p}\ R^2 63$. Thus their multiplication comes to $R^2\ \underline{R^2 405\ \bar{p}\ 63}$.

Eight further worked examples follow, including the following:

Whoever would multiply $R^2\ \underline{R^2 7\ \bar{m}\ 2}$ by 3 $\bar{p}$ $R^2 5$. It is necessary first to multiply and reduce 3 $\bar{p}$ $R^2 5$ to a bound root of the nature of $R^2\ \underline{R^2 7\ m\ 2}$ by multiplying it by itslef; it comes to 14 $\bar{p}$ $R^2 180$ of which the second root is $R^2\ \underline{14\ \bar{p}\ R^2 180}$ which must be rearranged and thus one will have $R^2\ \underline{R^2 180\ \bar{p}\ 14}$, which one should multiply by $R^2\ \underline{R^2 7\ \bar{m}\ 2}$ in this manner. First, it is necessary to multiply $R^2 180$ by $R^2 7$, coming to $R^2 1260$. Then $R^2 7$ must be multiplied by plus 14 reduced to a second root, which is $R^2 196$, coming to $R^2 1372$. Afterwards $R^2 180$ must be multiplied by $\bar{m}$ 2 reduced to a second root, which is $R^2 4$, coming to $R^2 720$. Then after, plus 14 must be multiplied by $\bar{m}$ 2 coming to $\bar{m}$ 28. Thus this multiplication comes to $R^2\ \underline{R^2 1260\ \bar{p}\ R^2 1372\ \bar{m}\ R^2 720\ \bar{m}\ 28}$, which is to be understood as the second root of 1260, of 1372 and of 720 added together with $\bar{m}$ 28. And then further to take the second root of all of it; this is what the multiplication comes to.

(Here Chuquet appears to have overlooked the negative attached to $R^2 720$.)

Whoever should wish to multiply $R^3\ \underline{R^2 7\ \bar{m}\ 2}$ by $R^2 5$. It is necessary first to reduce $R^2 5$ to a third root and one will have $R^6 125$. Then $R^3\ \underline{R^2 7\ \bar{m}\ 2}$ must be reduced to a second root by multiplying it by itself in this manner. $R^2 7$ by $R^2 7$ comes to $R^2 49$, then $R^2 7$ by $\bar{m}$ 2 comes to $\bar{m}$ $R^2 28$, and further, $\bar{m}$ 2 by *plus* $R^2 7$ *comes to* $\bar{m}\ R^2 28$, *then after, it is necessary to multiply* $\bar{m}\ 2$ *by* $\bar{m}\ 2$, *coming to* $\bar{p}\ 4$.

Thus it comes to $R^3\ R^2 49\ \bar{m}\ R^2 28\ \bar{m}\ R^2 28\ \bar{p}\ 4$ *which simplifies by adding* $\bar{m}\ R^2 28$ *with* $\bar{m}\ R^2 28$, *coming to* $\bar{m}\ R^2 112$, *and so this calculation comes to* $R^3\ R^2 49\ \bar{m}\ R^2 112\ \bar{p}\ 4$.

Whoever should wish to multiply $R^3\ R^2 7\ \bar{m}\ 2$ *by* $R^3\ R^2 7\ \bar{m}\ 2$. *It is necessary to multiply* $R^2 7$ *by* $R^2 7$ *coming to* $R^2 39$, *then after* $R^2 7$ *must be multiplied by* $\bar{m}\ 2$ *reduced to a second root which is* $R^2 4$, *coming to* $\bar{m}\ R^2 28$. *Then* $\bar{m}\ 2$ *must be further multiplied by* $R^2 7$ *coming to* $\bar{m}\ R^2 28$: *Afterwards, multiply* $\bar{m}\ 2$ *by* $\bar{m}\ 2$ *coming to* $\bar{p}\ 4$. *Thus this multiplication comes to* $R^3\ R^2 49\ \bar{m}\ R^2 28\ \bar{m}\ R^2 28\ \bar{p}\ 4$ *which, simplified by addition of* $\bar{m}\ R^2 28$ *with* $\bar{m}^2 28$, *comes to* $R^3\ R^2 49\ \bar{m}\ R^2 112$ *plus 4. Alternately one can do this multiplication: multiply* $R^2 7$ *by* $R^2 7$ *coming to 7. Then whoever multiplies* $R^2 7$ *by* $\bar{m}\ 2$ *and then* $\bar{m}\ 2$ *by* $R^2 7$ *and adds everything together, it comes to* $\bar{m}\ R^2 112$. *Afterwards, whoever multiplies* $\bar{m}\ 2$ *by* $\bar{m}\ 2$, *it comes to* $\bar{p}\ 4$ *which, added with 7, is plus 11. Thus this multiplication comes to* $R^3\ R^2 112\ \bar{p}\ 11$. *And thus other differences of root should be understood.*

The other five worked examples are: $\sqrt{(\sqrt{7} - 2)} \times \sqrt{3}$; $[\sqrt{(\sqrt{7} - 2)}]^2$; $\sqrt[3]{(\sqrt[3]{7} - 2)} \times 4$; $\sqrt[3]{(\sqrt{7} - 2)} \times \sqrt[3]{5}$. Chuquet next discusses how to determine which of two bound roots is the greater.

There are two rules to know which of two numbers compounded in the same way is the greater or the lesser

Whoever multiplies two numbers one by the other, if this product is equal to one of them multiplied by itself, these two numbers are equal necessarily.

Whoever multiplies two or more numbers, each by itself, if the products are equal, the numbers that one has multiplied are equal. If they are unequal, the numbers are unequal.

Example: whoever should wish to know if $R^2 8\ \bar{p}\ R^2 7$ *is equal to* $R^2 20\ \bar{p}\ 1$, *and if they are unequal which is the greater. To do this multiply* $R^2 8\ \bar{p}\ R^2 7$ *by itself making 15 plus* $R^2 224$. *And then multiply* $R^2 20\ \bar{p}\ 1$ *by itself, which comes to 21* $\bar{p}\ R^2 80$. *Now, whoever consideres well these two multiplications will find that 15* $\bar{p}\ R^2 224$ *is nearer to 30 than is 21* plus $R^2 80$, inasmuch as $R^2 224$ is closer to 15 than 80 is to 9. Thus $R^2 7\ p\ R^2 8$ is the greater.

(Chuquet has here omitted R^2 before 80.)

Also, whoever multiplies $R^2\underline{8\ \bar{p}\ R^2 60}$ by itself, it comes to $8\ \bar{p}\ R^2 60$. And whoever multiplies $R^2 5\ \bar{p}\ R^2 3$ by itself, it comes to $8\ \bar{p}\ R^2 60$, by which it appears that $R^2\underline{8\ p\ R^2 60}$ and $R^2 5$ plus $R^2 3$ are equal.

Likewise, whoever should wish to know which of $2\ \bar{m}\ R^2 2$ and $R^2\underline{2\ \bar{m}\ R^2 2}$ is the greater, if the one and the other of these two numbers are each multiplied several times by themselves, when the multiplications are skillfully considered, one will find that $R^2\underline{2\ \bar{m}\ R^2 2}$ is greater than $2\ \bar{m}\ R^2 2$. This is a great marvel, for at first sight the opposite seems to be true.

The sixth and last chapter, which is of the division of roots

If the divisor and the number to be divided are not of one nature, one should reduce them, and then divide one by the other, and it will be done. And one should know that the quotient is always of the nature of the number divided and of the divisor too, that is, if the number divided and the divisor are numbers, the quotient will be a number, and if they are second, third, fourth or other roots, the quotient will likewise be a root of a number of the same species as the number divided and the divisor.

Example: whoever should wish to divide $R^2 12$ by 2, it is necessary first to reduce 2 to a second root, which is $R^2 4$. Now divide 12 by 4; you will have $R^2 3$ and that is what the division comes to.

Whoever should wish to divide 5 by $R^2 12$, it is necessary to divide 5, reduced to a second which is 25 by 12, and one will have $R^2 2\frac{1}{12}$...

Whoever should wish to divide $R^2 20$ by $R^2 5$, 20 must be divided by 5; it comes by the division to $R^2 4$, which is 2.

Three similar worked examples now follow for cube roots and a further three for fourth roots. There is one fifth root example and then the following three where the divisor and the dividend are both roots but of different order.

Whoever should wish to divide $R^2 12$ by $R^3 16$, it is necessary to reduce $R^2 12$ to R^3, which is $R^6 1728$; then afterwards $R^3 16$ must be reduced to a second root, which is $R^6 256$. Now divide 1728 by 256, you will have $R^6 6\frac{3}{4}$...

Whoever should wish to divide $R^4 96$ by $R^2 3$; it is necessary to reduce $R^2 3$ to a fourth root which is $R^4 9$. Now divide 96 by 9, you will have $R^4 10\frac{2}{3}$ for the quotient.

Whoever should wish to divide $R^3 7$ by $R^4 5$, it is necessary first to reduce $R^3 7$ to a fourth root, which is $R^{12} 2401$. Then $R^4 5$ must be reduced to R^3, which is $R^{12} 125$. Now divide 2401 by 125, you will find from the division $R^{12} 19\frac{26}{125}$. And thus must the other simple roots be understood. But to come to compound roots, it is first necessary to know what follows.

Whoever divides plus by plus and minus by minus, it comes to plus. And whoever divides plus by minus or minus by plus, it comes to minus.

Example: whoever should wish to divide $R^2 245 \bar{p} 21$ by 7, it is necessary to reduce 7 to a second root which is $R^2 49$. Now divide 245 by 49. You will have $R^2 5$. Then divide plus 21 by 7 and you will find 3. Thus the division comes to $R^2 5$ plus 3.

Next, $\sqrt{637} - 14$ is divided by 7, $48 + \sqrt{320}$ by 8, $84 - \sqrt{245}$ by 7, $\sqrt{108} + \sqrt{21}$ by $\sqrt{3}$, and $\sqrt{108} - \sqrt{21}$ by $\sqrt{3}$. There follows:

Whoever should wish to divide $R^2 108 \bar{p} R^2 21$ by $6 \bar{p} R^2 7$: it is necessary for doing such problems and the like to simplify its divisor and reduce it to a non-compound number in this way. The divisor must be multiplied by a number which is to be equal to it in number and unlike in plus or minus. As for example, with the above divisor which is $6 \bar{p} R^2 7$, its equal and unlike is $6 \bar{m} R^2 7$. And by such number by which the divisor is multiplied, the number to be divided must be multiplied by the same. And in this manner one will have a simple divisor, where it was compound before, and by this one should divide the number to be divided in the form and manner aforesaid, and it will be done.

Now, therefore, let $6\ \bar{p}\ R^2 7$ be multiplied by $6\ \bar{m}\ R^2 7$, and one will find that the multiplication comes to 29 for divisor. Then let $R^2 108\ \bar{p}\ R^2 21$ be multiplied by $6\ \bar{m}\ R^2 7$. The multiplication comes to $R^2 3888\ \bar{m}\ R^2 147$. Now, whoever divides $R^2 3888\ \bar{m}\ R^2 147$ by 29 as shown above will find from the division $R^2 4\frac{524}{841}\ \bar{m}\ R^2\ \frac{147}{841}$. Simplified, this comes to $R^2 3$, And thus whoever divides $R^2 108\ \bar{p}\ R^2 21$ by $6\ \bar{p}\ R^2 7$, the division comes to $R^2 3$.

Or otherwise, before one divides $R^2 3888\ \bar{m}\ R^2 147$, one can simplify them by adding $\bar{m}\ R^2 147$ with $\bar{p}\ R^2 3888$, and one will have $R^2 2523$. Now, whoever divides $R^2 2523$ by 29 times 29 will find $R^2 3$, as above.

Another way of doing it. Divide $R^2 108$ by 6 times 6 and you will find $R^2 3$. Divide also $R^2 21$ by $R^2 7$, and you will likewise have $R^2 3$. Now take whichever quotient you like, and you will have $R^2 3$ as above.

Whoever divides also $R^2 108\ \bar{m}\ R^2 21$ by $6\ \bar{m}\ R^2 7$, it is necessary as above to simplify the divisor and multiply the number to be divided also as abovesaid, and one will have 29 for divisor and $R^2 3888\ \bar{m}\ R^2 147$, which simplified is $R^2 2523$, for the number to be divided. Now divide $R^2 2523$ by 29 times 29; you will have $R^2 3$, and the division comes to such.

$(\sqrt{108} + \sqrt{21}) \div (6 - \sqrt{7})$ is discussed similarly.

Whoever should wish to divide $6\ \bar{p}\ R^2 7$ by $6\ \bar{p}\ R^2 7$ should simplify the divisor and multiply the number to be divided in the aforesaid manner, and one will find 29 for the divisor and 29 for the number to be divided. Now divide 29 by 29, and you will find 1. And so much comes for quotient.

Or otherwise, it is necessary to know that whoever divides a number by another which is equal to it and like, it always comes to 1 by the division. And therefore whoever divides $6\ \bar{m}\ R^2 7$ by $6\ \bar{m}\ R^2 7$, he finds 1 by the division.

Whoever should wish to divide $R^2 6\ \bar{p}\ R^2 7$ by $R\ 5\ \bar{m}\ R^2 3$. It is necessary first to simplify the division and do according to the manner aforesaid and one will find 2 for divisor and $R^2 30\ \bar{p}\ R^2 35\ \bar{p}\ R^2 18\ \bar{p}\ R^2 21$ for the number to be divided...

(In the manuscript this last number has been corrected

– seemingly in another hand – from $R^2 150\ \bar{p}\ R^2 175\ \bar{p}\ R^2 18\ \bar{p}\ R^2 21$.)

Now divide everything by 2 times 2 and you will find for the division $R^2 7\frac{1}{2}\ \bar{p}\ R^2 8\frac{3}{4}\ \bar{p}\ R^2 4\frac{1}{2}\ \bar{p}\ R^2 5\frac{1}{4}$.

Three examples now follow, namely: $(\sqrt{13} - \sqrt{7} + \sqrt{6}) \div (\sqrt{5} - \sqrt{2})$; $(\sqrt{65} - \sqrt{35} - \sqrt{39} + \sqrt{21}) \div (\sqrt{5} - \sqrt{3})$; $(\sqrt{117} - \sqrt{63} + \sqrt{54}) \div (\sqrt{13} - \sqrt{7} + \sqrt{6})$; then:

Whoever should wish to divide $R^2 117\ \bar{m}\ R^2 63\ \bar{p}\ R^2 54\ \bar{p}\ R^2 65\ \bar{m}\ R^2 35\ \bar{p}\ R^2 30$ *by* $R^2 13\ \bar{m}\ R^2 7 + R^2 6$. *It is necessary first to simplify the divisor and to multiply the number to be divided by the aforesaid manner and one will find* $R^2 168$ *for divisor and for the number to be divided* $R^2 702\ \bar{p}\ R^2 845\ \bar{m}\ R^2 245\ \bar{p}$. $R^2 1512\ \bar{p}\ R^2 840\ \bar{m}\ R^2 180$. *Which it is necessary to divide by* $R^2 168$, *and one will find by the division* $R^2 4\frac{30}{168}\ \bar{p}\ R^2 5\frac{5}{168}\ \bar{m}\ R^2 1\frac{77}{168}\ \bar{p}\ 3\ \bar{p}\ R^2 5\ \bar{m}\ R^2 1\frac{12}{168}$. *Which simplified must be equal to* $3\ \bar{p}\ R^2 5$.

Chuquet has carefully chosen his numbers so as to ensure simplification of the divisor, which is in the form $(\sqrt{a+b} - \sqrt{a} + \sqrt{b})$, and when multiplied by $(\sqrt{a+b} + \sqrt{a} - \sqrt{b})$ reduces to $\sqrt{4ab}$. He also suggests an alternative method, which further reveals his choice of numbers:

...divide $R^2 117$ *by* $R^2 13$ *and* $R^2 63$ *by* $R^2 7$ *and* $R^2 54$ *by* $R^2 6$ *and you will find for each* $R^2 9$ *which is 3. Afterwards, divide the other three orders, that is to say,* $R^2 65$ *by* $R^2 13$, *and* $R^2 35$ *by* $R^2 7$ *and* $R^2 30$ *by* $R^2 6$ *and you will find plus* $R^2 5$ *for each division.*

There are two further worked examples: $(12 + \sqrt[3]{320}) \div 4$ and $(\sqrt[3]{12} - \sqrt[3]{5}) \div \sqrt[3]{4}$.

Bound roots can be divided in the manner given above. As, for example, whoever should wish to divide $R^2 \underline{180\ \bar{p}\ R^2 3888}$ *by 6, it is necessary first to reduce 6 to a second root, which is 36. Now divide 180 by 36 and you will find* $R^2 5$. *Further, it is necessary again to reduce 36 to a second root, which is 1296. Now divide* $\bar{p}\ R^2 3888$ *by* $\bar{p}$ *1296, you will have* $\bar{p}\ R^2 3$ *joined with* $R^2 5$; *you will have* $R^2 \underline{5\ \bar{p}\ R^2 3}$. *And the division comes to so much.*

Twelve further worked examples are given, including the following:

Whoever should wish to divide $R^3\underline{86\ \bar{p}\ R^2 12}$ by $R^3\underline{12\ -\ R^2 3}$.

To do this, it is necessary to simplify the divisor by multiplying it by $R^3\underline{12\ \bar{p}\ R^2 3}$, coming to $R^3 141$ for divisor. Then the number to be divided must be multiplied by $R^3\underline{12\ \bar{p}\ R^2 3}$ coming, simplified, to $R^3\underline{1038\ \bar{p}\ R^2 36300}$, which it is necessary to divide by $R^3 141$ and one will find $R^3 7\frac{17}{47}\ \bar{p}\ R^2 1\frac{5473}{6627}$.

Whoever should wish to divide $R^4\underline{35\ \bar{p}\ R^2 10}$ by $R^4\underline{5\ \bar{m}\ R^2 3}$, it is necessary to simplify the divisor by multiplying it by $R^4\underline{5\ \bar{p}\ R^2 3}$; the divisor comes to $R^4 22$. Then the number to be divided must be multiplied by $R^4\underline{5\ \bar{p}\ R^2 3}$; the multiplication comes to $R^4\underline{175\ \bar{p}\ R^2 250\ \bar{p}\ R^2 3675\ \bar{p}\ R^2 30}$, which it is necessary to divide by $R^4 22$ and one will find $R^4 7\frac{21}{22}\ \bar{p}\ R^2\frac{125}{242}\ \bar{p}\ R^2 7\frac{287}{484}\ \bar{p}\ R^2\frac{15}{242}$...

Afterwards, whoever should wish to divide $R^2\ \underline{R^2 48\ \bar{p}\ 12}$ by 2, it is necessary first to reduce 2 to a fourth root, which is 16. Then divide 48 by 16, coming to $R^2 3$. Afterwards divide plus 2 by 2 times 2, you will have plus 3, which must join with $R^2\ R^2 3$; you will have $R^2\ \underline{R^2 3\ p\ 3}$...

Whoever should wish to divide $R^2\ \underline{R^2 48\ \bar{m}\ 2}$ by $R^2 3\ \bar{p}\ 2$, it is necessary first to reduce $R^2 3\ \bar{p}\ 2$ to a bound root of the same kind as the number to be divided by multiplying it by itself, coming to $7\ \bar{p}\ R^2 48$ of which the second root is $R^2 7\ \bar{p}\ R^2 48$, which rearranged is $R^2\ \underline{R^2 48\ \bar{p}\ 7}$.

So divide now $R^2\ \underline{R^2 48\ \bar{m}\ 2}$ by $R^2\ \underline{R^2 48\ \bar{p}\ 7}$ in the manner aforesaid by simplifying the divisor and multiplying the number to be divided in the aforesaid manner, and one will find by the division $R^2\ \bar{m}\ \underline{R^2 2304\ \bar{p}\ R^2 192\ \bar{p}\ R^2 2352\ \bar{m}\ 14}$, which abbreviated comes to $R^2\ \underline{R^2 192\ \bar{p}\ R^2 2352\ \bar{m}\ 62}$.

Whoever should wish to divide $R^3\ \underline{R^2 96\ \bar{m}\ 5}$ by $R^3\ \underline{R^2 13\ \bar{m}\ 2}$, it is necessary to simplify the divisor by multiplying it by $R^3\ \underline{R^2 13\ \bar{p}\ 2}$ and the number to be divided equally. The divisor comes to $R^3 9$ and the number to be divides comes to $R^3\ \underline{R^2 1248\ \bar{m}\ R^2 325\ \bar{p}\ R^2 384\ \bar{m}\ 10}$. Now perform this division, you will find for quotient $R^3\ \underline{R^2 15\frac{41}{27}\ \bar{m}\ R^2 4\frac{1}{18}\ \bar{p}\ R^2 4\frac{20}{27}\ \bar{m}\ 1\frac{1}{9}}$.

The remaining worked examples are: $\sqrt{(35 - \sqrt{147})} \div \sqrt{7}$; $\sqrt{(5 + \sqrt{7})} \div \sqrt{(5 + \sqrt{7})}$; $(5 + \sqrt{7}) \div \sqrt{(5 + \sqrt{7})}$; $\sqrt{(35 + \sqrt{75} + \sqrt{98} + \sqrt{6})} \div \sqrt{(5 + \sqrt{2})}$; $\sqrt{(\sqrt{48} + 12)} \div \sqrt{6}$; $\sqrt{(\sqrt{48} - 2)} \div \sqrt{(\sqrt{3} + 2)}$; $\sqrt[4]{(\sqrt{96} - 5)} \div \sqrt[4]{(\sqrt{13} - 2)}$. Chuquet concludes with the familiar phrase:

And thus it should be understood of the others.

Chapter 5

THE TRIPARTY - THIRD PART

It is in the third part of the *Triparty* that Chuquet presents his algebra, which he calls *"la rigle des premiers"*. Strictly, this translates as "the rule of firsts", the word "first" designating an unknown quantity to the power 1, written conventionally today as x. However, in order to distinguish clearly between "first numbers" (encountered earlier in the manuscript) and "firsts" in the context of the algebra, *"premiers"* is here translated as "first terms". Chuquet does not explicitly claim originality for his algebra, and it might appear that his singling out of the rule of intermediate numbers alone as his own invention suggests that the algebraic material is derived from earlier works of other writers. Certainly, Chuquet makes an explicit claim for his expression "rule of first terms", commenting that the unknown had earlier simply been designated as "thing"; but, as this third part unfolds, its originality becomes more and more obvious and justifies the title for its author "father of French algebra", at one time accorded to de la Roche on the strength of almost the same algebraic material - but not strictly the same, since de la Roche did not include some of the more sophisticated aspects of Chuquet's writings.

Chuquet himself explains how the third part of the work is subdivided, after a mention of the rules already discussed in the earlier parts, rules which he sees as being surpassed in "profundity" and usefulness by the algebraic methods of this final part of the *Triparty*. Nowhere, however, does the word "algebra" appear at any point in the *Triparty*.

The third and last part of this book, which deals with the rule of first terms

As Boethius says in his first book and in the first chapter, the science of numbers is very great, and among the sciences of the quadrivium it is the one in the pursuit of which every man ought to be diligent. And elsewhere he says that the science of numbers ought to be preferred as an acquisition before all others, because of its necessity and because of the great secrets and other mysteries which there are in the properties of numbers. All sciences partake of it, and it has need of none.

Chuquet's reference here is to the *De institutione arithmetica*, in which Boethius expressly claims that, amongst the disciplines of the *quadrivium*, the study of numbers holds the principal place as "mother" of the others.

And because this is a science of great utility and also of great necessity, inasmuch as it is necessary and propitious to clerks and laymen, many wise men have studied therein; and to attain the great and marvellous subtleties thereof a number of rules have been made therein, of which one is the rule of three, which is the ruler and mistress of proportions of numbers, and of such great recommendation that some philosophers have called it the golden rule. Likewise the rule of one position, by which more beautiful and delectable calculations are made than it would be possible to estimate. Also the rule of two positions which assists enquiry into deep matters and of such great subtlety that none of the rules named above can attain to them. And likewise there is the rule of apposition and remotion. There is also the rule of intermediate numbers of which I was at one time the inventor, by means of which I have made certain calculations which I could not do by two positions. Mention is made of all these rules in the first part of this book. But above all these rules named above in its marvellous excellence is this rule of first terms, which does what the others do, and also over and above them does innumerable calculations of inestimable profundity. This rule is the key, the entrance, and the door of the depths which are in the science of numbers.

This part is divided into three main parts, the first of which is an introduction for the others.

The second deals with the manner of comparing and reducing one part composed of several diversities of number against another part, simple or compound, together with the general canons of this rule.

The third part contains their application.

The first part contains five chapters, of which the first is of the order of numbers and of their diversities and interpretation.

The second teaches how one should add together one or more diversities of numbers.

The third deals with how one should subtract one diversity of number from another.

The fourth deals with the manner in which one can multiply one diversity of number by itself or by another whether like to it or unlike.

And the fifth gives the style of dividing one diversity of number by another, like or unlike.

Chuquet uses the word "diversity" to refer to a term of the general form $a\ x^n$, that is, a number where the unknown is raised to an integer power (positive, negative, or zero) which may also be multiplied by a coefficient. Thus he provides a generic term which subsumes "number", "thing", "root", "power", etc. used in writing equations in the rhetorical algebra of his day. In the first chapter he presents his own algebraic notation which replaces words with numerical symbols.

De la Roche mentions in his work that Chuquet adopts the expression "rule of first terms", and he describes its "marvellous excellence" in almost the same words. However, he calls it the "rule of the thing or the rule of 1", and adds that "other nations call it algebra and yet others almucabala", thus showing that he was unaware that in Arab writings, in the *Liber Abaci* of Leonardo of Pisa, and in

Pacioli's *Summa* both words are used and have their own distinctive meanings. *Al-jabr* referred to transferring negative terms to the other side of an equation, and *muqâbalah* to cancelling identical terms on either side, described in some Western works as "restoration" and "opposition" respectively. Since it is clear that Chuquet was aware of the writings of Leonardo, we can infer that his omission of mention of Arab technical terms was a deliberate act and that in coining the expression "rule of first terms" he was intending to provide a general title which related specifically to his notation rather than to algebraic operations in general.

Of the order of numbers, and of their diversities and interpretation

Number, insofar as it is expedient for our purpose, is taken here in the broad sense, not only as something that is a collection of several units, but also there is 1, or part and parts of 1, such as any fraction. Any number, whatever it may be, is understood and interpreted in many ways. The one and the first way is that one can interpret any number as a discrete quantity or as a number taken simply without any denomination, or whose denomination is 0, and so from henceforth these numbers shall have 0 above them for their denomination in this way: 12^0 and 13^0, etc.. Secondly, each number is interpreted as a first number of a continuous quantity, which is otherwise called a linear number. Such numbers are marked by placing a unity above them in this manner: 12^1, 13^1, 20^1, etc.. Thirdly, every number is contemplated as a second number or simply as a square number, and such numbers are marked by 2 in this way: 12^2, 13^2, 19^2. Fourthly, all kinds of numbers can be understood as third numbers, which are otherwise called cubic numbers, which one can mark thus: 12^3, 15^3, 1^3, etc.. One can also understand them to be fourth numbers, or squares of squares, which are written thus: 12^4, 18^4, 30^4, etc. And likewise one may interpret there to be fifth, sixth, seventh, or eighth numbers, and so continuing as far as one wishes to go, putting for each diversity of number its denomination above it in the aforesaid manner. And thus the numbers whose denomination is 0 are occupying the first place in the order of diversities. The first terms, that is, those whose denomination is 1, are of the second order. The second numbers are in the third place; the third numbers follow next after them, and then the fourths, and after them the fifths, and so for the others according to the natural progression of numbers.

The ancients called "things" that which I name first terms, of which the figure is thus: ρ . They named the second terms fields, of which the character is ƒ . The third terms are called cubes, of which the sign is □ . And the fourth terms they call fields of fields, of which the character is ƒƒ . And there they rested and hardly ever went beyond these. Such denominations are insufficient to provide for all diversities of numbers since these are innumerable.

One should also understand that each of the diversities aforesaid can be a second, third, fourth or fifth root, and so for the others, which may be written in this way: $R^2 12^1$, $R^2 12^2$, $R^2 12^3$, $R^2 12^4$, etc., $R^3 12^1$, $R^3 12^2$, $R^3 12^3$, $R^3 12^4$, etc., $R^4 12^4$, $R^6 12^6$, etc..

One should also know that each one of the diversities aforesaid, be they simple, first, second or other numbers, or second, third, fourth or other roots, are always understood to be plus unless they be expressly marked with this appellation <u>minus</u>, as $\bar{m}\ 12^0$, or $\bar{m}\ 12^1$, or $\bar{m}\ R^2 12^3$, etc..

Also it sometimes happens that the denominations are marked and understood to be minus, although their number may be plus and sometimes minus. Thus, the number may be plus and at the same time its denomination plus, like 12^1, or 12^2, or 12^3, etc.. Although the number is not marked with plus and neither is its denomination, yet it is considered to be plus, and its denomination also. Sometimes one may be plus and the other minus, or vice versa, like 12 first minus, which one may write as $12^{1\bar{m}}$ or minus 12 first, which one may write as $\bar{m}\ 12^1$. And sometimes the one and the other are minus, like minus 12 second minus, which one can write thus: $\bar{m}\ 12^{2\bar{m}}$. And just *as it has been said for 12, so should it be understood for all other numbers.*

And because it is said above that any diversity of number is always interpreted and contemplated as being plus or minus, and since, also, plus and minus presuppose something, as when one says plus 12 or minus 12, to be informed of this one should know that minus and plus are one towards the other as giving and taking or as debt and having, of which 0 is the common disposition preceding the one and the other. Like minus 12₶, which may thus be put: 0 minus 12₶ is to be understood that if a person had 0 m̄ 12₶ he would have nothing, and over and above this he would owe 12₶. And if he had 0 p̄ 12₶ we would have 12₶ over and above 0. And thus should it be understood for all other numbers.

There are a number of points of interest in this chapter, of which the most striking is Chuquet's algebraic notation. The concept of a number is further extended, beyond that stated in the second part of the *Triparty*, to include expressions involving powers and (later) roots of unknown quantities. Chuquet makes his notation sufficiently general to include powers and roots of the unknown and, by allowing the exponent 0, 'constant' terms also. The notation pre-empts the possibility of raising constants as 'given numbers' to a power; thus 12^2 means 12 times the square of the unknown and never 144. The figures which Chuquet describes in the second paragraph are Arabic characters, derived from earlier Hindu sanskrit letters. Similar characters can be found in the writings of Regiomontanus, and also in a number of European manuscripts of the mid-15th century. Chuquet rightly points out the greater generality of his system of exponents. The introduction of negative exponents is a daring innovation in view of the status of negative numbers not only in his day but for centuries afterwards. Although fractional exponents are not used, the notation does allow for roots of powers of the unknown by a combination of integer exponents and integer roots, such as $R^3 12^4$. Amongst the examples quoted in the manuscript $R^2 12^2$, that is $(12\ x^2)^{\frac{1}{2}}$, appears in the margin as a later insertion; though the hand seems to be Chuquet's. The negative exponents appear to be included as it were for completeness, that is, in order to provide an analogy with the positive and negative integers with zero; their interpretation is not covered by the explanation in terms of "debt". The interpretation as reciprocals does not appear until later - Part 2, Chapter 2.

The symbol  which occurs in the last translated paragraph above is very similar to an Arabic symbol for the unknown quantity which can also be found in the works of Johann Müller (Regiomontanus). However, because of the context in which the symbol appears, the possibility of it being a shorthand form denoting a particular denomination of coinage should not be overlooked.

The second chapter: how one should add one diversity of number with another or several.

The diversities of number which are like in plus and in minus as also in denomination may be added as one is accustomed to add numbers or the roots of numbers. As whoever should wish to add 6^1 with 10^1, they come to 16^1. And 8^2 with 12^2 come to 20^2, and thus for other like ones.

And if the diversities of number that one wishes to add are like in denomination and unlike in plus and minus, then one should subtract the lesser from the greater. As whoever should wish to add 8^1 with $\bar{m}\ 5^1$, it all comes to 3^1. Or 10^1 with $\bar{m}\ 16^1$ all comes to $\bar{m}\ 6^1$.

If the diversities added are of different denominations and like in plus or in minus, then such diversities of number should be added together by the appellation plus. As whoever should wish to add 5^1 with 4^2 would have $5^1\ \bar{p}\ 4^2$. Or 12^0 with 7^2, he would have $12^0\ \bar{p}\ 7^2$. And whoever would add 12^2 with $\bar{m}\ 8^0$, the addition would amount to $12^2\ \bar{p}\ \bar{m}\ 8^0$. And generally the remarks set down in the book of roots in the chapter on addition should be applied, observed and kept in due course and accordingly as the matter requires it.

The third chapter: how one should subtract one diversity of number from another

All diversities of numbers can be subtracted from their like in plus and in minus and also in denomination, just as one subtracts a number from a number. As whoever should subtract 5^1 from 13^1, there would remain 8^1. Or 6^2 from 21^2, there would remain 15^2. Or 20^3 from 13^3, there would remain $\bar{m}\ 7^3$. And so for the others. And if one of the diversities is unlike the other in plus or in minus and like in denomination, then the numbers should be added together, and then one should subtract plus from minus or minus from plus, and there always remains the like of the numbers from which the subtraction is made, and unlike the number subtracted. As whoever should take $\bar{m}\ 8^1$ from plus 12^1, there would remain plus 20^1. Or should take $\bar{m}\ 16^2$ from plus 12^2, there would remain $\bar{p}\ 28^2$. Or should take $\bar{p}\ 7^3$ from $\bar{m}\ 12^3$, there would remain $\bar{m}\ 19^3$. Or should take $\bar{p}\ 10^4$ from $\bar{m}\ 8^4$, there would remain $\bar{m}\ 18^4$. And so for the others.

If the diversities of number that one wishes to subtract are unlike in denomination and like or unlike in plus or in minus, such subtraction cannot be done except by means of this appellation minus. As whoever should wish to take 5^3 from 12^2, there would remain $12^2\ \bar{m}\ 5^3$...

The remainder of the chapter consists of further examples and a general reference to the material on roots, following the pattern of the previous chapter on addition. It includes the statement that $12^2\ \bar{m}\ \bar{m}\ 16^0$ has the value $12^2\ \bar{p}\ 16^0$.

The fourth chapter: how one can multiply one diversity of number by itself or by another, like it or unlike.

The remarks and rules for plus and for minus set down in the second part of this book in the chapter on multiplication of roots should be held and here consigned to memory in that place. And with them this. It is necessary to multiply number by number and denomination must be added to denomination.

Example: whoever multiplies 12^0 *by* 12^0, *it makes 144, then whoever adds 0 with 0, it makes 0, so this multiplication comes to* 144^0.

Further, whoever multiplies 12^0 *by* 10^2, *one must firt multiply 12 by 10, coming to 120, and then 0 must be added with 2. Thus the multiplication will come to* 120^2. *By this same method, whoever multiplies* 5^1 *by* 8^1, *the multiplication comes to* 40^2.

A further example follows in similar vein, then two examples each involving a negative exponent, of which the second is:

....whoever would multiply 8^3 *by* $7^{1\bar{m}}$, *it is first necessary to multiply 8 by 7 coming to 56, then he must add the denominations, that is to say* $3\bar{p}$ *with* $1\bar{m}$ *coming to 2. Thus this multiplication comes to* 56^2, *and so should others be understood.*

To understand the reason why denomination of number is added to denomination, and to have knowledge of the order of numbers which was mentioned in the first chapter, it is necessary to set down several proportional numbers beginning with 1 and arranged in a continuous sequence, like 1, 2, 4, 8, 16, 32, etc. or 3, 9, 27, etc..

Numbers	Denomination
1	0
2	1
4	2
8	3
16	4
32	5
64	6
128	7
256	8
512	9
1024	10
2048	11
4096	12
8192	13
16384	14
32768	15
65536	16
131072	17
262144	18
524288	19
1048576	20

Now it is necessary to know that 1 represents and is in the place of numbers, whose denomination is 0. 2 represents ...the first terms, whose denomination is 1. 4 holds the place of the second terms, whose denomination is 2. And 8 is in the place of the third terms, 16 holds the place of the fourth terms, 32 represents the fifth terms, and so for the others. Now whoever multiplies 1 by 1, it comes to 1, and because 1 multiplied by 1 does not change at all, neither does any other number when it is multiplied by 1 increase or diminish, and for this consideration, whoever multiplies a number by a number, it comes to a number, whose denomination is 0. And whoever adds 0 to 0 makes 0. Afterwards, whoever multiplies 2, which is the first number, by 1, which is a number, the multiplication comes to 2; then afterwards, whoever adds their denominations, which are 0 and 1, it makes 1; thus the multiplication comes to 2^1. And from this it comes that when one multiplies numbers by first terms or vice versa, it comes to first terms. Also *whoever multiplies 2^1 by 2^1, it comes to 4 which is a second number. Thus the multiplication amounts to 4^2. For 2 multiplied by 2 makes 4 and adding the denominations, that is, 1 with 1, makes 2. And from this it comes that whoever multiplies first terms by first terms, it comes to second terms. Likewise whoever multiplies 2^1 by 4^2, it comes to 8^3. For 2 multiplied by 4 and 1 added with 2 makes 8^3. And thus whoever multiplies first terms by second terms, it comes to third terms. Also, whoever multiplies 4^2 by 4^2, it comes to 16 which is a fourth number, and for this reason whoever multiplies second terms by second terms, it comes to fourth terms. Likewise whoever multiplies 4 which is a second number by 8 which is a third number makes 32 which is a fifth number. And thus whoever multiplies second terms by third terms or vice versa, it comes to fifth terms. And third terms by fourth terms comes to 7th terms, and fourth terms by fourth terms, it comes to 8th terms, and so for the others. In this discussion there is manifest a secret which is in the proportional numbers. It is that whoever multiplies a proportional number by itself, it comes to the number of the double of its denomination, as, whoever multiplies 8 which is a third number by itself, it comes to 64 which is a sixth. And 16 which is a fourth number multiplied by itself should come to 256, which is an eighth. And whoever multiplies 128 which is the 7th proportional by 512 which is the 9th, it should come to 65536 which is the 16th.*

It is on the basis of this chapter that claims have been made by Marre and others that Chuquet has anticipated Napier in the invention of logarithms. Such claims can hardly be substantiated from an examination of the manuscript, where all that appears is the rule for exponents applied only to integers and illustrated by a table consisting of an arithmetical and a geometrical progression placed side by side. The relation between the progressions 0, 1, 2, 3, 4,... and 1, 2, 4, 8,... are in fact to be found in the writings of Archimedes. Chuquet makes no attempt to 'fill the gaps' in the geometric series nor does he suggest that this is possible, hence it is difficult to see even the 'germ' of logarithms in his juxtaposition of the two progressions. Had he been prepared to include non-integral powers and suggested also that bases other than 2 could be adopted, some of the more enthusiastic claims made on his behalf might be considered to have a sound basis of fact.

In his commentary on the *Triparty*, Marre makes a comparison between the 'germ' of logarithms said to be found in the manuscript and that said to be found in the *General trattato di numeri et misure* of Nicolas Tartaglia, printed in Venice in 1556. Tartaglia also assigns 0 to constant terms, 1 to the unknown, 2 to the square of the unknown, and so on. He then makes the correspondence between the two progressions, as does Chuquet, and remarks that "multiplication in the geometric progression corresponds to addition in the arithmetic", and relates division and subtraction in the same way. He does not, however, admit the possibility of negative exponents, remarking "if...the rank of the divisor is greater than that of the dividend...this division cannot be carried out", in which case he requires that the division is expressed in the form of a fraction. Thus, 72 years after the *Triparty*, Tartaglia fails to include the negative exponents which Chuquet allowed.

The fifth chapter. How one can divide one diversity of number by another, like it or unlike

The remarks and rules set in the treatise of roots in the second part of this book, in the chapter on division, should not be forgotten. For they are needed here. Further, it is necessary to know that the number should be divided by the number, and denomination taken from denomination.

Example: whoever should wish to divide 36^0 *by* 4^0 *should divide 36 by 4. The quotient comes to 9. Then 0, which is the denomination of 4, must be subtracted from 0 which is the denomination of 36, and there remains 0 for the denomination of 9. Thus the division comes to* 9^0.

Whoever should divide also 36^1 *by* 4^1, *the number divided by the number, the division comes to 9, then take denomination from denomination, that is to say 1 from 1 leaving 0 for the denomination of 9. Thus the division comes to* 9^0.

A number of further examples follow, of which a selection is now given. In each case Chuquet represents the calculation in the margin. Thus, for the two examples above, there appear in the margin:

$$\begin{array}{l} \underline{36^{0}} \\ \underline{9} \text{ —— } 0 \\ 4^{0} \end{array} \qquad\qquad \begin{array}{l} \underline{36} \text{ —— } 1 \\ \underline{9} \text{ —— } 0 \\ 4 \text{ —— } 1 \end{array}$$

For examples where "like is divided by like", he comments that the division comes to "a simple number whose denomination is 0".

Also, whoever should wish to divide 96^3 by 6^0, it is necessary to divide 96 by 6, the division comes to 16, then 0 must be taken from 3, 3 remains for the denomination of 16. Thus the division comes to 16^3.

Likewise, when 96^3 is divided by 6^2, the division comes to 16^1... . One should know also that the numbers divided, the divisors, and their quotients, together with the denominations of the terms in the divisions stated above are all understood to be plus.

Whoever should wish to divide 72^0 by 8^3 should divide 72 by 8, and the quotient comes to 9. The 3, which is the denomination of 8, must be subtracted from 0, which denominates 72. There remains $\bar{m}3$ for the denomination of 9, which one must put thus: $9^{3\bar{m}}$.

...And similarly, whoever divides 72^2 by 8^5 finds for the division $9^{3\bar{m}}$. And if the denominations of the number to be divided and the divisor were unlike, that is, if the one were plus and the other minus, then these denominations should be subtracted by adding the one to the other according to the nature of the plus and the minus.

Example: whoever should wish to divide 84^0 by $7^{0\bar{m}}$, the number divided by the number comes to 12, then $0\bar{m}$ must be taken away from 0 plus, $0\bar{p}$ remains. Thus the division comes to 12^0.

Whoever wishes also to divide 84^1 by $7^{1\bar{m}}$, the number divided by the number comes to 12, then it is necessary to subtract $1\bar{m}$ from 1 plus. Thus the division comes to 12^2.

Also, whoever divides 84^2 by $7^{3\bar{m}}$, the number divided by the number - he finds 12. Then after $3\bar{m}$ must be subtracted from 2 plus, there remains 5 plus for denomination of 12. Thus the division comes to 12^5.

Similarly, whoever wishes to divide $84^{2\bar{m}}$ by 7^3... Thus the division comes to $12^{5\bar{m}}$.

Whoever wishes to divide $84^{3\bar{m}}$ by $7^{2\bar{m}}$... Thus the division comes to $12^{1\bar{m}}$.

In a number of the later examples, there also appear in the margin (probably in a different hand) partial transcriptions of the calculations making use of the Arabic characters mentioned earlier in the first chapter of this part of the *Triparty*. These have been omitted in Marre's transcription. Two examples are given below.

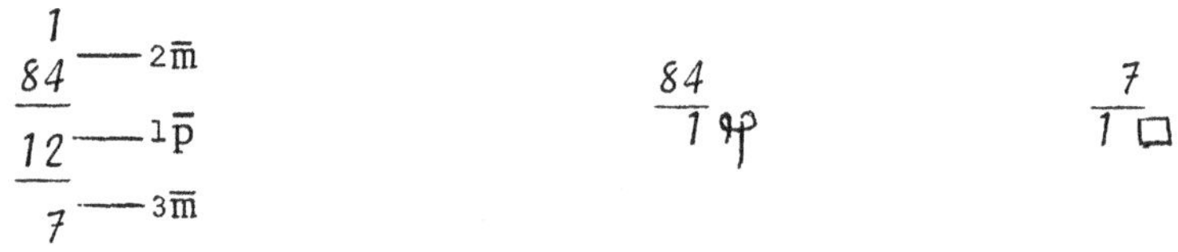

Chuquet now turns to a discussion of equations.

The second part of this third part of this book contains two chapters of which the first gives the manner of equating.

In the use of the rule of first terms one supposes that the thing one wishes to know be 1^1. And then this first term is added or subtracted, multiplied or divided, by some of its parts or by another number, as required by the calculation concerned. And everything on one side is equal or like to some other diversity of number on the other side. And although in general one always puts 1^1, yet one may also put 2^1 or 3^1 or 4^1 etc., as many as one wishes, and then manipulate as aforesaid. But if the assumption 2^1 is made, what will come out will be the subdouble of what one wishes to know. And if the assumption is 3^1, the answer will be the subtriple. Hence, if the assumption made is 2^1, double what comes out, or triple it if the assumption made were 3^1. And so must the other assumptions be understood.

In the rule of first terms, two parts are always required, one of which is equal to or like the other, not of equal quantity but of equal quality, or otherwise, of like computation and of the same intention. As, whoever should wish to find two numbers in triple proportion which added together make 12. To do this, one may suppose that one of them be 1^1. Thus the other will be 3^1, which added together make 4^1. These are like to 12, that is to say, of the same consideration as 12. For just as 12 is considered the assemblage of the triple with its subtriple, so 4^1 is produced by the addition of the triple which is 3^1 with 1^1 which is its subtriple.

These two aforesaid parts are sometimes like and sometimes unlike, simple or compound. Like, as first terms and first terms, seconds and seconds, and so on. Unlike, as numbers and first terms, or first terms and seconds, or seconds and fifths, etc.. Simple, as when in one of these parts there is a single diversity of number, as 17^0, or 17^1, or 12^2, etc. Compound, as when there are two or more diversities of number in one or the other of these parts, like 5^1 plus 17^0, or 3^2 plus 10^3, or 3^1 $\bar{p}$ 12^0 plus 5^2, etc.

When the two parts are like, if they are equal in number, like 12^1 and 12^1, or 15^2 and 15^2, etc., it is a sign that all numbers are of the nature and property that one is seeking or asking for, and that the question has infinitely many responses and not one single necessary one. If they are unequal, like 12^1 and 17^1, or 13^2 and 9^2, the calculation is impossible.

When these two parts are unlike and one of them is compound: if in the compound there is some diversity of number like to the simple part, this should be cut off and taken from that part, and likewise it should also be subtracted from the simple part which is like to it, be it greater or less. As, if 4^1 $\bar{p}$ 10^2 were equal to 24^2, 10^2 should be taken from 4^1 $\bar{p}$ 10^2 and also subtracted from 24^2. Thus one will have 4^1 on one side and 14^2 on the other. Or if 2^1 plus 8^0 is equal to 5^0, the 8^0 must be taken from 5^0 and $\bar{m}$ 3^0 remains. Or if 1^1 $\bar{p}$ 8^0 $\bar{p}$ 5^2 be equal to 12^2, one must take the 5^2 from its part and subtract it and take it also from 12^2. Thus one will have 1^1 $\bar{p}$ 8^0 for one part and 7^2 for the other part.

And if in the compound part there were some diversity of number which was marked by the appellation <u>minus</u>, whether it were like or unlike to the other part, it should be yielded up or given to that part by striking out this diversity and giving so much to the other side. For what has been done to one must be done to the other. And in the same way, even if neither on one side nor on the other were there any diversity which was not marked or understood as plus, yet, as said above, there may be need of this, in case it is necessary to take one diversity of number away from its like which is the lesser and which is simple not compound. And since the greater is subtracted from the less, the remainder is marked with <u>minus</u>. As, if 4^1 $\bar{p}$ 6^0 were equal to 4^0, then the 6^0 should be cut off and taken from its side and also subtracted from the other. Thus one will have 4^1 equal to $\bar{m}$ 2^0. Also, if 4^1 $\bar{m}$ 6^0 is equal to 3^0, the $\bar{m}$ 6^0 must be given to the one and to the other part.

Thus one will have 4^1 equal to 9^0. Similarly, if $4^1\ \bar{m}\ 6^0$ is equal to 3^1, the $\bar{m}\ 6^0$ must be given to the one and to the other side and one will have 4^1 for one part and $3^1\ \bar{p}\ 6^0$ for the other part. Then it is necessary to take the 3^1 from its part and similarly from the other. Thus one will have 1^1 for one part equal to 6^0 for the other part. Similarly, if $4^1\ \bar{m}\ 6^0$ be equal to 3^2, the $\bar{m}\ 6^0$ must be cut and taken from its part and added to the other. Thus one will have 4^1 equal to $3^2\ \bar{p}\ 6^0$.

(In the penultimate sentence of the above, the words translated here as "be equal to 3^2, the $\bar{m}\ 6^0$" are omitted in the Marre transcription.)

And if both one and the other sides be compound, one should do it in the aforesaid manner, by striking out, adding, and taking away from the one side and from the other, as many times and in such a way that one of the parts should contain no diversity of number which is like any of the diversities on the other side.

In describing how negative terms are transferred to the other side of an equation by addition, Chuquet presents precisely what in Arabic works is given the technical term *al-jabr*.

It also happens sometimes that the compound parts are unbound roots, at other times bound roots. The bound roots are made when two diversities are joined together or subtracted one from the other before either of them is marked to be a root. As, whoever would join 3^1 with 12^2 would have $12^2\ \bar{p}\ 3^1$. And whoever should wish to have the root when this addition has been done would have R $12^2\ \bar{p}\ 3^1$, which should be bound in this manner: R $\underline{12^2\ \bar{p}\ 3^1}$. Therefore such a root should be multiplied by itself, by removing the R, and so one will have $12^2\ \bar{p}\ 3^1$. And just as one of the parts has been multiplied by itself, so the other part should also be multiplied by itself. The unbound roots are produced when some diversity is added or subtracted to a root or from a root. As, whoever should wish to add or subtract 3^1 to or from R 12^2 would have R $12^2\ \bar{p}\ 3^1$ or $\bar{m}\ 3^1$, which must not be bound like the one above. And in such a case, if R $12^2\ \bar{m}\ 3^1$ were equal to 4^1, or to some other diversity of number, then the $\bar{m}\ 3^1$ should be cut off from its part by adding to it $\bar{p}\ 3^1$, and do as much to the other part, and thus one will have $R\ 12^2$ on one part and 7^1 on the other. Then each part is to be multiplied by itself if the root is a second root, or threefold if the root is R^3, or otherwise as the nature of the root requires. And thus the two parts will be in one same calculation and will not be roots.

Also when the one and the other part are unlike roots, that is, one may be R^2 and the other R^3 or some other, one should reduce them so that they may be like, and then multiply each one by itself or according to the nature of the root until they have been converted into non-roots, as is aforesaid.

Again it is necessary to know that if in one of the two aforesaid parts, after they have been equated in the above manner, if it should have any denomination that is minus, one should then multiply the one part and the other by 1 whose denomination will be like to the aforesaid minus denomination, but unlike to it in being plus. As, for example, suppose that $20^0\ \bar{p}\ 2^1$ be equal to $480^{1\bar{m}}$. Because the 480 are minus first terms, it is necessary in this case to multiply them by 1^1, and the multiplication will come to 480^0. And likewise the $28^0\ \bar{p}\ 2^1$ must be multiplied by 1^1, and one will have $28^1\ \bar{p}\ 2^2$. Thus one will have $28^1\ \bar{p}\ 2^2$ for one part and 480^0 for the other part. Also, if $12^{1\bar{m}}$ be equal to 3^0, one should multiply $12^{1\bar{m}}$ by 1^1 and one will have 12^0. And similarly 3^0 must be multiplied also by 1^1 and one will find 3^1. Likewise, if $12^{2\bar{m}}$ be equal to 3^5, the $12^{2\bar{m}}$ must be multiplied by 1^2 and one will have 12^0, and also the 3^5 must be multiplied by 1^2 and one will find 3^7, by which one will have 12^0 equal to 3^7 or vice versa 3^7 equal to 12^0. And thus it is necessary to understand other similar examples.

(There is an error in the manuscript at the beginning of the next paragraph which is retained in Marre's transcription. The appropriate correction is made here in square brackets.)

Or, otherwise, since $12^{1\bar{m}}$ [$12^{2\bar{m}}$] *comes from the division of* 12^0 *by* 1^2, *as happens when one wishes to make such divisions, one may set the divisor whatever it may be underneath the number to be divided, similarly whatever it may be, and in this way one will have* 12^0, *divisor* 1^2, *which one may set thus:* $\frac{12^0}{1^2}$ *equal to* 3^5. *And since* $\frac{12^0}{1^2}$ *is an unknown number, to clarify it one should know whatever the divisor may be, when it is multiplied by its quotient it always produces the number divided, and so let* $\frac{12^0}{1^2}$ *which is the quotient be multiplied by* 1^2 *and one will have* 12^0. *And just as one of the parts has been multiplied by the divisor which is* 1^2, *so also the other side which is* 3^5 *should be multiplied by* 1^2, *and one will have* 3^7 *equal to* 12^0 *as before.*

Also, in this manner, whoever should wish to divide $30\ \bar{m}\ 1^1$ *by* 1^2 *plus* 1^1 *would have* $30\ \bar{m}\ 1^1$, *divisor* $1^2\ \bar{p}\ 1^1$, *which one may put thus:* $\frac{30\ \bar{m}\ 1^1}{1^2\ \bar{p}\ 1^1}$. *Now suppose this diversity of number were equal or like to* 3^0. *It would be necessary to multiply the one and the other part in the aforesaid manner, and one will have* $30\ \bar{m}\ 1^1$ *for one part and* $3^2\ \bar{p}\ 3^1$ *for the other. Then afterwards, it is necessary to give* 1^1 *to one part and to the other on account of* $\bar{m}\ 1^1$ *which is in one of them, and one will have* 30^0 *for one part and* $3^2\ \bar{p}\ 4^1$ *for the other, which are indeed equated and simplified.*

This last paragraph is quite remarkable in that it includes the rational function written today as

$$\frac{30 - x}{x^2 + x}$$

This is, however, the only place where such an expression occurs in the manuscript. It is further evidence that Chuquet's thinking was advanced for his day, not only in respect of accepting negative exponents and interpreting them as reciprocals or denominators, but also in the extent to which he was prepared to manipulate algebraic expressions, including the "compound parts" which he had already accepted might include roots.

Of equipollences of numbers

It is necessary to mark and to understand that when two diversities of numbers are equal or like to another diversity or vice versa, if it happens that one or more of these said diversities are roots, in order to grasp the nature of the balanced parts one should know the equipollences and properties of numbers, that is, that the root of any number whatever and a number are equipollent and of one same rank. Likewise R^2 of second terms, R^3 of third terms, R^4 of fourth terms, R^5 of fifth terms, etc. are all equipollent to first terms. And R^2 of fourth terms, R^3 of sixth, R^4 of eighth and R^5 of tenth are equipollent to second terms. And R^2 of sixth terms, R^3 of ninth and R^4 of twelfth, equipollent to third terms. The cause and reason for which the root of a number, whatever it be, is equipollent to a number is because of extraction from it, if it can be extracted, there comes a number. Also, by extraction of the second root of second terms, there come first terms, and this is the reason why R^2 of second terms is equipollent to first terms. Also equally by extraction of R^3 from third terms, of R^4 from fourth, of R^5 from fifth, etc., there come always first terms, and they are therefore equipollent to first terms. Likewise by extraction of R^2 from fourth terms, of R^3 from sixth, of R^4 from eighth, of R^5 from tenth, there come always second terms, and they are hence equipollent to second terms. And by extraction of R^2 from sixth terms, of R^3 from ninth, of R^4 from twelfth, there come always third terms, and they are thus equipollent to third terms. Example: suppose a root of a number and first terms were equal to second terms, it would be as if numbers and first terms were equal to second terms. Or if R^2 of second terms and a number were equal to second terms it would be as if first terms and a number were equal to second terms. Or if $R^2\ 12^2\ \bar{p}\ 4^2$ be equal to $R^2 180^0$, it would be as if first and second terms were equal to a number.

And to understand better what is said above, that is to say, the art and style of simplifying and equating the parts and reducing them to two simple parts as far as one can, there are set down after this some example, the first of which is as follows. I wish to simplify $R^2\,\underline{4^2\ \bar{p}\ 4^1}\ \bar{p}\ \ 2^1\ \bar{p}\ 1$ equal *to 100. First, I take $2^1\ \bar{p}\ 1$ from each of the two parts, and I am left with $R^2\ 4^2\ \bar{p}\ 4^1$ for one part and $99\ \bar{m}\ 2^1$ for the other. And because one of the parts is a bound second root, it is necessary to multiply it by itself, and one will have $4^2\ \bar{p}\ 4^1$ for that part. And likewise $99\ \bar{m}\ 2^1$ must be multiplied by itself, and one will have $9801\ \bar{m}\ 396^1\ \bar{p}\ 4^2$ for the other part. Now these parts must be simplied again by removing 4^2 from the one side and from the other. And then give to each of these 396^1, and thus one will have 400 for the one part and 9801 for the other.*

In order to see the considerable advance which Chuquet's notation represents, it is useful to place the working of this last example side by side with the same working in modern notation:

$\underline{R^2 4^2\ \bar{p}\ 4^1}\ \bar{p}\ 2^1\ \bar{p}\ 1$ equal to 100	$\sqrt{(4x^2 + 4x)} + 2x + 1 = 100$
$\underline{R^2 4^2\ \bar{p}\ 4^1}$ equal to $99\ \bar{m}\ 2^1$	$\sqrt{(4x^2 + 4x)} = 99 - 2x$
$4^2\ \bar{p}\ 4^1$ equal to $9801\ \bar{m}\ 396^1\ \bar{p}\ 4^2$	$4x^2 + 4x = 9801 - 396x + 4x^2$
400^1 equal to 9801	$400x = 9801$

It should be noted that in this example the required simplification involves removing both roots and negative terms. This is generally the case, as can be seen from the examples which follow.

... I wish to simplify and equate $12\frac{1}{2}\ \bar{m}\ R^2\ 156\frac{1}{4}\ \bar{m}\ 25^2$ which is equal or similar to 9. And to do this, it is necessary to take away $12\frac{1}{2}$ from one part and the other, and one will have $\bar{m}\ R^2\ 156\frac{1}{4}\ \bar{m}\ 25^2$ for one part and $\bar{m}\ 3\frac{1}{2}$ for the other. Now ... it is necessary to multiply the one part and the other by themselves and one will have $\bar{p}\ 156\frac{1}{4}\ \bar{m}\ 25^2$ for the one part and plus $12\frac{1}{4}$ for the other. Now it is further necessary to give 25^2 to each part and one will have $12\frac{1}{4}\ \bar{p}\ 25^2$ for one part and $156\frac{1}{4}$ for the other. Again it is necessary to take away $12\frac{1}{4}$ from each part and one will have 144 for one part and 25^2 for the other, which is the end of this equating.

The remaining examples are given worked out in Chuquet's notation.

$R^2 6^1\ \bar{p}\ 1^1$ equal to 12
$R^2 6^1$ equal to $12\ \bar{m}\ 1^1$

6^1 *equal to* $144\ \bar{m}\ 24^1\ \bar{p}\ 1^2$
30^1 *equal to* $144\ \bar{p}\ 1^2$

$R^2\underline{12^1\ \bar{m}\ 1^2}\ \bar{p}\ 1$ *equal to* $R^2\underline{36\ \bar{m}\ 1^2}$
$12^1\ \bar{m}\ 1^2\ \bar{p}\ R^2\underline{48^1\ \bar{m}\ 4^2}\ \bar{p}\ 1$ *equal to* $36\ \bar{m}\ 1^2$.
$12^1\ \bar{p}\ R^2\underline{48^1\ \bar{m}\ 4^2}\ \bar{p}\ 1$ *equal to* 36
$12^1\ \bar{p}\ R^2\underline{48^1\ \bar{m}\ 4^2}$ *equal to* 35
$R^2\underline{48^1\ \bar{m}\ 4^2}$ *equal to* $35\ \bar{m}\ 12^1$
$48^1\ \bar{m}\ 4^2$ *equal to* $1225\ \bar{m}\ 840^1\ \bar{p}\ 144^2$
48^1 *equal to* $1225\ \bar{m}\ 840^1\ \bar{p}\ 14^2$
888^1 *equal to* $1225\ \bar{p}\ 148^2$

Chuquet then concludes:

All these things being equated and simplified in the form and manner aforesaid one must then manipulate and perform the calculation by the canons here following.

The examples given above include equations of the first order, $bx = c$, and also of the second order in the forms:

$$ax^2 = c$$
and $$bx = c + ax^2.$$
Two other quadratic forms were well-known in Greek, Hindu and Arabic algebra, namely

$$ax^2 + bx + c = 0$$
and $$ax^2 = bx + c.$$
Chuquet deals with examples of these in what follows, noting cases which in his terms were "not solvable" and cases where there were two solutions. This did not represent a significant advance, though there were later writers who disputed two solutions for a given equation.

In the margin of the manuscript alongside the discussion of the last of Chuquet's examples quoted above, there appears the word "lyee" written in a hand other than Chuquet's. This word also appears inserted into identical text in de la Roche's work. The inclusion of such marginal notes, of which this is only one example amongst several, would appear to provide conclusive evidence that Chuquet's manuscript was at one time in the hands of de la Roche, as Marre quite rightly points out.

The second and last chapter of this second part containing the canons and general rules of the science of numbers, or of the rule of first terms

Before one may understand the canons which follow, one should know that in the order of numbers there are precedents and sequents and sometimes intermediates. The precedents are those diversities of number whose denominations precede the denomination of the others, as numbers precede first terms, and first terms precede second terms, and second terms precede third terms, and so for the others. The sequents are the diversities of number whose denomination comes after the precedent, as first terms are sequents of numbers, and second terms are sequents of first terms and also of numbers; and third terms are sequents of second terms, first terms, and likewise of numbers. The intermediates are those which are between the extremes; as between numbers and second terms there are first terms, and between first and fifth terms there are second, third and fourth terms as intermediates. etc.. Of three or more diversites of number, when one or two or more are equal to the other or to the others, these diversities, that is, their denominations, are sometimes neighbouring, like numbers, first and second terms, or like first, second and third terms, or like second, third and fourth terms, etc.. Sometimes they are not neighbouring, and this in two ways: for sometimes they are equally distant from each other, or the intermediate denomination is equally distant from its extremes, as numbers, second and fourth terms, or as first, third and fifth terms, or as numbers, third, sixth and ninth terms, and so for the others. And often it happens that the denominations are unequally distant from each other, like numbers, second and fifth terms, or like first, second and fourth terms, or like numbers, second and fifth terms; and so for the others.

(The repetition is exactly as in the manuscript.) The next paragraph below is a marginal insertion in Chuquet's hand, after which the main text continues.

One should know that there are like numbers and unlike numbers. Like numbers are those whose denominations are alike, as numbers are like to numbers, and first terms to first terms, second terms to second terms, etc.. The unlike numbers are those whose denominations are unlike, as numbers and first terms, or first and third terms, or second and third terms, etc..

Again it is necessary to know that when the sides have been equated and abbreviated in the form and manner aforesaid, and when one has added, subtracted, multiplied or divided, considering the nature and denomination of the numbers according to what is described above in this third part, now and hereafter, when they must be added, subtracted, multiplied or divided according to the following canons, all denominations are annulled, for all diversities of number will be interpreted as numbers and treated as numbers or roots of numbers, as appears in full in the third part of this third book.

The first canon of the rule of first terms is thus:

Of two unlike numbers when one is equal to the other: the precedent should be divided by the sequent, for the quotient is that which one requires. And if the denominations are neighbouring, then the quotient is a number. If they are not neighbouring, it is a root of a number, whose denomination is that by which the greater denomination exceeds the lesser.

The second canon

Of three diversities of number equally distant from each other, when the two precedents are equal to their sequent or vice versa, then the two precedents should be divided by their sequent, and then the half of the intermediate multiplied by itself and added to its precedent. The second root of this addition added to half the intermediate is what one requires, provided that the three diversesities are neighbouring. If they are not neighbouring, it is the bound root of all the number whose denomination is that by which the denomination of the intermediate exceeds the denomination of its precedent or is exceeded by that of its sequent.

The third canon

Of three diversities of number equally distant, when the two sequents are equal or like to their precedent, it is necessary to divide the two precedents by the sequent. And then if half of the intermediate be multiplied by itself and added to its precedent the second root minus the $\frac{1}{2}$ of the intermediate is that which one wants to know, provided that the three denominations are neighbouring. If not, it will be the bound root of all this number as is stated above in the second canon.

The fourth canon

Of three diversities of number equally distant, when the two extremes are equal to their intermediate it is always expedient to divide the two precedents by the sequent, and then to multiply half by the intermediate by itself, and from the multiplication to subtract its precedent. For the second root of the remainder added to or subtracted from the half of the intermediate is that which one requires provided that the three denominations are neighbouring. If not, it is the bound root of all the addition or subtraction whose denomination is as is stated above in the two preceding canons.

In these four canons Chuquet deals with each of four species of equation, all terms being assumed positive.

$$ax^m = bx^{m+n}$$

with solution

$$x^n = \frac{a}{b};$$

$$ax^m + bx^{m+n} = cx^{m+2n}$$

with solution

$$x^n = \frac{b + \sqrt{b^2 + 4ac}}{2c};$$

$$ax^m = bx^{m+n} + cx^{m+2n}$$

with solution

$$x^n = \frac{-b + \sqrt{b^2 + 4ac}}{2c};$$

and

$$ax^m + cx^{m+2n} = bx^{m+n}$$

with solution

$$x^n = \frac{b \pm \sqrt{b^2 - 4ac}}{2c}.$$

(The equations and solutions are given here in equivalent modern form.) In the fourth case, Chuquet subsequently shows that he is aware both that such equations may be unsolvable in his terms and that there may be two solutions. The remainder of the *Triparty* deals with the application of the canons.

The third part of this third part of this book contains the application and exposition of the aforesaid canons. And initially of the first which is as follows.

(The first canon is here repeated.)

For declaration of this canon one must know that when numbers are equal to numbers, or first terms of first terms, or second terms to second terms, and so on for others like to them, if the divisions are equal in quantity as 12^0 and 12^0, or as 12^1 being equal to 12^1, or 15^2 to 15^2, etc., this signifies that the calculation has no necessary solution, and that all numbers are included in the said answers.

Example: I wish to find a number such that when multiplied by 2, and the product multiplied by itself comes to as much as if it were first multiplied by itself and that which ensues further multiplied by 4. To find this number it is necessary to propose 1^1, which multiplied by 2 comes to 2^1, and then 2^1 multiplied by itself given 4^2 for one part. After, let us multiply 1^1 by itself, it comes to 1^2, and further by 4, coming to 4^2 for the other part. Thus we have 4^2 equal to 4^2. Now we can say that this problem has no necessary result, for one cannot state a number which is not according to this proposition. If the like parts are unequal, such as 12^0 and 17^0, or 12^1 and 8^1, or 13^2 and 20^2, this signifies that the calculation is impossible. Example: I wish to find a number which, multiplied by 3 and that which results multiplied by itself, comes to as much as if it were multiplied by itself and that which ensues by 5. I propose that the number that I seek be 1^1, which multiplies by 3 comes to 3^1, then 3^1 multiplied by itself comes to 9^2 for one part. After, let us multiply 1^1 by itself; it comes to 1^2, and again by 5, it comes to 5^2 for the other part. Thus we have 9 second terms equal to 5^2, which signifies that the calculation is impossible. And nevertheless for like numbers it does not necessarily follow, for this cause the first canon speaks of two unlike numbers of which a necessary conclusion follows.

The solution of $9x^2 = 5x^2$ is, of course, not "impossible"; it is $x = 0$.

And for a special declaration of this and also of the other canons and also to avoid too verbose a description, one must understand that when hereafter and henceforth one will find that a number can be reduced to a second term or multiplied by itself, which is all one, one should understand that the number that one wants to reduce to a second term is posed as 2 times the one below the other, and then multiply the one by the other, then it is reduced to a second term. To reduce to a third term is to propose a number three times the one below the other and then multiply the first by the second and that which ensues by the third, then it is reduced to a third term. Also to reduce to a fourth term is to put this number that one wishes to reduce in fourth terms in four places and then multiply the first by the second and that which ensues by the third, and again that which ensues by the fourth. Then it is reduced to fourth terms. And thus must one understand of fifth terms, and of those that one wishes to reduce to sixth terms, and so on for the others.

Further, one should know that when one divides numbers by first terms, or first terms by second terms, or second terms by third terms, or third terms by fourth terms, and generally all precedents when they are divided by their next sequent, that which ensues as quotient is a number. And whoever divides a number by second terms, or first terms by third terms, or second terms by fourth terms, etc., that which ensues is a second root of a number whenever the greater denomination exceeds the lesser by 2. And whoever divides numbers numbers by third terms, or first terms by fourth terms, or second terms by fifth terms, or third terms by sixth terms, etc., that which ensues is a third root, for the greater denomination exceeds the lesser by 3. Also, similarly, whoever divides numbers by fourth terms, and first terms by fifth terms, or second terms by sixth terms, or third terms by seventh terms, etc., the result is a fourth root since the greater denomination exceeds the lesser by 4, and thus must other diversities of number be understood.

Further, it is necessary to know that in infinitely many ways one diversity of number can be equal to another. For the numbers can be equal to number, to first terms, to second terms, to third terms, to fourth terms, to fifth terms, etc. And the first terms can be equal to numbers, to first terms, to second terms, ... And the second terms can be equal to numbers, to first terms, to second terms, ... And the third terms similarly ... And also the fourth terms, fifth terms, sixth terms, seventh terms, etc. can be equal one and all to numbers, first terms, second terms, third terms, fourth terms, etc..

In the first sentence of the next paragraph there is a curious repetition of the phrase "ou de premier" (translated in brackets), which might again suggest that Chuquet is copying from an earlier version of the manuscript.

Equally, a second root, or a third, or a fourth, or a fifth, etc. of a number or of a first term (of a first term), or of a second, or of a third, or of a fourth, etc. can be equal to a number, to a first term, to a second, to a third, and to a fourth, etc.. And it can further be equal to a second root, or a third, or a fourth, or a fifth, etc. of a number, of a first term, of a second, of a third, of a fourth, etc., which combinations are infinite and infinite in manners, as one can consider and understand each of these three following figures.

The "figures" are given below directly from the manuscript.

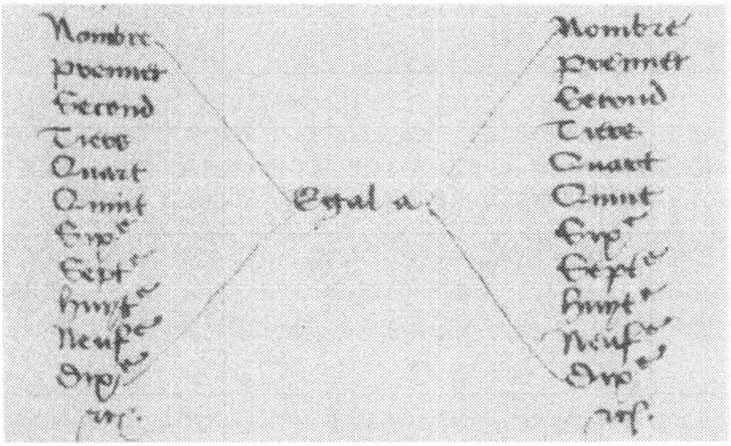

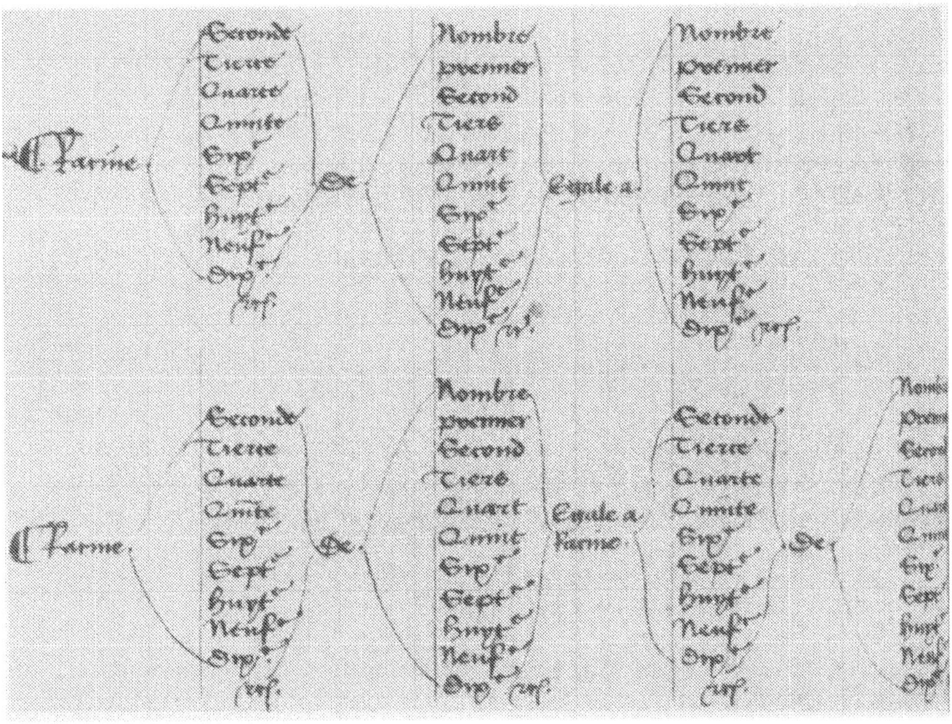

Now, to come to the application of this first canon. I wish to find a number such that when one will have added to it its third and to all the sum further added the fourth, and again 8 over and above, it all comes only to 12. For this I propose that this number which I wish to find 1^1, of which the third is $\frac{1}{3}^1$, added together amounts to $1^1\frac{1}{3}$, of which the fourth is $\frac{1}{3}^1$, which added with $1^1\frac{1}{3}$ comes to $1^1\frac{2}{3}$ which one should add to 8. Thus one will have 8 $\bar{p}$ $1^1\frac{2}{3}$ or $1^1\frac{2}{3}$ $\bar{p}$ 8 equal to 12. Now one must eqate these parts by taking 8 from one part and the other; thus there will remain $1^{\bar{1}}\frac{2}{3}$ equal to 4. Now divide 4 which is the precedent by $1^1\frac{2}{3}$ which is the sequent, and one will find $2\frac{2}{5}$ which is the number which one seeks.

I wish to find two numbers which added together make 17, and which subtracted the one from the other, that is to say, the lesser from the greater, the remainder will be 3. To do this I propose that the greater of these be 1^1 *and the lesser will be* $17\ \bar{m}\ 1^1$, *which subtracted from* 1^1 *leaves* $1^1\ \bar{m}\ 17\ \bar{p}\ 1^1$, *which simplified is* $2^1\ \bar{m}\ 17$ *equal to 3. Now simplify the parts; you will have* 2^1 *for one part and 20 for the other. Now divide the number by the first terms; you will have 10 for one of the numbers and consequently 7 for the other.*

Further, I wish to find two parts of 12 such that the greater divided by the lesser the quotient will be 12. I propose that the lesser be 1^1, *then the other will be* $12\ \bar{m}\ 1^1$, *which divided by* 1^1 *yields the part* $12^{1\bar{m}}\ \bar{m}\ 1$ *equal or alike to 12. And because one of the parts has minus first terms, multiply each by* 1^1; *it will be* $12\ \bar{m}\ 1^1$ *for one part and* 12^1 *for the other. Then give* 1^1 *to each of the parts; you will have 12 on one part and* 13^1 *on the other. Divide the number by* 13^1: *you will have* $\frac{12}{13}$ *for one of the numbers and consequently* $11\frac{1}{13}$ *for the other.*

The last two examples are found on folio 97r. The right margin of this folio contains three examples added subsequently, apparently in Chuquet's hand. These additional examples are to be found in de la Roche's work precisely as worded in Chuquet's manuscript but incorporated without comment in the main text. In all, Chuquet has 156 examples illustrating the application of the first canon. Only a representative selection are translated below, the content of the remainder being indicated by giving the equations involved in modern notation.

One number satisfying:

$$(5x)^2 + 5x = 80x$$

Further, I wish to find two numbers in double proportion and such that, multiplied the one by the other and that which ensues further by 5, this multiplication comes to as much as if the subdouble of the two numbers were reduced into third terms and further multiplied by $3\frac{1}{3}$. *To find these two numbers I propose that the subdouble be* 1^1. *Thus the double will be* 2^1, *which multiplied the one by the other comes to* 2^2. *Which multiplied further by 5 comes to* 10^2 *for one side.*

Afterwards it is necessary to reduce 1^1 to a third term and it comes to 1^3 which it is necessary further to multiply by $3\frac{1}{3}$, coming to $3^3\frac{1}{3}$. Now divide the precedent by the sequent, that is to say, 10^2 by $3^3\frac{1}{3}$, and you will find 3, which is the subdouble of the two numbers which one seeks. And hence 6, which is the double, will be the other number.

Three numbers in triple proportion, the equation being:

$$[x^4 + (3x)^4] \times 24 = (9x)^3 \times 8\tfrac{8}{81}$$

Two numbers in double sesquialter proportion, the equation being:

$$625x^5 = (2\tfrac{1}{2}x)^4 \times 64$$

Two numbers x, y in the proportion 3 : 2, the equation being:

$$x^5 \times 128 = y^6 \times 243$$

One number satisfying:

$$336 / 3x = 7x$$

Further, I wish to find two numbers in equal proportion such that the one multiplied by 12 and the other reduced into third terms and further multiplied by 5, the second product is the triple of the first. To do this calculation I propose that the one be 1^1, thus the other will be 1^1 ... And you will find $R^2 7\frac{1}{5}$.

One number satisfying:

$$9x^2 = x^4.$$

One number satisfying:

$$(\tfrac{3}{5}x)^3 = (\tfrac{2}{5}x)^5.$$

Two numbers x, y is sesquitier proportion such that x^6 is the sesquitier of y^4. The numbers found are

$$\sqrt{(7\tfrac{1075}{2187})},\ \sqrt{(4\tfrac{52}{243})}.$$

Chuquet then gives the laborious calculations involved in checking that the two numbers found satisfy the given conditions. He adds:

And thus it must be understood of fifth terms when they are equal to seventh terms, and of sixth terms equal to eighth terms, etc.. Thus it appears that when the precedents are divided by their second sequent that which ensues is a second root of a number.

I wish to find two numbers in such proportion as are 5 and 7 and such that the lesser being multiplied by itself and that which ensues further multiplied by the other number, the last multiplication comes to 40. To find these two numbers I propose that the lesser be 1^1*, and thus the greater will be* $1\frac{2}{5}$*...* $R^3 28\frac{4}{7}$ *is the first of the two numbers which one seeks. The other number can be investigated by the rule of three thus. If* $5/7/R^3 28\frac{4}{7}/$ *then multiply and divide as the rule of three requires and one will find* $R^3 78\frac{2}{5}$ *for the other number. Another way of doing this calculation. I propose that the lesser number of the two be* 5^1*, thus the other will be* 7^1*. Now whoever multiplies* 5^1 *by itself it comes to* 25^2 *and further by* 7^1 *yields* 175^3 *equal to 40. Now divide 40 by 175, you will have* $R^3 \frac{8}{35}$*. And as the position has been done with* 5^1*, for this reason* $R^3 \frac{8}{35}$ *is the subquintuple of the first number which one seeks, by which whoever multiplies* $R^3 \frac{8}{35}$ *by 5 reduced to a third term, he will find* $R^3 28\frac{4}{7}$ *as before.*

Or otherwise since 175^3 *is equal to 40, multiply then by 40 by 5 reduced to a third term, which is 125; the multiplication comes to 5000 which it is necessary to divide by* 175^3 *and you will find* $R^3 28\frac{4}{7}$ *as before. The other number can be sought by the rule of three as aforesaid.*

Two numbers x, y in proportion 5 : 7 such that $x^2y^2 = 6(x + y)$.

Two numbers x, y in triple proportion such that $x^3y^2 = xy$.

Three numbers x, y, z in proportion 1 : 2 : 3 such that $x^2y^2z^2 = x\ y\ z$. Chuquet adds:

And thus it must be understood of fourth terms when they are equal to seventh terms and vice versa, and of fifth terms equal to eighth terms, and of sixth terms equal to ninth terms, and so for the others. By which it appears that when the precedent is divided by its third sequent, the quotient is a third root of a number which otherwise one calls a cubic root.

Four numbers w, x, y, z in proportion 2 : 3 : 4 : 5 such that $wxyz - 12 = 10$.

Two numbers x, y in sesquialter proportion such that $x^3y^2 = 2(x + y)$.

One number satisfying

$$(\frac{2}{5}x)^4 \times (\frac{3}{5}x)^2 = 3x^2$$

Two numbers x, y in proportion 4 : 3 such that $x^3y^4 = x^3$. Chuquet adds:

And thus it is necessary to understand of fourth terms when they are equal to eighth terms and vice versa. Or of fifth terms equal to ninth terms. And thus of the others by which it appears that when the precedent is divided by its fourth sequent the quotient is a fourth root of a number which otherwise one calls a root of a root.

Two numbers x, y in proportion 4 : 7 such that $x^3y^2 = 30$.

Two numbers x, y in proportion 2 : 3 such that $x^3y^3 = (x + y)^4$.

Two numbers x, y in double proportion such that $x^4y^3 = xy$. Chuquet adds:

And thus it is necessary to understand of third terms when they are equal to eighth terms, and of fourth terms equal to ninth terms, and of fifth terms when they are equal to tenth terms, and thus for the others. By which it appears that when the precedent is divided by its fifth sequent that which ensues is a fifth root of a number.

Three numbers x, y, z in proportion 3 : 4 : 5 such that $x^2y^2z^2 = 18$.

Two numbers x, y in double sesquialter proportion such that x^4y^3 = subtriple $(x + y)$. Chuquet adds:

And thus it must be understood of second terms when they are equal to eighth terms, and of third terms when they are equal to ninth terms, and of fourth terms equal to tenth terms, and thus for the others. By which it appears that when a precedent is divided by its sixth sequent that which ensues is a sixth root of a number. And similarly it must be understood that when the precedent is divided by its seventh sequent that which follows is R^7 of a number. And if it is divided by its eighth sequent, the quotient is an eighth root of a number. And thus it must be understood of the ninth sequent and of the others following.

For fuller explanation of this first canon of the rule of first terms one must know that there are some calculations or questions for which it is necessary to make two positions or several, of which one or two or more are of several fixed numbers such as one wishes, or of several fixed numbers, and the other position must be 1^1. Such questions have as many answers as one wishes accordingly as one varies the number or the fixed numbers of its position, and thus they have an infinity of examples.

Against this last paragraph in the margin, apparently in Chuquet's hand, there appears:

Note of questions which have several answers.

Chuquet has here proposed a combination of his algebraic method (take 1^1 as one unknown) and the rules of false position (suppose one unknown is n). This is interesting in that he is bringing out the parallel between the two methods. The main text then continues with further examples.

Example: I wish to find three numbers of which the second added to the first, the addition is the triple of the third and the addition of the third with the first is the quintuple of the second... I propose that the first be 12 and that the second be 1^1, which added with 12 is 12 $\bar{p}$ 1^1, *of which the third part, which is $4\ \bar{p}\ \frac{1}{3}^1$ is the third number. Afterwards, whoever adds $4\ \bar{p}\ \frac{1}{3}^1$ with 12 gets $16\ \bar{p}\ \frac{1}{3}^1$ similar to 5 times 1^1 which is 5^1. Now it is necessary to simplify these parts and one will have $4\frac{2}{3}$ for divisor and 16 for the number to be divided. Divide then 16 by $4\frac{2}{3}$; you will have $3\frac{3}{7}$ for the second number, which added with 12 which is the first comes to $15\frac{3}{7}$ of which the third is $5\frac{1}{7}$ for the third number.*

Another answer: I propose 1^1 for the first number and 8 for the second number and hence $\frac{1}{3}^1$ plus $2\frac{2}{3}$ will be for the third. Then, whoever adds 1^1 with $\frac{1}{3}^1$ plus $2\frac{2}{3}$ gets $1^1\frac{1}{3}\ \bar{p}\ 2\frac{2}{3}$ equal to 5 times 8, which is 40. Now simplify the parts; you will have $1\frac{1}{3}$ for divisor and $37\frac{1}{3}$ for the number to be divided. Now divide $37\frac{1}{3}$ by $1\frac{1}{3}$ and you will find 28 for the first number, 8 for the second, and consequently 12 for the third. Thus it appears that the number proposed alone determines the varying answer.

Further, one should know that when the parts of a calculation are equated and the divisor is minus, it is always a sign that such a problem is impossible. Example: I wish to find a number such that, multiplied by 20 and to the product 7 added and set down, then further this same number multiplied by 30 and 9 taken away from the product, the first product and the second are in such proportion as 3 : 10.

In the margin against this last paragraph there appears in Chuquet's hand:

Note for the impossible problems.

The calculation then continuous:

To find this number I propose 1^1, which multiplied by 20 and 7 added comes to $20^1\ \bar{p}\ 7$ for one part. Then afterwards whoever multiplies 1^1 by 30 and takes from it 9 gets $30^1\ \bar{m}\ 9$, of which the $\frac{3}{10}$, which is $9^1\ \bar{m}\ 2\frac{7}{10}$ is like to $20^1\ \bar{p}\ 7$. Now simplify the parts giving $2\frac{7}{10}$ to each of the two parts and you will have $20^1\ \bar{p}\ 9\frac{7}{10}$ for one part and 9^1 for the other. Now take 20^1 from one part and from the other; you will have $9\frac{7}{10}$ for the number to divide and $\bar{m}\ 11^1$ for the divisor. And as the divisor is minus it *is a sign that this calculation is impossible.*

Here, Chuquet totally rejects a negative solution. However, as can be seen elsewhere, he is not consistent in this. "More questions" now follow.

Two numbers satisfying $x = y + 1$ and $\sqrt{(xy)} + x + y = 100$.

Two numbers x, y in sextuple proportion satisfying $x + y = \sqrt{21}$.

Two numbers x, y in proportion 3 : 2 satisfying $8xy = \sqrt{24}$.

Five problems involving one number x:

$$5x^3 = \sqrt{75}$$

$$7x = \sqrt{(147x)}$$

$$\sqrt{\frac{5x}{3}} = 8$$

$$\sqrt{(6x)} = 3x$$

$$\sqrt{(12x)} = 3x$$

Two numbers x, y in proportion 2 : 3 such that $\sqrt{(12x)} = y^3$.

Two numbers x, y in double proportion such that $\sqrt{(10x)} = y^4$.

One number satisfying $\sqrt{(12x^2)} = 6$.

Further, I wish to find two numbers in triple proportion and such that, when the subtriple shall be multiplied by itself and further by 8, the R^2 of this product will be as much as the triple when it shall be multiplied by 7. To do this, I propose 1^1 for the subtriple, which multiplied by itself comes to 1^2, and further by 8 comes to 8^2, of which the R^2 8^2 is equal to 21^1 which is the triple 1^1 multiplied by 7. Now multiply R^2 8^2 by itself; you will have 8^2 for one part and similarly 21^1 by itself comes to 441^2

for the other part. Thus we have second terms equal to second terms and unequal numbers, for the one is 8 and the other 441, which signifies that the problem is impossible.

The possibility of the zero solution of $8x^2 = 441x^2$ is, in accordance with earlier stated principles, not considered. In the margin, Chuquet has written "Impossible problem".

Two numbers x, y in "superbiparciens tierces" proportion such that $\sqrt{(10x^2)} = \frac{1}{2}y$.

Two numbers x, y in proportion 5 : 7 satisfying $\sqrt{(6x^2)} = xy^2$.

One number x satisfying $\sqrt{(6x^2)} = x^4$.

One number x satisfying $\sqrt{(1\frac{19}{125}x^3)} = 2$.

Two numbers, $x : y = 2 : 3$ and $\sqrt{(7x^3)} = 5y$.

One number: $\sqrt{x^3} = 5x^2$.

Two numbers x, y in sesquialter proportion such that $\sqrt{y^3} = x^3$. The same proportion but satisfying $\sqrt{(2y^3)} = x^4$.

One number: $\sqrt{(3x^4 - 24)} = 8$.

One number: $\sqrt{(\frac{1}{3}x)^4} = \frac{2}{3}x$.

Further, I wish to find a number of which the third of it, when it shall be reduced to a fourth term and further multiplied by 5, the R^2 *of this product comes to as much as* $\frac{2}{3}$ *of it when they are multiplied by themselves. To do this I propose* 1^1, *of which the third is* $\frac{1}{3}^{1}$, *which reduced to a fourth term comes to* $\frac{1}{81}^{4}$, *which multiplied by 5 comes to* $\frac{5}{81}^{4}$. *Afterwards, the two third of* 1^1, *which is* $\frac{2}{3}^{1}$, *multiplied by itself comes to* $\frac{4}{9}^{2}$ *equal to* $R^2\ \frac{5}{81}^{4}$. *Now multiply each part by itself; you will have* $\frac{5}{81}^{4}$ *for one part and* $\frac{16}{81}^{4}$ *for the other part. And since the two are like and unequal, it follows that the problem is impossible.*

Again a zero solution is, according to earlier principles, rejected.

Two numbers x, y in double proportion such that $\sqrt{y^4}$ = subdouble x^3. The same proportion but satisfying $\sqrt{(x^2y^2)} = x^4$.

One number: $\sqrt{[(\frac{1}{4}x)^3(\frac{3}{4}x)^2]} = 12$.

Two numbers x, y in double proportion such that $\sqrt{(x^3y^2)} = x + y$. The same proportion but satisfying $\sqrt{(x^4y)} = 3xy$.

One number: $\sqrt{\frac{x^5}{6}} = \frac{x^3}{2}$.

One number: $\sqrt{(5x^5)} = \frac{1}{3}x^4$.

Further, I wish to find two number in triple proportion and such that added together they come to $R^3 10$. *To do this I propose* 1^1 *for the subtriple and* 3^1 *for the triple, which added together come to* 4^1 *equal to* $R^3 10$. *Now multiply each part in third terms; you will have* 64^3 *for one part and 10 for the other. Divide therefore the number by the third term; you will have* $R^3 \frac{5}{32}$ *for the subtriple, and consequently* $R^3 4\frac{7}{32}$ *for the other number.*

Two numbers x, y in triple proportion such that $xy = \sqrt[3]{10}$. The same proportion satisfying $x^2y = \sqrt[3]{10}$.

Two numbers x, y in double proportion such that $x^2y^2 = \sqrt[3]{7}$. The same proportion satisfying $x^3y^2 = \sqrt[3]{6}$.

One number: $\sqrt[3]{\frac{x}{7}} = 5$.

One number: $\sqrt[3]{x} = \frac{3}{4}x$.

One number: $\sqrt[3]{(10x)} = \frac{1}{2}x^2$.

Two numbers x, y in sesquialter proportion such that $\sqrt[3]{(10y)} = x^2$.

Further, I wish to find two numbers in the proportion aforesaid and such that both added together and that which ensues multiplied by 10, the R^3 *of this product comes to as much as these two numbers when they are multiplied each by itself, and then these two products multiplied further the one by the other. To find these two numbers I propose* 1^1 *for the subsesquialter and* $1^1\frac{1}{2}$ *for the* *sesquialter which, added together, are $2^1\ \frac{1}{2}$, which multi- by 10 comes to 25^1. Afterwards it is necessary to multiply 1^1 by itself, coming to 1^2, and $1^1\ \frac{1}{2}$ by itself coming to $2^2\ \frac{1}{4}$, which one should further multiply by 1^2; this last multiplication comes to $2^4\ \frac{1}{4}$ equal to $R^3 25^1$. Now reduce each part to third terms; you will have 25^1 for one part and $11^{12}\ \frac{25}{64}$ for the other. Now the first term must be divided by the twelfth term; you will have $R^{11}2\frac{142}{729}$ for the subsesquialter, and consequently $R^{11}189\frac{27}{32}$ for the sesquialter for the completion of the problem.*

Whoever should wish to prove and test this problem must first add $R^{11}2\frac{142}{729}$ with $R^{11}189\frac{27}{32}$; it all comes to $R^{11}52327\frac{18869}{23328}$, which multiplied by 10 reduced to an 11th term, this multiplication comes to the sum of $R^{11}52327808856310001\frac{271}{729}$, of which the third root is $R^{33}52327808856310001\frac{271}{729}$, which simplified by extraction of R^3 comes to $R^{11}173611\frac{1}{9}$. Afterwards it is necessary to multiply $R^{11}2\frac{142}{729}$ by itself, and likewise $R^{11}189\frac{27}{32}$, and that which ensues further the one product by the other and one will find $R^{11}183611\frac{1}{9}$ as above. Thus the problem is true and well tested. Also, whoever divides $R^{11}189\frac{27}{32}$ by $R^{11}2\frac{142}{729}$, he comes to the part $R^{11}86\frac{1019}{2048}$, which simplified by extraction of the eleventh root comes to $1\frac{1}{2}$, which signifies that the number divided and the divisor are in sesquialter proportion.

This example immediately above illustrates that Chuquet was not afraid of the most laborious arithmetical calculations.

Two numbers x, y in triple superbipartiens quintes such that $^3\surd(xy) - 26 = 5$.

One number: $^3\surd(6x^2) = \dfrac{x}{4}$

One number: $^3\surd(7x)^2 = \frac{1}{2}x \cdot \frac{2}{3}x$

Two numbers x, y in proportion $3 : 4$ such that $^3\surd(10xy) = x^2y$.

One number: $^3\surd x^2 = 2x^4$

One number: $^3\surd(4x^3) = 12$

One number: $^3\surd[(2x)^3] = 4x$. Here the zero solution is again rejected.

One number: $\sqrt[3]{(2x^3)} = (\frac{1}{3}x)^2$

Three numbers x, y, z in proportion 1 : 2 : 3 such that $\sqrt[3]{(xyz)} = 3xyz$.

Two numbers x, y in double proportion such that $\sqrt[3]{(x^2y^2)} = 5$. The same proportion satisfying $\sqrt[3]{(x^2y^2)} = y - x$. The same proportion satisfying $\sqrt[3]{(x^2y^2)} = xy$.

Two numbers x, y in proportion 2 : 3 such that $\sqrt[3]{(x^3y^2)} = 10$. The same proportion satisfying $\sqrt[3]{(x^3y^2)} = x + y$. The same proportion satisfying $\sqrt[3]{(x^3y^2)} = xy$.

Two numbers x, y in the ratio 7 : 2 such that $\sqrt[4]{\frac{x}{y}} = 5$. The zero solution is rejected.

Further, I wish to find two numbers in triple proportion and such that, the triple divided by its subtriple, the R^4 *of the quotient is the subdouble of the two numbers when they are added together. For this I propose* 1^1 *for the subtriple and* 3^1 *for the triple. Now divide* 3^1 *by* 1^1*; you will have 3, of which* $R^4 3$ *is equal to* 2^1*, which is the half of the sum of* 1^1 *with* 3^1*. Now multiply each part into fourth terms; you have 3 for one part and* 16^4 *for the other. Divided, the number ensues for the fourth term; you will have* $R^4 \frac{3}{16}$ *for the subtriple. And consequently* $R^4 5\frac{3}{16}$ *for its triple. And whoever divides the triple by the subtriple finds the part* $R^4 81$*, of which the* R^4 *is* $R^{16} 81$*, which simplified by extraction of the second root comes to* $R^8 9$*. Which further simplified from the beginning by extraction of* R^2 *comes to* $R^4 3$*. Thus the* R^4 *of the quotient is* $R^4 3$*. Afterwards, whoever adds* $R^4 \frac{3}{16}$ *with* $R^4 15\frac{3}{16}$ *finds* $R^4 48$*, of which the half is* $R^4 3$*. Thus the calculation is good, for the quotient is the subdouble of the addition.*

Two numbers x, y in triple proportion such that $\sqrt[4]{(\frac{y}{x})}$ = subquadruple (xy).

Four numbers w, x, y, z in proportion 1 : 2 : 3 : 4 such that $\sqrt[4]{(wxyz)} = 5$.

One number: $\sqrt[4]{(\frac{1}{3}x^2 + \frac{1}{2}x^2)} = x$. Zero solution is rejected.

One number: $\sqrt[4]{[(\frac{2}{3}x^2)^2]} = (\frac{1}{5}x)^2$.

Two numbers x, y in double sesquialter proportion such that $\sqrt[4]{(xy)} = 10$. The same proportion satisfying $\sqrt[4]{(xy)} = x + y$. The same proportion satisfying $\sqrt[4]{(xy)} = y^2$.

Two numbers x, y in double proportion such that $\sqrt[4]{(x^2y)} = 10$. The same proportion satisfying $\sqrt[4]{(x^2y)} = y$.

Two numbers x, y such that $x = \frac{2}{3}y$ and $\sqrt[4]{(x^2y^2)} = 10$. The same first relationship but satisfying $\sqrt[4]{(x^2y^2)} = \frac{1}{2}(x + y)$. The same first relationship but satisfying $\sqrt[4]{(x^2y^2)} = xy$.

And thus it must be understood of the fifth, sixth, seventh, eighth, etc. roots of a number or of first, second, third, fourth, etc. terms. When it is equal to a number or to a first, second, third, fourth, fifth, etc. term. It is always necessary to multiply each part by itself according to the nature of the root, and then afterwards it is necessary to divide the precedent by the sequent and it will be done.

Further, I wish to find a number such that multiplied by 12 and then divided by this number multiplied by 3, the R^2 of the quotient is $R^2 5$. For this I propose 1^1, which multiplied by 12 comes to 12^1, which it is necessary to divide by 3 times 1^1, which is 3^1, and arrive at the quotient 4, of which $R^2 4$ is equal to $R^2 5$. And since the parts are like and unequal, it signifies that the number which one seeks is 0.

Here, Chuquet has broken with his previous practice and accepted the zero solution, though here he is incorrect in so doing.

One number: $\sqrt{\frac{12x}{4}} = \sqrt{5}$.

One number: $\sqrt{(6x^2)} = \sqrt{3}$.

Two numbers x, y in double proportion such that $\sqrt{(x^2y)} = \sqrt{5}$. The same proportion satisfying $\sqrt{(x^2y^2)} = \sqrt{12}$.

One number: $\sqrt[3]{(6x)} = \sqrt{10}$.

One number: $\sqrt[3]{(6x^2)} = \sqrt{10}$

Two numbers x, y in triple proportion such that $\sqrt[3]{(x^2y)} = \sqrt{7}$.

Two numbers x, y in double proportion such that $\sqrt[4]{(x + y)} = \sqrt{5}$. The same proportion satisfying $\sqrt[4]{(xy)} = \sqrt{5}$. The same proportion satisfying $\sqrt[4]{(y^2x)} = \sqrt[4]{5}$.

Two numbers x, y in proportion 2 : 3 such that $\sqrt[5]{(y - x)} = \sqrt{3}$. The same proportion satisfying $\sqrt[5]{(2xy)} = \sqrt{3}$. The same proportion satisfying $\sqrt[5]{(x^2y)} = \sqrt{3}$.

One number: $\sqrt{x} = \sqrt[3]{7}$. One number: $\sqrt{[(2x)^2]} = \sqrt[3]{7}$.

Two numbers x, y in proportion 3 : 4 such that $\sqrt{(x^2y)} = \sqrt[3]{7}$.

Two numbers x, y in proportion 2 : 1 such that $\sqrt[3]{(x + y)} = \sqrt[3]{5}$. The same proportion satisfying $\sqrt[3]{(xy)} = \sqrt[3]{5}$.

One number: $\sqrt[3]{(x^2 \cdot \frac{1}{2}x)} = \sqrt[3]{5}$.

$\sqrt[4]{\frac{x}{5}} = \sqrt[3]{7}$

$\sqrt[4]{(2x^2)} = \sqrt[3]{7}$

$\sqrt[4]{(3x^2.x)} = \sqrt[3]{7}$

$\sqrt[5]{x} = \sqrt[3]{6}$.

Two numbers x, y in double proportion such that $\sqrt[5]{(xy)} = \sqrt[3]{6}$. The same proportion satisfying $\sqrt[5]{(xy.3x)} = \sqrt[3]{4}$.

One number: $\sqrt{(7x^2)} = \sqrt{(5x)}$.

Two numbers x, y in triple proportion such that $\sqrt{(x^2y)} = \sqrt{(x + y)}$. The same proportion satisfying $\sqrt{(x^2y^2)} = \sqrt{(x + y)}$.

One number: $\sqrt[3]{(12x)} = \sqrt{(8x)}$
$\sqrt[3]{(5x^2)} = \sqrt{(6x)}$
$\sqrt[3]{(5x^3)} = \sqrt{x}$
$\sqrt[3]{(5x^2)} = \sqrt[3]{(7x)}$
$\sqrt[3]{(5x^3)} = \sqrt[3]{(2x)}$

Two numbers x, y in double proportion such that $\sqrt[3]{[(xy)^2]} = \sqrt[3]{(x + y)}$.

Two numbers x, y in proportion 3 : 5 such that $\sqrt{(x^2y)} = \sqrt{(4xy)}$. The same proportion satisfying $\sqrt{(x^2y^2)} = \sqrt{(4xy)}$. The same proportion satisfying $\sqrt{(x^3y^2)} = \sqrt{(4xy)}$.

Further, I wish to find a number such that multiplied to a third term and further by 5, the R^3 of this product is equal to R^2 of this number when it is multiplied by itself and further by 6. To find this number I propose 1^1, which reduced to a third term comes to 1^3 and further by 5 comes to 5^3, whence $R^3 5^3$ is equal to $R^2 6^2$, which is 1^1 multiplied by itself and further by 6. Now reduce the two roots to one like root and then to no root; you will have 25^6 on one side and 216^6 on the other, and since the two parts are alike and unequal it signifies that such a number is irretrievable.

Chuquet here once more rejects the zero solution.

One number: $\sqrt[3]{[(3x^2)^2]} = \sqrt{(8x^2)}$.

Two numbers x, y in double proportion such that $\sqrt[3]{(x^3y^3)} = \sqrt{(8xy)}$.

Two numbers x, y in proportion 2 : $\sqrt{3}$ such that $xy = \sqrt{7}$.

Two numbers x, y in proportion $\sqrt{2}$: $\sqrt{3}$ such that $\frac{y}{x} = \sqrt{5}$. The same proportion such that $xy = \sqrt{5}$.

Two numbers x, y in proportion $\sqrt{3}$: $\sqrt[3]{5}$ such that $\frac{x}{y} = \sqrt{7}$. The same proportion such that $xy = \sqrt{7}$. The same proportion such that $xy = 4$.

Two numbers x, y in proportion $\sqrt[3]{5}$: $\sqrt[3]{6}$ such that $xy = \sqrt[4]{7}$.

One number:
$$\sqrt{(5x + 6)} = 10$$
$$\sqrt{(4x - 7)} = \sqrt{21}$$
$$4x + \sqrt{6} = 12$$
$$4x + \sqrt{6} = \sqrt{12}$$
$$\sqrt{(3x)} + 6 = 18$$
$$\sqrt{(3x)} + \sqrt{6} = \sqrt{15}$$
$$\sqrt{5x} + \sqrt{7} = \sqrt{20}.$$

Further, I wish to find a number such that multiplied by $R^2 5$ and $R^2 7$ taken away from this product the remainder is $2\ \bar{p}\ R^2 3$. To do this I propose 1^1, which multiplied by $R^2 5$ comes to $R^2 5^2$, from which it is necessary to remove $R^2 7$, thus $R^2 5\ \bar{m}\ R^2 7$ remains, which is alike to $2\ \bar{p}\ R^2 3$. Now to equate these parts it is necessary to add $R^2 7$ to the two parts and thus one will have $R^2 5^2$ for one side and $2\ \bar{p}\ R^2 3\ \bar{p}\ R^2 7$ for the other. Now multiply each side by itself, you will have 5^2 for one side and $14\ \bar{p}\ R^2 48\ \bar{p}\ R^2 112$ plus $R^2 84$ for the other side. Now divide this number by 5^2, you will have $R^2 2\frac{4}{5}\ \bar{p}\ R^2 1\frac{23}{25}\ \bar{p}\ R^2 4\frac{12}{25}\ \bar{p}\ R^2 3\frac{9}{29}$; which is a bound root which simplified comes to $R^2 \frac{4}{5}\ \bar{p}\ R^1 2\frac{3}{5}\ \bar{p}\ R^2 1\frac{2}{5}$, which is the number that I wish to find. Another way of doing: add $R^2 7$ with $2\ \bar{p}\ R^2 3$ and you will have $2\ \bar{p}\ R^2 3\ \bar{p}\ R^2 7$, which it is necessary to divide by $R^2 5^2$, and one will find $R^2\ \frac{4}{5}\ \bar{p}\ R^2\ \frac{3}{5}\ \bar{p}\ R^2 1\frac{2}{5}$ as before.

Further, I wish to find a number such that multiplied by 8 and then this multiplication to keep aside. Then afterwards this number multiplied by itself and further by 2, and then this product added to the first one set aside, the R^2 of this sum is 10. To do this calculation I propose 1^1, which multiplied by 8 comes to 8^1. Afterwards it is necessary to multiply 1^1 by itself coming to 1^2, and further by 2 coming to 2^2, which one must add with 8^1, and it comes to $8^1\ \bar{p}\ 2^2$ of which $R^2\ \underline{8^1\ \bar{p}\ 2^2}$ is equal to 10. Now multiply each side by itself, you will have $8^1\ \bar{p}\ 2^2$ for one part and 100 for the other. And as there are here three diversities of numbers, that is to say, number, first terms and second terms, for this reason this first canon cannot do this calculation. But it can be done by the others.

Further, I wish to find a number such that multiplied by itself and further by 5, the R^2 of this product is equal to $R^2 12$ minus the triple of this number. To do this I propose that this number be 1^1, which multiplied by itself comes to 1^2 and further by 5 comes to 5^2 of which the second root which is $R^2 5^2$ is equal or alike to $R^2 12\ \bar{m}\ 3^1$. Now give 3^1 to the one and other part; you will have $R^2 5^2\ \bar{p}\ 3^1$ for one part and $R^2 12$ for the other part. And thus $R^2\ 5^2$ and 3^1 are one same kind, for R^2 of second terms and first terms are equipollent. Thus we have here first terms equal to numbers. But as the divisor, which is $R^2 5^2\ \bar{p}\ 3^1$ is a compound number, it is necessary to *simplify by multiplying it by $R^2 5^2$ $\bar{m}$ 3^1, coming to $\bar{m}$ 4^2 for divisor. It is necessary also to multiply $R^2 12$ by this nu-number and one will have $R^2 60$ $\bar{m}$ $R^2 108$. Now divide $R^2 60$ $\bar{m}$ $R^2 108$ by $\bar{m}$ 4, reduced also to R^2, you will have $\bar{m}$ $R^2 3\frac{3}{4}$ $\bar{p}$ $R^2 6\frac{3}{4}$, which one must thus rearrange $R^2 6\frac{3}{4}$ $\bar{m}$ $\bar{m}$ $R^2 3\frac{3}{4}$, which is the number that I wish to find. Which multiplied by itself comes to $10\frac{1}{2}$ $\bar{m}$ $R^2 101\frac{1}{4}$ which it is necessary further to multiply by 5, coming to $52\frac{1}{2}$ $\bar{m}$ $R^2 2531\frac{1}{4}$, of which the second root, which is $R^2 52\frac{1}{2}$ $\bar{m}$ $R^2 2531\frac{1}{4}$ comes to as much as $R^2 12$ $\bar{m}$ $R^2 60\frac{3}{4}$ $\bar{p}$ $R^2 33\frac{3}{4}$, which is the triple of this number taken away from $R^2 12$. Afterwards, whoever simplifies the $R^2 52\frac{1}{4}$ $\bar{m}$ $R^2 2531$ by extraction of a root, finds $R^2 33\frac{3}{4}$ $\bar{m}$ $R^2 18\frac{3}{4}$. Also whoever simplifies $R^2 12$ $\bar{m}$ $R^2 60\frac{3}{4}$ $\bar{p}$ $R^2 33\frac{3}{4}$ by adding $\bar{p}$ $R^2 12$ with $\bar{m}$ $R^2 60\frac{3}{4}$ finds $\bar{m}$ $R^2 18$, which added with $R^2 33\frac{3}{4}$ comes to $R^2 33\frac{3}{4}$ $\bar{m}$ $R^2 18$.*

Other combinations, diversities and varieties of two parts are innumerable, as was stated above at the beginning of this first canon, and may be understood likewise. For all the calculations and countings preceding, it appears that the precedent should be divided by the sequent, which if it is next the quotient is a number. If it is not next it is the root of a number as this present canon states.

There follows the explanation and application of the second canon of the rule of first terms, which is thus:

The second canon is repeated here.

Example: I wish to find a number such that multiplied by 3 and to the number added 12, it comes to as much as if it be multiplied by itself and then by 4. To find this number, I propose 1^1, which multiplied by 3 comes to 3^1, which one must add with 12 to give 12 plus 3^1 for one part. And afterwards it is necessary to multiply 1^1 by itself coming to 1^2 and further by 4 coming to 4^2 for the other part. Thus we have 12 $\bar{p}$ 3^1 equal to 4^2. Now divide the precedent by the sequent, that is to say 12 $\bar{p}$ 3^1 by 4^2; you will have 3 for the precedent and $\frac{3}{4}$ for the intermediate, of which the half is $\frac{3}{8}$, which multiplied by itself comes to $\frac{9}{64}$, of which R^2 $3\frac{9}{64}$ $\bar{p}$ $\frac{3}{8}$, which is the half of the intermediate, is the number that one seeks.

Two numbers x, y in double proportion such that $x^2 + y = xy^2$. The same proportion satisfying $x^2 + y^2 = x^2y^2$.

Three numbers x, y, z in proportion 2 : 3 : 4 such that $x^2y + x^2y^2 = xyz^3$.

One number: $\sqrt{(x^4 - x^3)} = x$.

And thus it is necessary to understand of fourth terms and fifth terms when they are equal to sixth terms. And of fifth terms and sixth terms equal to seventh terms. And of the other diversities of numbers of which their denominations are neighbouring. Of non-neighbouring denominations, there follows here several examples of which the first is as follows.

I wish to find two numbers in such proportion as are 3 and 5, and such that the lesser multiplied by itself and further by 2 and this product added to 12, this addition comes to as much as if these two numbers be multiplied each by itself and further the one by the other. To do this calculation I propose 1^1, *which multiplied by itself comes to* 1^2, *which it is necessary to multiply by 2, coming to* 2^2, *which added to 12 comes to* $12\ \bar{p}\ 2^2$ *for one part. Afterwards it is necessary to multiply* 1^1 *and* $1^1\ \frac{2}{3}$, *which are in such proportion as 3 and 5, each by itself and then the one product by the other and one will have* $2^4\ \frac{7}{9}$ *equal to* $12\ \bar{p}\ 2^2$. *Now divide the number and the second term by the fourth term; you will have* $4\frac{3}{25}$ *and* $\frac{18}{25}$ *for the intermediate, of which the half which is* $\frac{9}{25}$ *multiplied by itself and added to* $4\frac{8}{25}$ *comes to* $4\frac{281}{625}$, *of which the* R^2 *added to* $\frac{9}{25}$ *comes to* $\frac{9}{25}\ \bar{p}\ R^2 4\frac{281}{625}$. *And since from number to second terms or from second terms to fourth terms there are two degrees of difference, for this reason 2 will be the denomination of the root of this number in this manner* $R^2\ \underline{\frac{9}{25}\ \bar{p}\ R^2 4\frac{281}{625}}$ *which is a bound root for the lesser number. And thus* $R^2\ \underline{1\ \bar{p}\ R\ 34\frac{1}{3}}$ *will be greater.*

Two numbers x, y in triple proportion such that $(x^2y \times 8\frac{8}{9}) + y = x^4y$.

One number: $256x^2 + (2x)^4 = 2x^6$.

Two numbers x, y in double proportion such that $y^3 + 16 = x^6$.

One number: $12x + 161\frac{5}{9}x^4 = 6x7$

$$7x^2 + 16x^5 = 2\frac{1}{8}x^8$$

Two numbers x, y in quadruple proportion such that $y^4 + 6 = x^8$.

One number: $x^9 = (x^5 + x) \times 15\frac{1}{17}$

$$x^{10} = (x^2 + x^6) \times 80\frac{1}{82}$$

$$x^{10} = (x^5 + 36) \times 211\frac{20}{31}$$

And thus it is necessary to understand first terms and sixth terms when they are equal to 11th terms, and second terms and seventh terms equal to 12th terms, and all other numbers of which the intermediate is equally distant from its extremes, and of which the first and precedent extreme together with the intermediate is equal to the other extreme sequent. Thus must this second canon be understood.

There follows the explanation and application of the third canon of the rule of first terms, which is thus:

The third canon is repeated here.

Example: I wish to find a number such that it is multiplied by 3 and kept apart. Then this number multiplied by itself and again by 6 and then added with the first product, all this addition comes to 30. I propose 1^1, which multiplied by 3 comes to 3^1, then after one must multiply 1^1 by itself coming to 1^2, and again by 6 coming to 6^2, which added with 3^1 gives 3^1 plus 6^2 equal to 30. Now divide the two precedents, that is to say, 30 and 3^1 by 6^2 which is the sequent. You will have 5 for the precedent and $\frac{1}{2}$ for the intermediate of which the half is $\frac{1}{4}$, which multiplied by itself comes to $\frac{1}{16}$, which it is necessary to add with 5, and one will have $5\frac{1}{16}$, of which $R^2 5\frac{1}{16}$ minus $\frac{1}{4}$, which is the half of the intermediate, is the number which one seeks, which simplified comes to 2.

Two numbers x, y in double proportion such that $30x = x^2 + y^3$. The same proportion satisfying $60x^2 = x^3 + y^4$.

Two numbers x, y in triple proportion such that $6\frac{4}{9}y^3 = y^4 + 3x^5$.

Further, I wish to find a number such that added with $R^2 6$ and then this addition multiplied by this number, the multiplication comes to 24. To do this I propose 1^1, which added with $R^2 6$ comes to $R^2 6\ \bar{p}\ 1^1$, which multiplied by 1^1 comes to $R^2\ 6^2$ plus 1^2 equal to 24. And since the $R^2\ 6^2\ \bar{p}\ 1^2$ is not a bround root one must take away the $\bar{p}\ 1^2$ from each part, thus one will have $R^2\ 6^2$ for one part and $24\ \bar{m}\ 1^2$ for the other. And as one of the parts is a second root, it is necessary to multiply each of them by itself and one will have 6^2 for one side and $576\ \bar{m}\ 48^2\ \bar{p}\ 1^4$ for the other. Simplify or equate the parts; you will have 54^2 for one of the parts and $576\ \bar{p}\ 1^4$ for the other. And as the two extremes are equal to their intermediate, that is to say numbers and fourth terms are equal to second terms, this problem can be done according to the fourth canon, and thus one will find $R^2\ R^2 153\ \bar{p}\ 27$, or $R^2 27\ \bar{m}\ R^2 153$, which is the number that I wish to know.

Or otherwise, since in this calculation we have found that $R^2\ 6^2\ \bar{p}\ 1^2$ is equal to 24, as second root of second terms is equipollent to first terms, for this cause in this problem first terms and second terms are equal to a number. And continue the calculation accordingly as this third canon says in dividing 24 and $R^2\ 6^2$ by 1^2 and one will find 24 and $R^2 6$ for intermediate, of which the half which is $R^2\ 1\frac{1}{2}$ multiplied by itself comes to $1\frac{1}{2}$, which added with 24 is $25\frac{1}{2}$; $R^2 1\frac{1}{2}$ is taken away from R^2 of $25\frac{1}{2}$, and one will have $R^2 25\frac{1}{2}\ \bar{m}\ R^2 1\frac{1}{2}$ as above.

One number: $(x + 6)x = \sqrt{24}$
$(x + \sqrt{6})x = \sqrt{24}$

And thus it must be understood of fourth terms when they are equal to fifth terms and sixth terms. And of fifth terms equal to sixth terms and seventh terms. And of all other numbers whose denominations are neighbouring and of which the precedent is equal to its two neighbouring sequents or of which the two sequents are equal to their neighbouring precedents.

One number: $8x^2 + 2x^4 = 12$.

Two numbers x, y in double proportion such that $10y^3 + 2x^5 = 294(x + y)$. The same proportion satisfying $24(x^6 + y^4) = 1365\frac{1}{3}(x + y)^2$.

Two numbers x, y in triple proportion such that $12y^3 + 2x^6 = 36$. The same proportion satisfying $2y^4 + 6x^7 = 420(x + y)$.

One number: $6(x^5 + x^8) = 4536x^2$.

Two numbers x, y in proportion 2 : 3 such that $12x^4 + y^8 = 24$.

One number: $12x^5 + 8x^9$: $21384x$ in double sesquialter proportion.

One number: $8x^6 + 2x^{10} = 5(128x^2)$
$x^5 + x^{10} = 20$.

And thus it is necessary to understand of the first terms when they are equal to sixth terms and to eleventh terms, and of second terms equal to seventh terms and to twelfth terms, and of all other numbers of which the precedent is equal to its two sequents, neighbouring or not heibhouring. Or the two sequents are equal to their precedent. And this is what this third canon says.

There follows the fourth canon and its explanation by several examples, which is thus:

The fourth canon is repeated here.

One should know that the problems which are done by this canon generally have a double answer. For when the R^2 of the remainder is added to the half of the intermediate, it produces a number. And when it is subtracted from it, it produces another number, both of which have the properties which it is necessary to have, and consequently one can take whichever one wishes. Also, when the half of the intermediate is multiplied by itself and if this product is less than the precedent which must itself be subtracted, such calculations necessarily cannot be done.

Of the first two examples following, which illustrate this fourth canon, one has no real roots and the other has equal roots.

Example: I wish to find a number such that multiplied by itself and again by 3 and then this product added to 12, this addition comes to as much as if the number be multiplied by 9. To do this problem I propose 1^1, *which multiplied by itself and again by 3 comes to* 3^2, *which added to 12 comes to* $12\ \bar{p}\ 3^2$ *equal to 9 times* 1^1 *which is* 9^1. *Now divide the two precedents, that is to say, 12 and* 9^1 *by the sequent, that is by* 3^2; *you will have 4 and 3 for the intermediate of which the half which is* $1\frac{1}{2}$ *multiplied by itself comes to* $2\frac{1}{4}$, *from which it is necessary to take away 4 which is its precedent. And since* $2\frac{1}{4}$, *which is the multiplication of the intermediate, is less than the precedent, it follows that this problem is impossible.*

Further, I wish to find a number such that multiplied by itself and further by 3, and then this product added to 12, it comes to as much as if this number be multiplied by 12. To do this I propose 1^1, *which multiplied by itself and further by 3 comes to* 3^2, *added to 12 comes to* $12\ \bar{p}\ 3^2$ *equal to* 12^1, *which is* 1^1 *multiplied by 12. Now divide the two precedents by the sequent; you will have 4 and 4 for the intermediate of which the helf, which is 2, multiplied by itself comes to 4, from which it is necessary to take away its precedent, which is 4, leaving 0. Whence* $R^2 0$ *added or subtracted with 2 or from 2 comes to 2, which is the number that one requires.*

One number: $3x^2 + 12 = 30x$. Chuquet arrives at $5 + \sqrt{21}$ and $5 - \sqrt{21}$, the latter being selected as the number required.

Further, I wish to find two numbers such that added together they make 10 and multiplied the one by the other come to 10. To do this I proposed 1^1 *for one of the numbers, thus the other will be* $10\ \bar{m}\ 1^1$. *Which multiplied the one by the other comes to* $10^1\ \bar{m}\ 1^2$ *alike to 10. Now equate the parts; you will have* 10^1 *for one part and* $10\ \bar{p}\ 1^2$ *for the other. Then divide 10 and* 10^1 *by* 1^2; *you will have 10 and 10 for the intermediate, of which the half, which is 5, multiplied by itself comes to 25, from which it is necessary to take away the precedent, which is 10, leaving 15, whence* $R^2 15$ *added to 5 comes to* $5\ \bar{p}\ R^2 15$. *Now subtract from 5 leaving* $5\ \bar{m}\ R^2 15$ *for one of the numbers. The other is found by subtracting one of these from 10 and thus you will have* $5\ \bar{p}\ R^2 15$ *and* $5\ \bar{m}\ R^2 15$, *which are the numbers that I wish to ask for. Or otherwise, by a rule special to this purpose. Take the half of 10 which is 5, multiplied by itself coming to 25, from which it is necessary to take away 10 leaving 15. Now we may say that one of these numbers is* $5\ \bar{p}\ R^2 15$ *and the other* $5\ \bar{m}\ R^2 15$, *and thus one may do like things. Whoever by this means wishes to find two numbers which added together make* $R^2 24$ *and multiplied the one by the other come to* $R^2 24$. *He will find that one of these is* $R^2 6\ \bar{p}\ R^2 \underline{6\ \bar{m}\ R^2 24}$; *the other is* $R^2 6\ \bar{m}\ R^2 \underline{6\ \bar{m}\ R^2 24}$. *Also, whoever wishes to have two numbers which added together make* $R^2 72$ *and multiplied the one by the other make* $6\frac{1}{4}$. *As above, one may take the half of* $R^2 72$ *which is* $R^2 18$, *which multiplied by itself comes to 18. Take away* $6\frac{1}{4}$, *there remains* $11\frac{3}{4}$. *Now one can say that one of these numbers is* $R^2 18\ \bar{p}\ R^2 11\frac{3}{4}$ *and the other is* $R^2 18\ \bar{m}\ R^2 11\frac{3}{4}$.

Two numbers x, y such that $x + y = 12$ and $12x = y^2$.

Campanus, who was an esteemed geometer and Euclidean commentator, claims that such calculations cannot be done by numerical reasoning, as it appears in the commentary in several places, and specifically in the ninth book of Euclid at the end of the 16th proposition.

This is a reference of the thirteenth-century translation of Euclid's *Elements* from Arabic into Latin, which first appeared in printed form in 1482.

Two numbers x, y in double proportion such that $5x^3 + x = 6y^2$. The same proportion satisfying $x^2 y^2 + x^2 = 2\frac{21}{40}y^3$.

Three numbers x, y, z in proportion 2 : 3 : 4 such that $3x^3 + 2y^5 = 2\frac{93}{128}z^4$.

One number: $3x + \sqrt{(5x^2)} = 20$.

And likewise it must be understood of fourth and sixth terms when they are equal to fifth terms, and of fifth and seventh terms equal to sixth terms, and of all other numbers whose denominations are neighbouring and of which the two extremes are equal to their intermediate or of which the intermediate is equal to its next precedent and to its next sequent.

Further, I wish to find a number such that multiplied to a fourth term and further by 6 and then this product added to 24, this addition comes to as much as if this number be multiplied by itself and further by 2. To do this calculation I propose 1^1, which multiplied to a fourth term and further by 6 comes to 6^4 which it is necessary to add to 23, and one will have 24 $\bar{p}$ 6^4 equal to 2^2, which is 1^1 multiplied by itself and further by 2. Now divide the two precedents, that is to say, 24 and 2^2 by the sequent which is 6^4; you will have 4 for one part and $\frac{1}{3}$ for the intermediate, of which the half is $\frac{1}{6}$, which multiplied by itself comes to $\frac{1}{36}$, from which it is necessary to take the precedent which is 4. And because the precedent is greater than the multiplication of the intermediate, it signifies that this number is irretrievable.

One number: $6x^4 + 18 = 48x^2$.

Two numbers x, y in double proportion such that $8x + y^5 = 48xy^2$.

Two numbers x, y in triple proportion such that $6x^6 + 8y^2 = 11\frac{1}{6}x^2y^2$. The same proportion satisfying $(2x)^6 + 512 = (4y)^3$.

Two numbers x, y in double proportion such that $4x + y^7 = 864\frac{1}{27}x^2y^2$.

One number: $8x^2 + 2x^8 = 128\frac{1}{8}x^5$

$$6x^8 + 12 = (12x^2)^2$$

$$12x + 2x^9 = 162\frac{4}{27}x^5.$$

$$2x^{10} + 243 = 487x^5.$$

Two numbers x, y such that $x + y = \sqrt{72}$ and $xy = 6\frac{1}{4}$.

Two numbers x, y such that $x + y + \sqrt{20}$ and $xy = \sqrt{20}$.

And likewise it must be understood of first and 11th terms when they are equal to sixth terms. And of second and twelfth terms when they are equal to seventh terms.

And of all other numbers whose intermediate is equally distant from its extremes and of which the extremes are equal to their intermediate or vice versa. Always the precedents should be divided by their sequent, be it neighbouring or not, and then half of the intermediate multiplied by itself, and from this product it is necessary to take away the precedent and then add or subtract as it appears in the several examples given above. And this is what this fourth canon says.

The solution of quadratic equations of certain types is found in most ancient mathematical cultures, for example, the Babylonian, the Greek and the Indian. In Arab works it was customary to distinguish between the three types:

$$x^2 + bx = c$$
$$x^2 + c = bx$$
$$x^2 = bx + c$$

where all the constant terms are to be taken as positive, and where negative roots are ignored. The case

$$x^2 + bx + c = 0$$

is usually ignored because the roots are complex. The three accepted standard types have solutions equivalent to the modern general formula but any proofs are, following the Greeks, geometrical. The same types of equation appear in Leonardo of Pisa's *Liber Abaci* and other Western works such as the *Treatise of Algorithm* by John of Seville. The latter, however, avoids the possibility of double roots by a careful selection of coefficients, though double roots were well known in Arab works. Chuquet's response to roots is, as can be seen from the many examples above, inconsistent. Usually, he is content to accept only one of two solutions, frequently he rejects obvious zero roots, and he fails to understand the discussion in Campanus from which he quotes. In fact, Campanus shows that certain problems do not have rational solutions, not that they are impossible of solution.

Like Pacioli, Chuquet is aware that cubic equations in general are beyond the immediate state of mathematics of his day. Indeed, he makes no claim to discuss cubics in his *Triparty*, nor does he attempt to present rules for three "diversities" of numbers unequally spaced. He is, however, prepared to be optimistic about the necessary research programme, for he begins the final paragraph of the *Triparty* as follows:

> *There remains still for the perfection and completion of this book to find rules and general canons for three diversities of numbers unequally distant. And again for four or more diversities of numbers being equally or unequally distant the one from the other. These are left for those who would wish to proceed more deeply.*

The paragraph ends with the usual ascription to the deity and the specific claim of authorship.

And thus to the honour of the glorious Trinity ends this work, which by reason of its three general parts I call Triparty. And also because it has been made by Nicolas Chuquet, parisian, bachelor in medecine, I call it the Triparty of Nicolas in the science of numbers, which was begun, continued and and finished at Lyon on the Rhone in the year of salvation 1484.

Explicit. Deo gracias.

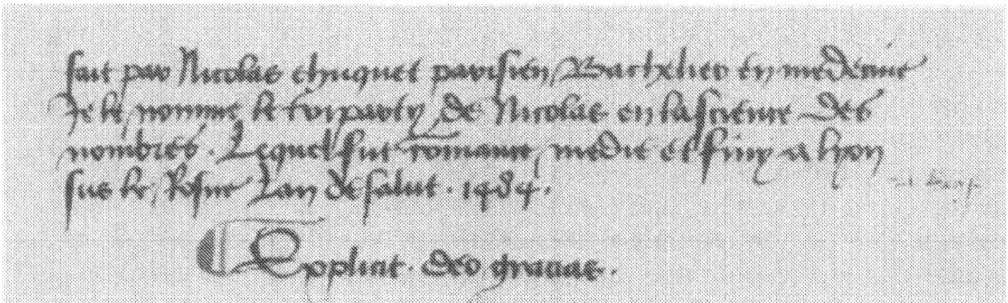

fait par Nicolas chuquet parisien, Bachelier en medicine
Je le nomme le triparty de Nicolas en la science des
nombres. Lequel fut commance, moyenne et finy a lyon
sus le Rosne Lan de salut. 1484.

Explicit. Deo gracias.

Chapter 6

THE PROBLEMS

Collections of mathematical problems, ranging from straightforward calculations in fancy dress to purely logical brainteasers, have a long history, and played a prominent role in the transmission of mathematical culture throughout the Middle Ages. From a tenth-century manuscript, sometimes attributed to Bede or to Alcuin, with the title *Propositiones ad acuendos juvenes*, to the best-known of the early published collections, the *Problemes Plaisans et Delectables* of Bachet de Meziriac (1612), the same puzzles are repeated time and again. Many of them predate Alcuin by centuries, and some are still found in popular paperbacks.

In Chuquet's manuscript, the prime purpose of the *Problems* is, as the title suggests, to illustrate the applications of his *Triparty*, and in particular of the rule of first terms. The *Triparty* itself contains no examples apart from purely numerical ones; even the rule of three, whose applications occupy such a large part of the *Commercial Arithmetic*, is presented in purely abstract terms in the *Triparty*. The problems are not original; but whereas in medieval manuscripts the reader was expected to solve similar problems either by guesswork or by the rule of three or the rule of false position, Chuquet's solutions are presented in the rudimentary algebraic terminology and notation of the rule of first terms. His algebraic methods, and the breadth of his concept of number, led him to present solutions to some problems which his contemporaries would have called insoluble. If his method led to a negative number, this number was the answer, and had to be interpreted - frequently an involved process. Given a problem solved with arithmetic progressions, he was quite content to set down as solution the sum of a progression with $8\frac{1}{3}$ terms, although, as we shall see below, it is not clear that this would be the correct model for the problem. Fractional answers to problems which should logically have integer solutions only (e.g. the number of children in a family) caused no anxiety to either Chuquet or his contemporaries; only in rare cases, where the author had not covered fractions, was there any special pleading, as in a problem about buying eggs to share, where the penny left over was spent on salt. But Chuquet goes beyond many of his contemporaries in that he also admits irrational solutions (and hardly ever gives a simple approximation). Chuquet also accepts 0 as a number, both in the formulation of equations and in their solution, though as we have seen in the *Triparty*, he occasionally missed a zero solution. He also sought to generalize problems dealing with geometric progressions, and it was in this context, in trying to discover how long a vessel would take to lose half its liquid if it lost one tenth of its contents each day, that he came closest to perceiving the need for logarithms (Problem 94).

The manucript contains 166 problems, two of which are identical (numbers 74 and 76 according to Marre's numbering which is adopted in this chapter). There are a few complete problems and solutions written in the wide margins - see page 226 for an example. L'Huillier (1979, p. 78) has identified two different hands, and provisionally ascribed some of these to Chuquet, the rest to de la Roche, in whose book (1520) some of these marginal problems appear. A number of problems, including the first 17, are purely numerical, and another 26 use the theme of a sum of money divided between two or more men as a thin disguise for what we should now regard as simultaneous equations. Many are traditional problems which may be solved by a variety of algebraic and semi-algebraic rules. Groups of problems are distinguished as solved by finding proportionals, involving progressions, and using the method of apposition and remotion, though the latter, as Chuquet pointed out, tends to arise in connection with problems having many possible solutions. There are other problems which are very well-known but harder to classify - the problem of the weights which is sometimes ascribed to Bachet (142), a foretaste of modular arithmetic in the guise of an unlikely story about some broken eggs (143, 144), the problem of measuring 4 pints with only a 5 pint and a 3 pint measure (165), and brainteasers like the story of the wolf, the goat and the cabbage, that are still used, as they have been since the time of Alcuin, to sharpen the wits of the young.

There is some overlap in subject-matter with the *Commercial Arithmetic*: both sections of the manuscript contain problems about buying and selling, mixtures of different qualities of a substance, and masters and servants. On the other hand, the *Commercial Arithmetic* uses almost no mathematics beyond the rule of three, and the *Problems* shows a lack of concern for practical detail which does not encourage the reader to infer a social history from the very brief stories (one sentence at most) leading up to the questions. The question of what, if anything, can be inferred from these settings will be discussed in the chapter on the *Commercial Arithmetic*.

Aristide Marre's edition (1881) contains the statements of all the problems, and also the answers, but not the detailed algebraic solutions. Marre also documented those problems which were reproduced by de la Roche (1520) - about a quarter of the total - and twelve problems which bear an affinity to another French fifteenth-century manuscript which did not cover algebra. A complete history of all Chuquet's problems is beyond the scope of this chapter, and the reader is referred to the new edition of Tropfke (1970) for copious details. The selection of problems below has been devised to represent each type of problem which appears in the collection, and to give enough detailed solutions to indicate Chuquet's methods and incidental remarks. Where the solution is not included, the answer is given in brackets.

The numerical problems with which the collection begins are treated as applications of the first canon of the rule of first terms; twelve yield linear equations, and the other five may be expressed as quadratic equations with one term omitted. They are rather simpler than the examples on this canon given in the *Triparty*, and de la Roche used seven of them in his own chapter on applications of the first canon.

<u>There follow several other discoveries of numbers in general which are achieved by the rule of first terms</u>

[1] *And firstly from 10 I wish to make three parts such that the first shall be the triple $\bar{m}$ 1 of the second and the $\frac{3}{4}$ plus 4 of the third ...* $(5\frac{5}{8},\ 2\frac{5}{24},\ 2\frac{1}{6})$

[9] *Again, from 10 I wish to make two parts such that when the greater is divided by the lesser the quotient shall be 29. To do this I propose 1^1 for the lesser part. Thus the greater will be 10 $\bar{m}$ 1^1. Dividing this by 1^1,*

there comes from it $10^{1m}\ \bar{m}\ 1$ *equal to 29. Balance the sides and you will have* 10^{1m} *on one side and 30 on the other. And because one of the sides is first minus, one should multiply that side and the other also by* 1^1*. Thus one will have 10 on one side for the number to be divided and* 30^1 *on the other side for divisor. Or otherwise, whoever divides* $10\ \bar{m}\ 1^1$ *divided by* 1^1*, which may be set down thus:*

$$\frac{10\ \bar{m}\ 1^1}{1^1}$$ *equal to 29. Now multiply each side by* 1^1 *and you will have* $10\ \bar{m}\ 1^1$ *on one side and* 29^1 *on the other. Now give* 1^1 *to each side and you will have 10 and 30 as above. So divide the number by the first terms and you will have* $\frac{1}{3}$ *for the lesser part of 10, and in consequence* $9\frac{2}{3}$ *for the greater. Or otherwise I suppose that the greater be* 1^1*. Thus the lesser will be* $10\ \bar{m}\ 1^1$*. The division comes to* $\frac{1^1}{10\ \bar{m}\ 1^1}$ *equal to 29. Now let us multiply each side by* $10\ \bar{m}\ 1^1$*, and we shall have* 1^1 *on one side and* $290\ \bar{m}\ 29^1$ *on the other. Let us lend to each side 291, and we shall have* 30^1 *on one side and 290 on the other. Let 290 be divided by 30 and you will have* $9\frac{2}{3}$ *for the greater part as above, and in consequence* $\frac{1}{3}$ *for the other. Or otherwise by a special rule. Add 1 onto 29 which is the quotient, which comes to 30. Now divide 10 by 30 and you will have the same as above. And thus should one do with like problems.*

Chuquet has obviously gone to particular trouble over this solution. It is the first problem in the collection to require algebraic division, and the first method involves a negative power of the unknown. The second and third methods show how this can be avoided by multiplying out to clear the fractions, whichever term is taken as the unknown, so that the student faced with a similar problem in future would be more likely to recognize a precedent. Chuquet does not, however, follow this example with a batch of similar questions for practice; instead, he quotes a "special rule" which he says should be applied in "like" cases, but expresses it in a form which applies only to this question, rather than in the more general proposition:

To find two numbers, given their sum and quotient: add 1 to the quotient and divide the result into the sum; the answer is the divisor, or algebraically:

$$\text{If } \frac{a}{b} = N, \text{ then } \frac{a+b}{b} = N + 1, \text{ hence } b = \frac{a+b}{N+1}.$$

Consideration of the "special rules" introduced by Chuquet later in this part of the manuscript, and of his own account of his method of finding special rules following problem 124, leads to the conclusion that he worked on the basis of intuition and induction rather than abstract reasoning. Occasionally (e.g. in problem 93) this intuition led him astray.

Chuquet also introduces a special rule as an alternative method of solution in the next problem, which is reproduced on folio 54v of de la Roche (1520). De la Roche does not give Chuquet's special rule, but appends four lines of working in a primitive algebraic notation to his verbal solution by the rule of first terms (see below).

¶ Plus partiz .20. en deux parties telles q̃ multipliee chascũe en soy. Et puis soustraire la mineur de la maieur. La reste soit .50. Responce. Pose que la maieur partie soit .1. ſ. Ainsi la mineur sera .20. m̄ .1. ſ. multiplie chascune partie en soy/et auras .1. ℭ. pour la maieur partie/⁊ .400. m̄ .40. ſ. p̄ .1. ℭ. pour la mineur. Ores de .1. ℭ. soustraiz .400. m̄ .40 ſ. p̄ .1. ℭ. reste .1. ℭ. m̄ .400. p̄ .40. ſ. m̄ .1. ℭ. Qui abbreuiez sont .40. ſ. m̄ .400 egaulx a .50. donnez .400. dung coste/⁊ daultre. Et auras .40. ſ. egales a .450. Partiz .450. par .40. ⁊ en vient .11. $\frac{1}{4}$. pour la maieur partie. Et par consequẽt .8. $\frac{3}{4}$. pour la mineur.

1. ſ. —— 20. m̄ .1. ſ. | 1. ℭ. m̄ .400. p̄ .40. ſ. m̄ .1. ℭ.
1. ſ. 20. m̄ .1. ſ. | 40. ſ. m̄ .400. —— 50
1. ℭ. — 400. m̄ .40. ſ. p̄ .1. ℭ. | 400 / 450

1
450 | 11
40

[10] *Again, from 20 I wish to make two parts such that when each is multiplied by itself, and then the lesser is subtracted from the greater, the remainder shall be 50. To do this, I suppose that the smaller part be* 1^1. *Thus the larger will be* $20\ \bar{m}\ 1^1$. *Now, let us multiply each one by itself and we shall have* 1^2 *for the first multiplication and* $400\ \bar{m}\ 40^1\ \bar{p}\ 1^2$ *for the second. Then subtract* 1^2 *from* $400\ \bar{m}\ 40^1\ \bar{p}\ 1^2$ *and there remains* $400\ \bar{m}\ 40^1$ *equal to 50. Now give* 40^1 *to the one and the other side and you will have* $50\ \bar{p}\ 40^1$ *on one side and 400 on the other. Next take 50 away from each side and you will have 350 for the number to be divided and* 40^1 *for the divisor. So divide 350 by 40 and you will*

have $8\frac{3}{4}$ for the first and lesser part, and in consequence $11\frac{1}{4}$ for the other. Or otherwise by a special rule: divide 50 by 20 and the division comes to $2\frac{1}{2}$. Taken from 20, there remains $17\frac{1}{2}$, the half of which, which is $8\frac{3}{4}$, is one of the parts, and thus the other will be $11\frac{1}{4}$ as above.

[14] *Further, from 20 I wish to make two parts such that when the $\frac{1}{4}$ and the $\frac{1}{5}$ of the first are added to the second, and then whoever takes away the $\frac{1}{3}$ and the $\frac{1}{4}$ of this sum, the remainder shall be 10. And if this third and quarter are added to what is left of the first, all together shall be 10. To do this I propose 1^1 for one part, and hence $20\ \bar{m}\ 1^1$ for the other. Then from 1^1 I take the $\frac{9}{20}$ which makes $\frac{9}{20}^{1}$ and add it to $20\ \bar{m}\ 1^1$ which comes to $20\ \bar{m}\ \frac{11}{20}^{1}$ in all. From this sum I remove the $\frac{7}{12}$ which makes $8\frac{1}{3}\ \bar{m}\ \frac{11}{48}^{1}$ equal to 10. Now give $\frac{11}{48}^{1}$ to each side and you will have $8\frac{1}{3}$ on one side and 10 plus $\frac{11}{48}^{1}$ on the other. Again, subtract 10 from the one side and from the other and you will have $\bar{m}\ 1\frac{2}{3}$ for the number to be divided and $\frac{11}{48}^{1}$ for the divisor. Now divide the one by the other and you will have minus $7\frac{3}{11}$ for the first part and in consequence $27\frac{3}{11}$ for the other. Now whoever should wish to prove this calculation, it is first necessary to take the $\frac{9}{20}$ of the first part, which makes $\bar{m}\ 3\frac{3}{11}$, which added to the second makes 24. From this, whoever takes away the $\frac{7}{12}$ which is 14, the remainder is 10. And whoever adds 14 to the remainder of the first part when the $\frac{9}{20}$ have been taken away, that is to the $\frac{11}{20}$ of that part which is $\bar{m}\ 4$, the whole comes to 10. Thus this calculation is true which some hold to be impossible.*

Notice that the two sentences in the statement of the problem are in fact equivalent. Chuquet includes the second statement to provide an additional check on his solution. Those unidentified mathematicians who regarded the problem as impossible presumably would have rejected negatives.

[17] *Further, I wish to find a number such that, when it is multiplied by itself and this multiplication taken away from 4, and the square root taken of what remains, and then 1 is taken away, and what remains is taken away from 2, and then what is left is multiplied by itself and then the multiplication is added to the first number, it all comes to 4. To determine what this number is...* $(R^2\ 1\frac{3}{4})$

This is the most intricate of these numerical problems. In modern notation, $(2 - (\sqrt{4 - x^2} - 1))^2 + x = 4$.

Chuquet now moves on to more concrete applications of the rule of first terms - in other words, to the solution of traditional problems. There is no further heading between problems 18 and 82, but there are several groups of problems relating to a common situation: trading at fairs, buying and selling fruit, and determining how much money each man has, though the last, and largest, of these groups is really a very thin disguise for numerical equations.

How the rule of first terms may be applied in special cases and in several ways.

[18] *There is a man who says thus. If I were as many years old again as I am, and* $\frac{1}{2}$ *and* $\frac{1}{3}$ *and yet again* $\frac{1}{4}$ *of those years, all added together, I should be 50 years old. To determine how old this man is. To do this I suppose that the number of his years be* 1^1, *onto which there must be added another* 1^1 *and again* $\frac{1}{2}{}^1$, $\frac{1}{3}{}^1$ *and* $\frac{1}{4}{}^1$. *All added together come to* $3^1\ \frac{1}{12}$ *equal to 50 years. Now divide 50 by* $3^1\ \frac{1}{12}$ *and you will have 16 years and* $\frac{8}{37}$ *of a year. And that is how much time this man has lived.*

There is a version of this problem in the *Propositiones ad acuendos juvenes*, and, no doubt, in many earlier and later collections. Its solution could be regarded, in the absence of more sophisticated algebraic techniques, as in application of the rule of three, and it is also easily solved by the rule of false position.

[19] *A spear is $\frac{1}{2}$ and $\frac{1}{3}$ in a sheath and 9 feet outside the sheath. To determine how many feet long this spear is...* *(54 feet)*

[20] *A piece of cloth is dyed black for $\frac{1}{3}$ and $\frac{1}{4}$ (of its length) and the 8 aulnes which are left over are dyed grey. To determine how many aulnes long the piece is...* *($19\frac{1}{5}$ aulnes)*

[21] *In a vessel full of wine there are three taps, one larger than the other in such manner that, whoever should draw out of the largest, all the wine will have been emptied in 3 hours. Whoever should draw out of the middle one, the vessel will be empty in 4 hours, and by the smallest tap the vessel will be emptied in 6 hours. To determine, if the three taps were opened together, in how long that vessel will be emptied...* *($1\frac{1}{3}$ hours)*

[22] *Two men depart on one day and in one hour, that is to say, one leaves Paris to go to Lyon which is a hundred leagues by road, and the other man leaves Lyon to go to Paris and makes the journey in 7 days. And the one who is going from Paris to Lyon makes the journey in 9 days. To determine after how many days they will meet each other...*

Chuquet's answer to this problem is incorrect. Letting x be the number of days (in modern notation), he claims that the men have covered $7x$ and $9x$ leagues when they meet, instead of $\frac{100}{7}$ and $\frac{100}{9}$ leagues. Problem 127, which is identical except that the two men take 7 and 8 days, is solved correctly. The problem of two men going on a journey, either in opposite directions or with one man pursuing the other, is a common one, and Chuquet has three further groups of such problems: 98 to 100 and 103 to 108 illustrate the case where one or both men travel not at a constant rate, but increasing their daily journey in arithmetic progression, while the other problems between 125 and 128 introduce the implausible complication of travelling backwards at night, familiar to most readers in the guise of the frog in the well.

[23] *A man makes his will and leaves his wife pregnant. He disposes of 100 ecus in such a way that if his wife has a daughter he wishes the mother to take twice as much as the daughter. If she has a son, he ordains that he take twice as much as his mother. It happened that the mother had a son and a daughter. To determine how many of these 100 ecus the mother, the son and the daughter should have, respecting the intention of the testator.*
($28\frac{4}{7}$ ecus for the mother, $57\frac{1}{7}$ and $14\frac{2}{7}$ for the son and daughter)

This problem, which goes back to the Roman emperor and legislator Justinian, was popular in medieval times, and far more implausible multiple births are postulated in some versions. One Arab writer described the interpretation of wills as one of the major applications of mathematics, and Pacioli tells this story as if he had heard it from someone in a shop only the other day. Robert Record's scholar [1552] opined: "if somme cunning lawyers had thys matter in scanning, they wold determine this testament to be quite voide, and so the man to die untestate", but the master replied "the mind of the testatour is to be taken fauorably, for the ayde of ye lagatories, when ther ryseth suche doubtes". An *ecu* was a gold or silver coin, whose exact relationship to the livre-sol-denier system (corresponding to the English pounds, shillings and pence) would appear to have fluctuated, for a number of different values are given in the *Commercial Arithmetic*. There is not reason to suppose that it could have been divided exactly into sevenths, but every reason to think that Chuquet would not have cared.

[24] *A merchant bought three pieces of cloth which cost him 60 ecus, and does not know how much each cost by itself. But he does know that the second cost him twice as much as the first, and the third cost him three times as much as the second. To determine how much each piece cost him...* *($6\frac{2}{3}$ ecus, $13\frac{1}{3}$ ecu, 40 ecus)*

[25] *An apothecary laid out 100 ecus, in pepper, ginger and grains as well as in cloves. And he does not know how much each of his spices cost him by itself, although he does know that the ginger cost twice as much as the pepper, and the grains cost three times as much as the pepper, and the cloves cost 4 times as*

much as the pepper. To determine what each spice cost him by itself... (*10 ecus, 20 ecus, 30 ecus, 40 ecus*)

Grains may refer to cochineal or to the West African spice known as grain of paradise or Guinea grain. Lyon was a centre for the trade in spices.

[26] *A man spent $\frac{1}{4}$ of his money, and again, of what remained he spent $\frac{1}{5}$. At the end he had 100 ecus left. To determine how much money he had at the beginning...* (*$166\frac{2}{3}$ ecus*)

The next three problems are similar, though the equations are slightly more complicated. The following four differ only in the setting: the man becomes a merchant trading at fairs, and consequently he makes money instead of just spending it. The problem reproduced below includes 0 in its statement, and, probably for this reason, was not taken up by de la Roche. There is some marginal calculation on the page, probably de la Roche's working.

[30] *Further, a merchant was at three fairs, and at each fair he doubled his money and spent 12 deniers. And at the end he had 0 left. To determine how much money he had at the beginning...* (*$10\frac{1}{2}$ deniers*)

[35] *Further, a merchant bought 15 pieces of cloth for the sum of 160 ecus, of which some had cost him 11 ecus each and the others 13 ecus. To determine how many there are at each price...*

This follows a similar problem in which the cheaper lengths of cloth cost only 9 ecus each. Here, however, the "answer" is that $17\frac{1}{2}$ of the 15(!) pieces were at the cheaper price, the remaining $-2\frac{1}{2}$ at 13 ecus. The following explanation fails to be entirely clear or convincing:

> *However, it seems according to general custom that such reasoning as this is impossible. The reason is that if we divide 160 ecus by 15 pieces, the division comes to 10 ecus $\frac{2}{3}$. Thus, taking one with the other, they cost 10 ecus $\frac{2}{3}$. And in as much as $10\frac{2}{3}$ is not intermediate between 11 and 13, from this the impossibility arises, for in such problems the common price which is $10\frac{2}{3}$ should be greater than 11 and less*

than 13. Hence whoever should wish to make it possible should prescribe that one of the prices be less than $10\frac{2}{3}$ and the other greater. Nevertheless according to the strict rule of first terms it becomes possible insofar as it gives $17\frac{1}{2}$ at the price of 11 ecus and $\bar{m}$ $2\frac{1}{2}$ at the price of 13. Which one may understand thus. This merchant bought 17 pieces $\frac{1}{2}$ for ready money and at 11 ecus each, which amounts to 192 ecus $\frac{1}{2}$, and for 2 pieces $\frac{1}{2}$ he bought and took in credit at the price of 13 ecus, which are worth 32 ecus $\frac{1}{2}$. Thus this merchant has only 15 pieces of cloth which would properly be his, and he has (spent) only 160 ecus, since the 32 ecus $\frac{1}{2}$ are to be deducted.

Put more simply, it would seem that he bought $17\frac{1}{2}$ pieces at 11 ecus and sold $2\frac{1}{2}$ pieces at 13 ecus, ending up with 15 more pieces of cloth and 160 fewer ecus than he started with. Compare the negative answer given to problem 43.

The next two problems deal with the payment of one of the taxes or customs duties which beset the medieval merchant, though if we believe Chuquet's formulation it would appear that payment in kind was acceptable.

[36] *Two merchants come from the fair. One of them has 20 bags of wool, for which he paid one bag of wool as duty and was given back 2 livres in change. And the other had 60 bags, for which he paid two bags of wool with 6 livres. To determine how much each bag of wool is valued...* *(12 livres)*

[38] *Further, a merchant has laid out 10 livres on wheat and oats, of which the wheat cost him 10 sols the sestier and the oats 5 sols. It happens that he resells the wheat for 12 sols and the oats for 4 sols, and gains 10 sols on the whole. To determine how many sestiers of wheat and how many of oats he bought...* *($12\frac{1}{2}$ of wheat, 15 of oats)*

Problem 39 is more of a disguised equation (he "spent so much money that the number of ecus and the number of sestiers add up to 60"), and the following additional problem is inscribed in the margin, in the same hand as the manuscript.

[39] *A merchant bought 66 sestiers $\frac{2}{3}$ of wheat and 33 sestiers $\frac{1}{3}$ of rye, making 100 sestiers in all which cost him 100 florins. But he said that the wheat cost 10 per cent more than the rye. One asks how much a sestier of wheat and of rye cost. Answer: a sestier of wheat cost 12 groschen $\frac{3}{8}$ and a sestier of rye 11 groschen and $\frac{1}{4}$.* (For the relationship between florins and groschen see Chapter 8).

[40] *A merchant finds a quintal of new wax at 14 ecus and a quintal of old wax at 9 ecus. Now in fact he only wants to use 11 ecus and he wishes to have a quintal of old and new wax. To determine how many pounds of new wax and how many pounds of old wax he must buy in order to have a quintal of wax which comes to him for 11 ecus...*
(40 pounds of new and 60 of old; a quintal is 100 pounds)

[41] *A pedlar bought a certain number of apples. One does not know how much they cost him. If he sells them again at 3 for a denier, he gains 2 deniers on them all, and if he sells them at 4 for a denier he gains 1 denier. To determine how many apples he bought, and how much they cost him...* *(12 apples, 2 deniers)*

Problems 42 to 46 are also concerned with buying and selling fruit. In 43, the answer is that 12 apples cost him -11 deniers, while in 45 he spent - $1\frac{13}{14}$ deniers on - $15\frac{3}{7}$ apples. There is a numerical error in Chuquet's answer to 46. Chuquet does not explain the negative answer to 45 (let alone what sort of people buy and sell fractions of apples), while in 43 he suggests that the pedlar obtained the apples from someone to whom he owed 11 deniers, and the debt was cancelled on condition that he sold the apples at the specified price. Problems 48 to 50 are also simple problems about the price of apples.

[47] *There is a gold cup whose base weighs $\frac{1}{3}$ and $\frac{1}{4}$ of the cup and lid. And the lid weighs $\frac{1}{4}$ and $\frac{1}{5}$ of the cup and the base, and the goblet weighs 7 ounces. To determine what the base and the lid of this cup weigh...*
($8\frac{5}{177}$ and $6\frac{45}{59}$ ounces)

[51] *A carpenter has agreed to build a house in 30 days, and that each day for which he laboured he would earn 5 ecus and each day for which he rested he would lose 6 ecus. It happens that, at the end of the 30 days when the house was built, this carpenter had laboured so many days and rested so many days that he neither gained nor lost money. To determine how many days he laboured and how many he rested...*

(He worked $16\frac{4}{11}$ days)

The next three problems deal with the same subject. There are also three problems written in the margin (by de la Roche), two about carpenters who worked at different rates, and one about three men buying a house jointly. Again, problems like 51, although the setting seems highly improbable, appear in many arithmetics and collections of puzzles of this period.

[53] *There are two carpenters who each wish to build a house in equal and like manner. One says to his fellow that if he will help him for 8 days with his house, it will be ready in 20 days. And the second replies that if the first will help him for 10 days, he will complete his in 15 days. One asks, if each one makes his own house separately, in how many days will they have built the houses. To do this problem and the like, one should know that the day's labour of the second compared to that of the first is such that when it is multiplied by 8 and added to 20 it comes to as much as if it were multiplied by 15 and added to 10. Now let us therefore set down 1^1, which multiplied by 8 and added to 20 comes to 8^1 plus 20. After this let us multiply 1^1 by 15 and add it to 10, and we shall have 15^1 plus 10 equal to 8^1 plus 20. Simplify the sides and you will have 10 on one side and 7^1 on the other. Divide 10 by 7 and you will have $1\frac{3}{7}$ and that is how much a day's labour of the second man is worth compared to a day's labour of the first. Now multiply 8 by $1\frac{3}{7}$ and then add 20, and you will have 31 days $\frac{3}{7}$, and in so many days will the first man's house be built. After this, divide $31\frac{3}{7}$ by $1\frac{3}{7}$ and you will have 22 days, and in that time the second will have completed his house without help from the first. Or otherwise, subtract 8 days that the first asks from the second from his 15 days. There remain 7 for*

divisor. Then take away the 10 days which the second one asks from the 20 days of the first and you will have 10 for the number to divide. So divide 10 by 7 and you will have $1\frac{3}{7}$ as above.

(In modern notation, solving the two equations $20a + 8b = H = 10a + 15b$ gives the relationship $10a = 7b$ by subtraction.)

[55] *A merchant has bought 60 pounds of saffron which cost him, or for which he handed over, 137 ecus, and there remains 18 ecus and also the value of 13 pounds of saffron. To determine how much 1 pound of saffron is worth...* *($3\frac{14}{47}$ ecus)*

[56] *A man has hired a servant to whom to gives 10 ecus and a garment of a certain price agreed between them each year. It happened that at the end of 5 months the master discharged him, and so that he might have profited from those 5 months he gave him the garment at the price which had been promised. To determine what the price of the garment is...* *($7\frac{1}{7}$ ecus)*

At this point, someone (de la Roche?) notes in the margin that there are 25 problems like the next, which follow one after the other. Problem 69 is reproduced here because of a surprising and unexplained remark that Berthelemy de Romans regarded it as impossible. Also included are a note on Chuquet's methods of solution, two problems that he recognized as indeterminate, and a final remark on alternative methods.

[69] *Further, two men have deniers in such number that if the first had 7 of the second man's deniers he would have 5 times as many as him plus 1. And the second says that if he had 5 of the other's, he would have 7 times as many as him and 12 more. To determine how many deniers each one has...* *($5\frac{11}{34}$, $9\frac{9}{34}$)*
... This calculation is impossible according to Master Berthelemy de Romans in his second book. However, trial shows that the contrary is true.

[70] *There are three men who have deniers in such proportion that if the first had 12 deniers from the two others with his own, he would have twice as many as what they have left, plus 6. The second replies and says that if he had 13 deniers from the first and the third*

he would have the quadruple of what they have left plus 2. And the third says that if he had 11 deniers from the first and the second he would have the triple of what they have left plus 3...

Simultaneous equations of the type

$$x + 12 = 12(y + z - 12) + 6$$
$$y + 13 = 4(x + z - 13) + 2$$
$$z + 11 = 3(x + y - 11) + 3$$

cannot easily be reduced by substitution to equations in a single unknown, particularly when no suitable algebraic notation is at hand. Chuquet's notation is especially unsuitable, since there is no symbol representing the variable; and in the *Triparty* he has said nothing about equations in more than one unknown.

Two methods of solution are used. In this problem, he combines the rule of two positions with the rule of first terms, choosing an arbitrary numerical value for one of the unknowns and putting one of the others equal to 1^1. He solves for the algebraic unknown, then calculates some function of the variables (what the second man has after he has received 13 deniers from the others) in two different ways to obtain the "position" as their difference; this he then repeats for another numerical value of the first unknown.

The second method, which according to the marginal annotations is called the "rule of quantity", is a purely algebraic method. Chuquet uses the notation 1^2 not as the square of the unknown, but as a second unknown, which he then evaluates in terms of 1^1. In fact (though unnecessarily), he also uses 1^2 for the third unknown before solving in numerical terms. This method would be perfectly satisfactory but for the fact that it introduces an ambiguity into his notation; in any given question, it is always clear whether 1^2 means x^2 or y.

[78] *Yet again, there are four men who have deniers in such proportion that if the first and second had 100 deniers from the other two they would have three times as much as the others had left. And the second and third together say that if they had 106 deniers from the fourth and the first, they would have four times*

> *what remained to the others. The third and fourth say that if they had 145 deniers from the other two, they would have five times what they had left. And the fourth and first say that if they had 170 deniers from the other two, they would have six times what they had left. To determine how many deniers each of them had...*

We have here four equations in the four unknowns $a+b$, $b+c$, $c+d$, $d+a$. These have a unique solution (obtained by the rule of quantity as given in problem 71) of $a+b = 125$, $b+c = 230$, $c+d = 205$, $d+a = 190$; but this in turn will produce a solution for any reasonable value of a - any value of a at all if negative solutions are acceptable. In Chuquet's words:

> *Thus it appears that such problems have a necessary answer for two by two, but for one by one they have whatever answer one desires.*

[82] *There are two people who have found a purse in which there is a certain sum of money. The first of them says to the other that if he had the money in the purse he would have three times as much as him. And the other replies and says that if he had the deniers in the purse he would have, with his own, 5 times as much as [the first]. To determine how many deniers there are in that purse and also how many each of them has.*

> *Answer and special rule for such problems. Inasmuch as the first said that he would have three times as much, and the other 5 times as much, 3 should be multiplied by 5 making 15, from which 1 must be taken away, leaving 14, and there are 14 deniers in the purse. Afterwards add 1 to 3 and also to 5, and you will have 4 deniers, which the first has, and 6 deniers for the second. And so must all the others be done. And whoever should wish to do such problems as the above by the rule of first terms, it would be necessary to make two positions, that is to say, one for the purse and the other position for one of the men. The position for the purse may be whatever number one likes, and the second position should be 1^1, and thus such problems have infinite responses. Example: in the above problem I suppose that there be*

30 deniers in the purse, and that the first man had 1^1...

And thus it appears that such questions do not have a necessary answer. All such problems as the preceding, or most of them, may be done by the rule of two positions, which has been treated in the Triparty, *in the first part.*

After this collection of problems related to simple and simultaneous equations in linear algebra, Chuquet moves on to illustrate some more traditional arithmetic methods and concepts, starting with the rule of apposition and remotion and more indeterminate equations.

Reckoning which is done by the rule of apposition and remotion

Problems which are dealt with and solved by a rule which the ancients called appostion and remotion can be sought by the rule of first terms, and by this one may find as many answers as one wishes, inasmuch as one or more of the numbers may be set at pleasure and the others are sought by the rule of first terms as it appears in the question above, and also as it may appear hereafter in several examples.

[83] *Example. A merchant has bought 30 aulnes of cloth which cost him 30 ecus in all. Of these, there are some that cost 4 ecus the aulne, and some at 2 ecus, and again some at $\frac{1}{2}$ ecu. To determine how many there are at each price, that is to say, at 4 ecus, at 2 ecus and at $\frac{1}{2}$ ecu.*

Answer. I suppose at my pleasure, for the first position, that there be 3 aulnes at the price of 4 ecus, with a vlue of 12 ecus, which should be taken away from 30 ecus. There remain 18 ecus. Now one must subtract 3 aulnes from 30 aulnes, leaving 27 aulnes. Now, to proceed further, it is necessary to make from 27 two parts such that when the one is multiplied by 2 ecus and the other multiplied by $\frac{1}{2}$, the two products added together amount to 18 ecus. To do this I suppose that one of the parts *be 1^1, thus the other will be $27\ \bar{m}\ 1^1$. Now 1^1 must be multiplied by 2, which amounts to 2^1, then $27\ \bar{m}\ 1^1$ by $\frac{1}{2}$ which amounts to $13\frac{1}{2}\ \bar{m}\ \frac{1}{2}^1$, which added to 2^1 comes to $13\frac{1}{2}\ \bar{p}\ 1^1\frac{1}{2}$ like to 18 ecus. Now the sides should be simplified, and one will find $1^1\frac{1}{2}$ for divisor and $4\frac{1}{2}$ for the number to be divided. So divide $4\frac{1}{2}$ by $1\frac{1}{2}$ and you will find 3 for one of the parts of 27, and in consequence 24 for the other. Thus there are 3 aulnes at the price of 4 ecus, and 3 aulnes at the price of 2 ecus, and 24 aulnes at the price of $\frac{1}{2}$ ecu, which make in all 30 aulnes for 30 ecus.*

Another answer: I wish that there should be 4 aulnes at the price of 4 ecus, which are worth 16 ecus. Taken from 30, there remain 14 ecus. Also, 4 aulnes taken from 30, there remain 26 aulnes. Now from 26 two parts must be made such that when the one is multiplied by 2 ecus and the other by $\frac{1}{2}$ ecu, the two products make 14 ecus. Proceed by the rule of first terms as it was done above, and you will find $\frac{2}{3}$ of an aulne at the price of 2 ecus and in consequence 24 aulnes $\frac{1}{3}$ at $\frac{1}{2}$ ecu the aulne. And in this way one can make as many answers as one likes...

In this particular problem, and in the following one which deals with buying different qualities of silver, there is no reason why fractional solutions should not be acceptable; but more traditional problems solved by the same method relate to buying three kinds of bird at different prices, or determining how many men, women and children there are in a party given the total consumption of drink, and require a solution in positive integers which may well be unique for the data in question. It is just because these problems are so simple to formulate, yet do not admit of a unique solution by algebraic methods, that they are still found in children's puzzle books, and even, with larger numbers, in brainteaser columns of more sophisticated newspapers.

The next section takes as its theme proportional numbers; two disguised equations on the finding of mean proportionals, seven problems of compound interest, a topic almost entirely omitted from the *Commercial Arithmetic* - perhaps because Chuquet did not want to restrict himself to the limited range of problems on this subject which could be solved by simple arithmetic and the rule of three alone - and four assorted problems of considerable interest, in two of which Chuquet expresses dissatisfaction with his answer.

Of the finding of proportional numbers and the problems that are done by them

[85] *Such proportion as there is between 12 ecus and the value of 5 aulnes of cloth, that proportion is between 5 aulnes of cloth and 15 ecus. To determine what an aulne of cloth is worth. I suppose that the 5 aulnes are worth 1^1. Then I shall say by the rule of three, if 12 ecus are worth 1^1, what is 1^1 worth? Multiply and divide and you will find $\frac{1}{12}{}^2$ equal to 15. Now divide 15 by $\frac{1}{12}{}^2$ and you will have $R^2 180$. And so many ecus are the 5 aulnes worth. Next divide $R^2 180$ by 5 times 5 and you will have $R^2 7\frac{1}{5}$ which is the value of an aulne. Or otherwise, to find a mean proportional between 12 and 15 multiply 12 by 15 making 180, whose second root is the mean proportional between 12 and 15. Or otherwise, put 12 over 15 and you will have $\frac{12}{15}$ which is simplified to $\frac{4}{5}$. Now divide 12 by $R^2\frac{4}{5}$. You will have $R^2 180$ as above, and whoever divides $R^2 180$ by $R^2\frac{4}{5}$ will have $R^2 225$ which is 15. And in this manner one may find one mean proportional between any numbers.*

[86] *Such proportion as there is between 12 ecus and the value of 5 aulnes of brunette, the same proportion is between the value of 5 aulnes of brunette and the value of 3 aulnes of scarlet, and between the value of 3 aulnes of scarlet and 15 ecus. To determine what the 5 aulnes of brunette are worth, and also the 3 aulnes of scarlet...* ($R^3 2160$, $R^3 2700$)

After solving this problem in 4 slightly different ways, Chuquet appends a note on how to find three, four or five mean proportionals. There is a marginal note (by de la Roche?) on "the property and nature of proportional numbers".

[87] *One merchant has lent to another the sum of 20 ecus at a certain interest, where the interest at the end of the year earns like the principal. And at the end of two years the 20 ecus have become 30. One asks at what*

interest per cent the 20 ecus were lent... ($R^2 15000-100$)

[88] *Someone lent 3 ecus at a certain interest, which 3 ecus, on account of the interest, had become 7 ecus in three years. One asks how much they earned in the first year and also in the other years... And I wish to know at what interest per cent the 3 ecus were lent...*

(*First year:* $R^3 63=3$; *Second:* $R^3 147-R^3 63$
Third: $7-R^3 147$; *Interest:* $R^3 2333333\frac{1}{3}-100$)

[89] *A merchant has lent to another a sum of money at the interest of 10 per cent, and the interest earned like the principal at the end of every year. It happened that at the end of three years, the debtor is found to owe, as much in interest as in principal, the sum of 100 livres. To determine how much had been lent to him in the first year...* ($75\frac{175}{1331}$)

[90] *A man has lent to another 200 ecus at the interest of 10 per cent at the end of the year, that is to say that the interest at the end of the year earned like the principal. And the other man has leased him a house in such a way that if he does not pay at the end of the year he will incur the same interest. It happens that at the end of three years the rent for the house with the interest for the aforesaid three years comes to as much as the 200 ecus with their interest. To determine at what price the house was let...* ($80\frac{140}{331}$ *ecus*)

[91] *A money changer lent a merchant an unknown number of livres at the rate of 4 deniers in the pound interest per month. The first year he returned him 50 livres. And at the end of the second year he gave over 70 livres to him. And he said, You have paid both the principal and the interest. To know how many livres the money changer had lent to the merchant...*

($90\frac{5}{18}$ *livres*)

The next two problems also deal with transactions between a merchant and a money changer at compound interest, but settlement has to be made in the middle of a year, for which Chuquet advocates finding mean proportionals - eleven if settlement is required after one month. Although the interest in problem 91 is quoted as so much per month, the answer makes it clear that the interest was only compounded at the end of the year.

Some authors of early printed arithmetics felt it necessary to preface their treatment of compound interest with some reflections on the morality or otherwise of this practice. Chuquet, however, had no wider audience in mind than the mercantile community of Lyon, who probably could not afford such ethical niceties. 36 years later, de la Roche took his passage on the subject from Luca Pacioli, who, since his mathematical career was dependent on the goodwill of his Order, may have felt obliged to try to reconcile Church teaching with commercial practice. For an interesting comparison of a number of sixteenth-century French texts on the subject, see N. Z. Davis (1960).

[94] *Yet again, a vessel has a tap through which, if it is opened, the tenth part of its contents is lost every day. To know in how many days half the vessel will be emptied. Response and rule for such problems. Suppose the vessel holds a large number of measures, from which number we can easily subtract the tenth part several times, such as 10000000. And for each time let us count one day until one arrives at two consecutive days of which one is more than a half of 10000000 and the other less. This part of a day should be sought by the rule of three. And thus one will find that this vessel will be emptied in 6 days* $\frac{314410}{531441}$ *of a day, as far as half its contents.*

Or otherwise take the $\frac{9}{10}$ *of 1, which is* $\frac{9}{10}$ *for the first day, then take* $\frac{9}{10}$ *of that* $\frac{9}{10}$ *and you will have* $\frac{81}{100}$ *for the second day. Again, of* $\frac{81}{100}$ *take* $\frac{9}{10}$ *and you will have* $\frac{729}{1000}$ *for the third day. And continue thus until you come to two consecutive broken numbers one of which is greater than* $\frac{1}{2}$ *and the other less. By continuing in this manner until the sixth day you will find* $\frac{531441}{1000000}$ *which is greater than* $\frac{1}{2}$, *and on the seventh day one will find* $\frac{4732969}{10000000}$ *which is less than* $\frac{1}{2}$. *Now, from that which is greater than* $\frac{1}{2}$, *that is from 31441, make the numerator, and from* $\frac{1}{2}$ *with the surplus, that is 531441, make the denominator and you will have* $\frac{31441}{531441}$ *of a day, and thus in 6 days and* $\frac{31441}{531441}$ *of a day, half that vessel will have been emptied. And some people are content with this manner of solution. However, it seems that for the greatest truth, one should search for a certain proportional number between 6 and 7 which is, for the present, unknown to us.*

We have not been able to trace a similar problem, in which the answer is required to be a fractional number of days, in spite of Chuquet's closing remark which suggests that he has taken not only the problem but at least one of his methods of solution from other writers. His second method is completely wrong, and it can only be speculated that this is one occasion when his "rule of method" (see commentary following problem 124) misled him because of a chance coincidence of figures. His first method, which amounts to linear interpolation, gives 6.592 to 3 decimal places, as opposed to the true answer (obtained by using logarithms to solve $0.9^x = 0.5$) of 6.579. It may be said that by realizing the inadequacy of his method Chuquet took one step towards the discovery of logarithms; another step was his grasp of the relationship between the multiplication of powers of a number and the addition of the corresponding indices as evinced in several places in the *Triparty*. But these two steps neither constitute an anticipation of logarithms nor (since the manuscript remained unknown to scholars with the exception of de la Roche, who himself had little originality and only limited influence) did they do anything towards advancing the history of mathematics in that direction. One thing Chuquet lacked was adequate motivation, for he had little interest in the numerical value of a solution, preferring to give his answers as exact sums of roots etc.

The next three problems represent a perversion of the study of proportion, and in the first case Chuquet himself acknowledges the fact. In the other two, he applies the formula for the sum of the first n terms of a geometric progression, without comment, to finding the sum of a fractional number of terms; that is, he finds the $9\frac{1}{2}$th term or $5\frac{1}{2}$th term, subtracts 1, and divides by the common ratio minus - a bold generalization exercised with insufficient prudence. There is a fuller discussion of this point following problem 105, where he takes Berthelemy de Romans to task for using a method based on linear interpolation.

[95] A merchant has made as many journeys as he had pieces of gold, in such a way that on the first journey he made from 3, 4 and 1 more. On the second journey he made from 3, 4 and 2 more. On the third journey he did as above, and 3 more. And thus continuing, always making from 3, 4 and progressing by 1, so that at the end of his journey 15 pieces of gold are found. To determine how many pieces he had at the beginning...

$(R^2 50\frac{16297}{16384} - 4\frac{13}{28})$

... However, it seems to some people that between 3 and 4 journeys one should find some proportional numbers and take one of them as the answer to this question. But such an argument is 0. For the numbers in the three journeys are not proportional and do not constitute any progression, whence that which is beyond the third journey should not be either proportional or in progression. Thus this calculation remans true.

[96] There is a vessel holding $9\frac{1}{2}$ barrels which has a tap of such a size that through it the first barrel would be emptied in 1 hour, the second in 2 hours, the third in 4 hours, the fourth in 8, and so on, doubling the hours each time. To determine in how many hours the $9\frac{1}{2}$ barrels will be emptied... $(R^2 524288 - 1 = \sqrt{2^{19}} - 1)$

A *barrel* could be either a measure of liquid, as here, or a container. Authorities differ on the exact dimensions - one estimate is 1 barrel = 36 pints, but Cotgrave (1611) thought it was a vessel holding 72 Parisian pints, or about 72 English quarts.

[97] Again, a man goes one league the first day, 3 leagues the second day, 9 leagues the third day, and so on, tripling his day's journeys. To determine how many leagues he will have travelled in 5 days $\frac{1}{2}$...

$(R^2 44286\frac{3}{4} - \frac{1}{2})$

This manner of proceeding derives from the nature of numbers in triple proportion.

Questions that are solved by the nature of progressions of numbers.

Questions 98 to 108 resemble the last two problems above in their general setting and in the improbability of their assumptions. Two are concerned with wine coming out of a vessel, the remainder with one or two travellers at least one of whom increases his day's journey each day in arithmetic progression - a familiar source of problems on this topic. Again, Chuquet regards it as perfectly acceptable to use the formula for the sum of an arithmetic progression to cover the case of a fractional number of terms. One example, followed by a criticism of a less radical method, will suffice.

[105] *Again, there is a man who is fleeing and goes 14 leagues each day. And another follows him immediately, going after him, who travels 3 leagues on the first day, 6 leagues on the second day, 9 leagues on the third day, and so on, augmenting his day's journeys by 3 leagues. To determine in how many days he will have overtaken his man. To do this it is necessary to find a number of numbers such that if 1 is added and this addition is multiplied by the sesquialter of the number of numbers, this multiplication comes to as much as if that number of numbers were multiplied by 14. And to find this I suppose that this number be 1^{1}... And in 8 days $\frac{1}{3}$ he will have overtaken him. The proof: Add 1 to $8\frac{1}{3}$ making $9\frac{1}{3}$. Multiply this by $12\frac{1}{2}$, which is the sesquialter of $8\frac{1}{3}$, and the multiplication comes to $116\frac{2}{3}$. Thus this calculation is true.*

However, master Berthelemy de Romans, once of the Order of Friars Preacher at Valence and Doctor of Theology does these calculations otherwise, and in this manner. He divides 14 leagues by 3 which is the number of the progression, and from this comes 4 wholes, and he does not heed the fraction. This 4 he doubles, making 8. Now he considers whether, in 8 days, the man who is progressing and going after will have reached his man, putting 8 days' travel thus: 3, 6, 9, 12, 15, 18, 21, 24, coming to 108. Then he sees how many leagues the man who is fleeing has done, by multiplying 14 by 8, making 112. And thus he perceives that the fugitive has not been caught. 4 more leagues are needed, and therefore he adds one more day and finds that the progression comes to 126 *and 9 times 14 makes 126, hence between 8 and 9 days the fugitive is caught. And to know in what part of the day: of the 4 leagues that remain on the 8th day, he makes the numerator, and of what remains of the 27 leagues which the man who is progressing does in the 9th day when 14 has been subtracted he makes the denominator, which is 13, and thus he has $\frac{4}{13}$. And he says that in 8 days $\frac{4}{13}$ he is caught...*

(There follow details of Berthelemy's method of checking his answer.)

... But in this manner of proceeding the nature of the progression is not observed, whence it should be considered of no value.

Since it is impossible to take the problem seriously, any judgement between Chuquet's model (which assumes that the man who is pursuing the fugitive walks faster as the day goes on) and Berthelemy's (which assumes that he travels at a constant pace throughout any given day) must be arbitrary. Berthelemy's seems less unlikely than Chuquet's, but other theories are possible - the second man might have come to spend a greater part of each day on the road, hired faster horses, or found the terrain more favourable. Moreover, given Chuquet's approach to the questions of proportion discussed above, it is reasonable to conclude that he was not constructing a model at all, but simply applying a formula.

[109] *A merchant bought 10 measures, of oats as well as of wheat, which cost him 10 livres, that is to say, 100 sols in wheat and 100 sols in oats. And a measure of wheat cost him 2 sols more than a measure of oats. To determine how many measures of oats and also how many measures of corn that merchant bought ...*
($R^2 2525$ - 45 measures of oats, 55 - $R^2 2525$ of wheat)

[110] *A money changer has bought 12 ounces of gold, some of which is alloyed with silver, and this cost him 48 livres. The rest is alloyed with copper, and this likewise cost him 48 livres. However, each ounce of the gold alloyed with silver cost him 5 livres more than an ounce of the gold alloyed with copper. To determine how many ounces there were which were alloyed*

with copper ...
($R^2\ 128\frac{4}{25} - 3\frac{3}{5}$ ounces alloyed with copper,
$15\frac{3}{5} - R^2\ 128\frac{4}{25}$ with silver)

Three problems dealing with money shared between two or three men are omitted here.

[114] *Again, of $R^3 72$ I want to make two parts such that when one is multiplied by the other, the multiplication comes to $R^2 60$ ...* ($R^3 9 + R^6 \underline{81 - R^2 60}$, $R^3 9 - R^6 \underline{81 - R^2 60}$

There are several orthographic errors in the working of this question: the most serious is the writing of $R^2\ \underline{R^3 81\ \bar{m}\ R^2 60}$ instead of $R^2\ R^3 \overline{81\ \bar{m}\ R^2 60}$. More interesting is the marginal note in Chuquet's hand which begins as follows:

This calculation is impossible. And the reason is that whoever divides a number into two unequal parts and then multiplies one part by the other, the answer cannot be more than that of multiplying the half of that number by itself ...

The two equal parts of $R^3 72$ would be $R^3 9$, hence the product cannot be greater than $R^3 81$; but $R^3 60$ is greater than $R^3 81$. Another three problems about the sharing of money are omitted.

[118] *There is a nobleman who has a cup of gold at 13 carats $\frac{5}{7}$, in which there is so much copper that if the amount which it contains were multiplied in a cube and the number of marcs of gold were multiplied in a square, the two multiplications would be equal. One asks what this cup weighs ...*

This question is followed by two further problems written into the margin, probably in Chuquet's hand, one numerical question arising from 118, and one similar problem about a cup made from silver alloy.

Questions which are solved by special and extraordinary rules

Special rules, Chuquet says in his solution to the next problem, are for the most part easier than general rules –

easier to apply, perhaps, once they have been formulated and memorized, but harder to understand; and most of the questions in the section are extremely mundane compared to many other parts of the book. To of them must serve as a sample: 119 involves an unknown quantity not merely in its solution, but also in its formulation, while 124A, so numbered because it escaped the attention of Marre in (1881), is followed by Chuquet's rule of method, in which he describes how he discovers these special rules. Cardano formulates a similar principle in his *Ars Magna* (1545, Chapter 29).

[119] *From $1^1\frac{1}{2}$ I wish to make two parts such that when 13 is taken away from the second to be added to the first, this first part will be the quadruple of the second and 2 more. To do this I shall take an ordinary number like 20, and from it I shall make two parts, as stated above, by a special rule, for special rules are for the most part easier than general rules. And to do this I consider that when the first part is combined with 13, it will be $\frac{4}{5}$ of 20, for it will be the quadruple of the other; and inasmuch as it will be four times as much as the other plus 2, this plus 2 should be divided by 5, which is the denominator of $\frac{4}{4}$, and it will be $\frac{2}{5}$. And thus the first part by itself should be $\frac{4}{5}$ of 20 $\bar{m}$ 13 plus $\frac{2}{5}$ which is $3\frac{2}{5}$...And in the same manner as I have done with 20, I wish to do with $1^1\frac{1}{2}$ plus 15...*

$$(1^1\ \tfrac{1}{5} - \tfrac{3}{5},\ \tfrac{3}{10}{}^1 + 15\tfrac{3}{5})$$

This is not quite a solution by rule of false position; rather, it is an example of testing an algebraic method by trying the problem with numbers first. In the margin, Chuquet gives an alternative solution using the rule of first terms, in which the variable is denoted as 1^2, as in the rule of quantity (see comment on problem 70).

[124A] *Further, from 19 I wish to make two parts such that when one adds 8 to the second, and from this addition one takes away 13 to add it to the first, this first amounts to four times as much as the second plus 2. To do this I consider that whoever adds 8 to the second, and from this addition takes away 13 to add it to the first, this is nothing but taking 5 from the second. Also when I give 13 to the first it is four times as much as the remainder of the second plus 2. And therefore I take 2 from 13 leaving 11 to add to the first. By which this question is none other than to make two parts of 19 such that when the first is added to 11 and 5 taken from the second, these two numbers shall be in quadruple proportion. This question is solved above, in the preceding problem.*

<u>The style of finding and seeking the aforesaid special rules and others like them.</u>

Although such problems as the above may easily be done by the special rule of first terms, yet these special rules often serve in place of the rule of first terms in the inquiry into difficult matters which may be helped thereby, and therefore such special rules are propitious and necessary. The style and manner of seeking and making such special rules is that one should carry out, by the rule of first terms or by another, two or three problems of a similar nature with various numbers. And then syllogize and reflect on the known numbers, how they might have been found had they been unknown, by adding or subtracting, multiplying or dividing the whole or part with the whole or part in several and divers manners, as much by consideration of the properties of numbers as by experiments made with several and divers numbers, until one has found a means by which one might arrive at the numbers of the known numbers and also the unknowns constituted in like proportion. And let this be a general precept for finding new and marvellous things in the science of numbers.

The next group of problems is again concerned with one or two travellers, and although special rules are introduced, the only novelty is that these travellers appear to lose ground at night. They are followed by a large number of questions about a father sharing a treasure-chest of money with his children. One of these will serve as an example, though the answer may be noted here; there are $2\frac{5}{6}$ children; each of the two has $6\frac{2}{3}$ deniers and the $\frac{5}{6}$ of a child has $5\frac{5}{9}$ deniers.

[129] *There is a father of a family who has children - one does not know the number - and he has in his treasure chest a sum of deniers, whose amount one does not know. And this father says to the first of his*

children, Go to the treasure chest and take 1 denier and $\frac{1}{10}$ of the remainder. To the second he says, Go after your brother and after he has taken what I told him, take 2 deniers from the treasure chest and the $\frac{1}{10}$ part of the residue. To the third he says, Go after your brothers and after they each have their portion, take 3 deniers and the $\frac{1}{10}$ part of the rest. And thus he says to the others, always augmenting the portion of his children by 1 denier and taking the $\frac{1}{10}$ part of the rest, up to the last one, to whom he says, Go to the treasure chest and take all the deniers that your brothers have left to you. And when this is done, the children find that they each have as many deniers as the others. To determine how many children this man has, how many deniers were in the treasure chest, and also how many deniers each one of them had ...

(9 children, 9 deniers each)

Chuquet gives a general rule for the class of problems: if the children take first a sum of money in the progression a, $a+b$, $a+2d$, ... and then $\frac{1}{n}$ of the remainder, then each child has $(n-1)d$ deniers, and the total number of deniers is $((n-1)d-a)n+a)$. However, at the end of the section he gives alternative solutions using the rule of first terms.

More calculations, touching the case of weights

[142] *There is a man who has a balance with three weights, which weigh 10 pounds all together, and they are such that one can weigh from 1 pound to 10 pounds, neither more nor less than if he were provided with all sorts of weights. To determine how much each one weighs ...*

W. W. Rouse Ball, in his *Mathematical recreations and problems* (1892) describes a version of this problem as among the more difficult proposed by Bachet (1612). Although Chuquet formulated the problem in terms of weighing up to 10 pounds and gave the solution 1, 2, 7, his discussion extends up to and beyond Bachet's problem of weighing up to 40 pounds, with two solutions, one based on powers of 2 and the other on powers of 3, the latter being "more convenient", as fewer weights are required, though the former did not require a two-sided balance.

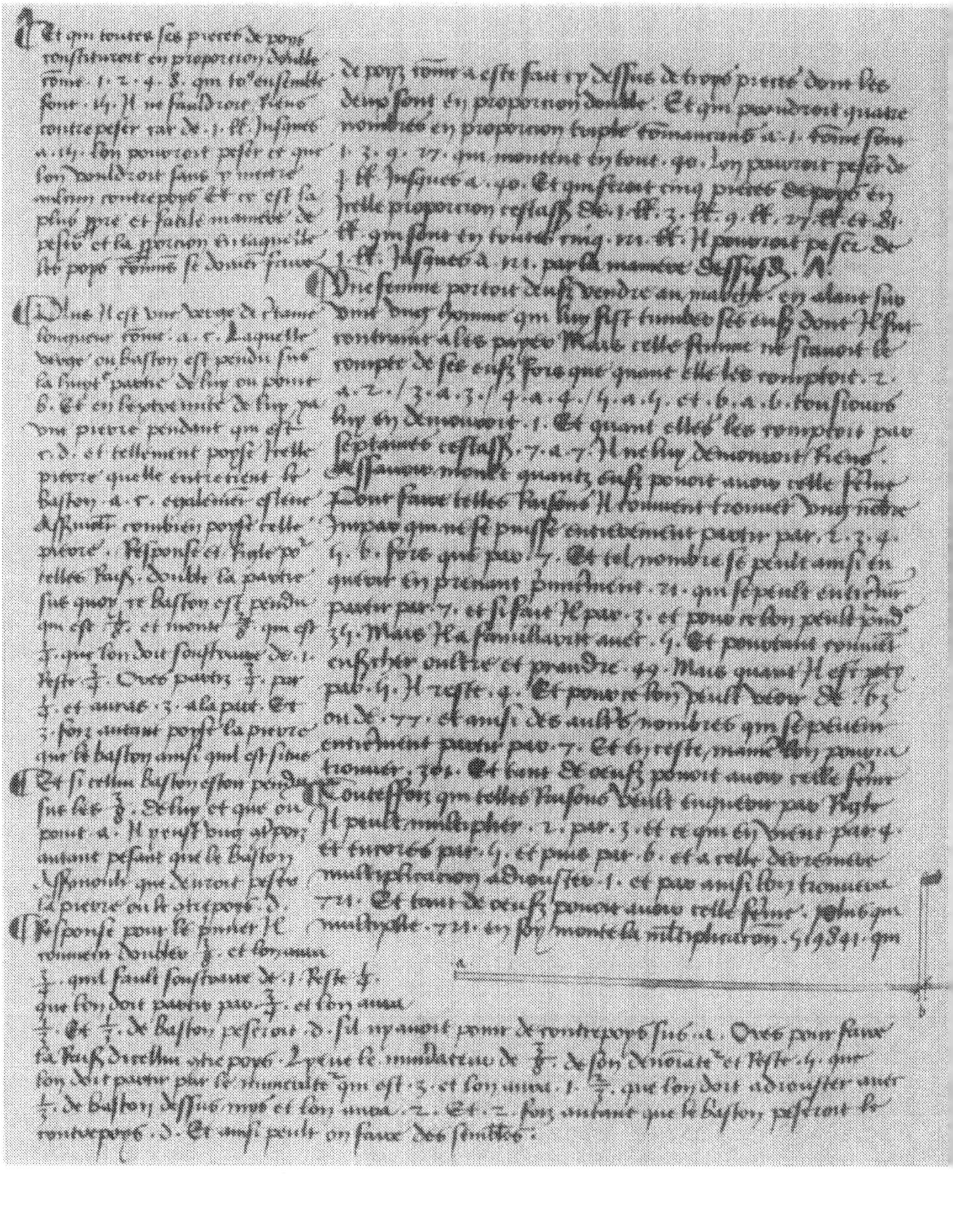

¶ Et qui toutes ses pieces de poys constitueroit en proporcion double comme .1. 2. 4. 8. qui tous ensemble font .15. il ne fauldroit riens contrepeser car de .1. lb. jusques a .15. l'on pourroit peser ce que l'on vouldroit sans y mettre aucun contrepoys. Et ce est la plus gente et subtile maniere de peser et la porcion en laquelle les poys comme se doivent faire de poys comme a este fait cy dessus de troys pieces dont les deux sont en proporcion double. Et qui prendroit quatre nombres en proporcion triple commencans a .1. comme sont 1. 3. 9. 27. qui montent en tout .40. l'on pourroit peser de 1 lb. jusques a .40. Et qui feroit cinq pieces de poys en telle proporcion c'est assavoir de .1. lb. 3. lb. 9. lb. 27. lb. et 81. lb. qui font en toutes cinq .121. lb. Il pourroit peser de 1. lb. jusques a .121. par la maniere dessusdicte.

¶ Plus il est une verge de chasne longueur comme .a. c. Laquelle verge ou baston est pendu sur la huytiesme partie de lui ou point b. Et en l'extremite de lui pa[r] une pierre pendant qui est c. d. et tellement poyse icelle pierre qu'elle entretient le baston .a. c. equalement estant. Assavoir combien poyse celle pierre. Responce et regle pour telles raisons. Double la partie sur quoy ce baston est pendu qui est $\frac{1}{8}$. et monte $\frac{2}{8}$ qui est $\frac{1}{4}$ que l'on doit soustraire de .1. reste $\frac{3}{4}$. Ores partis $\frac{3}{4}$ par $\frac{1}{4}$ et auras .3. a la part. Et 3 foiz autant poyse la pierre que le baston ainsi qu'il est situe.

¶ Et si cellui baston estoit pendu sur les $\frac{3}{8}$ de lui et que ou point .a. il pendist ung autre contrepoys autant pesant que le baston assavoir que devroit peser la pierre ou le contrepoys .d.

¶ Responce pour le premier il convient doubler $\frac{3}{8}$ et l'on aura $\frac{3}{4}$ qu'il fault soustraire de .1. Reste $\frac{1}{4}$ que l'on doit partir par $\frac{3}{4}$ et l'on aura $\frac{1}{3}$. Et $\frac{1}{3}$ de baston peseroit .d. s'il n'y avoit point de contrepoys sur .a. Ores pour faire la raison d'icellui contrepoys. prens le numerateur de $\frac{3}{8}$ de son denominateur et reste .5. que l'on doit partir par le numerateur qui est .3. et l'on aura .1 $\frac{2}{3}$. que l'on doit adiouster avec $\frac{1}{3}$ de baston dessus mis et l'on aura .2. Et .2. foiz autant que le baston peseroit le contrepoys .d. Et ainsi peult on faire des semblables.

¶ Une femme portoit des oeufz vendre au marche. en alant sur voie ung homme qui lui fist tumber ses oeufz dont il fut contraint a les payer. Mais celle femme ne savoit le compte de ses oeufz fors que quant elle les comptoit .2. a .2. / 3. a .3. / 4. a .4. / 5. a .5. et .6. a .6. tousjours lui en demouroit .1. Et quant elle les comptoit par septaines c'est assavoir .7. a .7. il ne lui demouroit riens. Assavoir moult quantz oeufz povoit avoir celle femme. Pour faire telles raisons il convient trouver ung nombre impar qui ne se puisse entierement partir par .2. 3. 4. 5. 6. fors que par .7. Et tel nombre se peult ainsi enquerir en prenant premierement .21. qui se peult entierement partir par .7. et si fait il par .3. et pour ce l'on peult prendre 35. Mais il a familiarite avec .5. Et pourtant convient enserchier oultre et prendre .49. Mais quant il est party par .5. il reste .4. Et pour ce l'on peult veoir de .63 ou de .77. et ainsi des autres nombres qui se peuvent entierement partir par .7. Et en reste, mais l'on pourra trouver .301. Et tant de oeufz povoit avoir celle femme.

¶ Toutesfois qui telles raisons veult enquerir par regle il peult multiplier .2. par .3. et ce qui en vient par .4. et encores par .5. et puis par .6. et a cette derreniere multiplicacion adiouster .1. et par ainsi l'on trouvera 721. Et tant de oeufz povoit avoir celle femme. Item qui multiplie .721. en soy monte la multiplicacion .519841. qui

Another problem concerning balances has been written in the margin at this point, probably by Chuquet; it relates to balancing a stick which is suspended one-eighth of the way along, and kept level by a weight hanging from one end, and is illustrated by a diagram.

[143] *A woman was carrying eggs to sell at market. On the way, a man came upon her and made her drop the eggs, for which he was obliged to pay. But this woman did not know how many eggs she had, except that when she counted them 2 by 2, 3 by 3, 4 by 4, 5 by 5 and 6 by 6, she was always left with 1. And when she counted them 7 by 7 there were none left over. To know how many eggs this woman could have had ...*

(301, by trial and error)

However, whoever wishes to investigate such calculations by a rule can multiply 2 by 3, and what comes from this by 4, and again by 5, and then by 6, and to this last multiplication add 1, and thus one will find 721. And this woman could have had so many eggs.

Further, whoever multiplies 721 by itself, the multiplication comes to 519841, which is also another true response. Thus it appears that such questions may have several and divers responses. Again, whoever multiplies 301 by itself will have 90601, which has the aforesaid property.

The next problem is identical except that the eggs were not counted two by two.

[145] *Three women are taking apples to sell at the market. One of them has 50, the second 30, and the third 10. And at such a price as one sells her apples, the others sell at the same price, that is, the one gives as many apples for a denier as the others. And after they have sold everything, they find that each has the same amount of money as the others. To know how they sold their apples.*

Chuquet's first "solution" - as he points out, others may be found by the rule of apposition and remotion - is that they each sold the apples at 7 for one denier, and those that were left over at 1 apple for 3 deniers. Dudney includes it in one of his collections, though expressing grave dissatisfaction with the answer.

[146] *There is a galley master who has in his care and charge 15 Christians and 15 Jews. This captain, because of the storming of the waves of the sea, is obliged to relieve the ship of half its occupants to save the other half, or otherwise all would perish, and the ship would sink in the sea. Now that captain desired to find means by which only the Jews would be thrown into the sea, and the Christians saved, but he suspects that the Jews would rebel and that they would refuse his command. In any case, to the end that he might not be seen to make distinctions between persons, favouring some more than others, he made them sit, one beside the other, in a circular figure, with the Christians intermingled with the Jews by a certain rule. And then he said, It is necessary that half of all you others be thrown into the sea, to save the rest. You see the great peril that we are in. I command that, counting round from 1 to 9 and then beginning again at 1 and continuing to 9 and continuing thus, that always the ninth be thrown into the sea. To this command they all consented. Then he began to count 1 2 3 4 etc. up to the ninth who was thrown into the sea, and then at the 10th he began again 1 2 3 etc. up to the ninth, who was likewise thrown into the sea. And this count was repeated so many times that 15 people were thrown into the sea. And this was done with such cunning that the 15 Christians were kept and saved, and the 15 Jews were suffocated in the sea. To know by what rule they were placed...*

This is sometimes known as the Josephus problem, after the Jewish historian of the first century AD, who tells the following story in *The Jewish War* (Translation, 1959, pp 202, 403). After the fall of the town of Jotapata, Josephus was anxious to yield to the Romans, but his companions were bent on mass suicide rather than surrender. Josephus "staked his life on one last throw", suggesting that they should draw lots to determine the order in which they should die. Then, according to one version of the manuscript, "he counted the numbers cunningly, and so deceived them all". In any case, he and one of his companions survived and surrendered to the Romans. The problem is comparatively rare

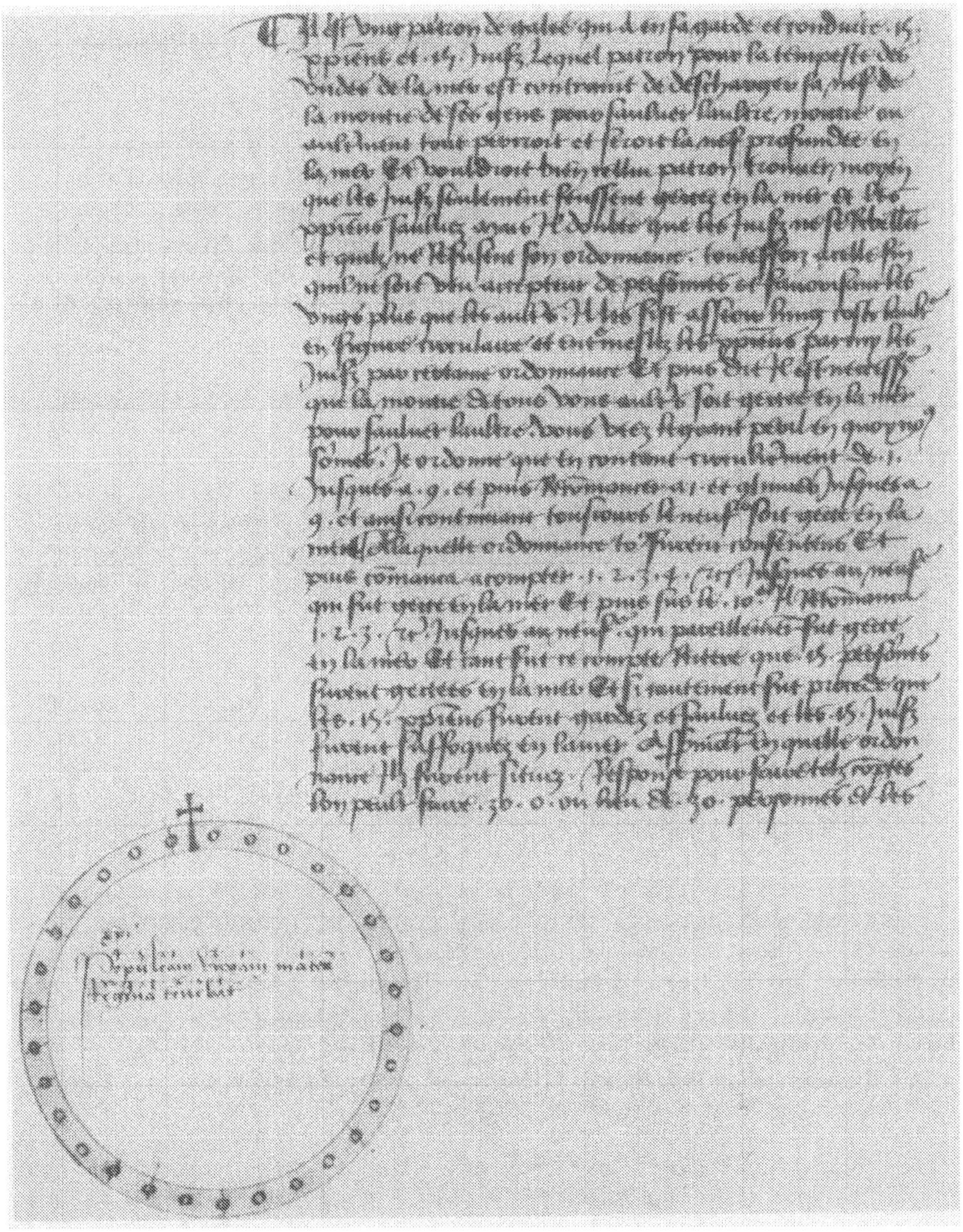

in fifteenth-century texts, though Chuquet is not the first Western writer of his time to mention it. The method of solution he describes involves drawing a picture and working out which positions will result in drowning, but his diagram is accompanied by the Latin slogan

4 5 21 3 1 1 2 2 3 1 2 2 1
Populeam virgam matre regina tenebat.

The numbers written over the vowels represent the position of that vowel in the alphabet, and the sentence should be interpreted as follows: 4 Christians, then 5 Jews, then another 2 Christians, then a Jew,... Chuquet also suggests generalizations with different numbers of Christian and Jews, and with throwing out the 10th or 11th instead of the 9th.

The next two problems (one about a man boiling down a large quantity of wine, the other about three men dicing for the money they have found in a purse) do not break new ground mathematically. They are followed by six problems about finding square and cube numbers with certain properties.

[*149*] *Rule for finding a square number which added to another number, whether square or not square, the addition shall be a square number. There is such a style, as, for example, I wish to find a square number which added to 7 shall be a square number. From 7, 1 must be taken away. There remain 6, of which the half is 3, which multiplied by itself comes to 9. And 9 is the square number which added to 7 makes 16 which is a square number.*

The collection of problems ends with an assortment of "Games and recreations which are done by the science of numbers". Most of these are variants on the theme of 'guessing' a number, or rather, deducing it from its simple mathematical properties. These must have enjoyed a popularity that is hard to imagine from our present level of mathematical sophistication, for they enjoy the same prominence in Bachet's *Problemes plaisans et delectables* (1612).

[*155*] *And first there follows the game of nine, which is this. If you want to know the number that someone has thought of, and holds in his mind, tell him to triple the number he has thought of, and from this triple to take the*

half. And make him triple this again, and from this triple make him take away one or several nines, like 18, 27, 36, etc. And so continue and repeat this several times until from his number he can no longer take away whole nines. Then again, tell him that from what remains he should take away 1 or 2 or 3 or some other number which is less than 9. For if he can take away 1 only, or 2, or some number, count it as 1, then secretly, for each nine which he has subtracted from his number, count 2. And thus you will know what number he has thought of.

Chuquet does append a numerical example but it is hard to see the fascination of this 'trick', though the sentence at the end of the example below may suggest an explanation.

[157] *The game of the circle: If you have a number of pieces of silver or other objects placed in a circle, and a man chooses one in his mind, and if you want to make him believe that you know which one he has though of, determine secretly the number of all the pieces. Then select one openly and tell him to begin to count at that one, in such a way that it shall be the first and the one that he has chosen shall be the last. Then know in which direction he counted, that is to say, whether he counted moving to the right or to the left. And count likewise in the same direction, over all the number of pieces and count as far as you like secretly, and the last one that you count is to be counted twice. After this, tell him to begin to count on his secret number where you left off counting, and to count backwards and in the direction opposite to that in which you and he have counted. And he should count quietly and secretly up to the number to which you yourself have counted. And tell him that where he reaches your number, that is the one he thought of.* (An example follows, and a simple diagram accompanies this example.) *And by this game one may know the lover or the beloved when they are in a dance, or assembled in order. And by the game of nine one can say on which day of the week or month the lover kissed his sweetheart.*

This suggests a party game for guessing information that is not exactly secret, but which people would prefer to reveal obliquely.

The other guessing games in this section will now be summarized briefly. In the game of equal things, the subject has the same number of coins in each hand. Tell him to transfer some number of coins from one hand to the other (if he cannot do this, choose a smaller number), and then put back into the first hand as many coins as remain there. You now amaze him by telling him how many coins he has in the other hand (i.e. twice the number you told him to transfer), but Chuquet suggests following this up with the game of nine on the coins in the first hand to discover how many coins he has in all.

In the game of two different things, the subject has gold in one hand and silver in the other. He is told that the gold is worth 4 and the silver 3, and asked to multiply what he has in his right hand by 3 and what he has in his left hand by 4, add the products together and say whether the result is odd or even. The answer could be reached, Chuquet says, by a single multiplication, but the trick would then be too obvious.

The game of three different things involves three men and 24 counters. The first man is given 1 counter, the second 2 and the third 3. Then whoever selected the first of the three different things is to take 1 counter for every counter he has been given, the man who selected the second thing is to take 2 for every counter he has been given, and the other should take 4 for every counter he has been given. It is now possible to work out, from the number of counters remaining on the table, who has what. Chuquet gives a table to memorize the six possible outcomes, while Bachet quotes a mnemonic sentence.

The game of the ring and the game of three dice each involve determining three numbers (in the former case, the number of the person who has the ring, which finger they have it on, and which joint of that finger). Ask for the result of the following calculation: double the first number, add 5, multiply by 5 and add the second number, then put the third number at the right hand of the answer. From the number so formed, subtract 250, and the three digits are the three numbers to be found.

The last game is another "think of a number" problem, in which what is found is not the number originally thought of, but the result of applying a series of calculations to it. Chuquet classifies the problem incorrectly in regarding it as equivalent to the game of nine.

The last four examples in this part of the manuscript are not games, but problems in recreational mathematics. The story of the wolf, the goat and the cabbage is too well known to be included here; it was so familiar in the Middle Ages that one twelfth-century manuscript claims that every little boy of five knows the answer. The tale of three husbands and wives crossing a river was generalized in the sixteenth century, with much discussion as to whether there was a solution for four couples; Rouse Ball (1892) mentions further extensions of the problem. The tavern-keeper's problem asked for measurement of 4 pints of wine using only a 3-pint and a 5-pint measure. Finally, a non-mathematical puzzle about two women who claimed that the babies they were carrying in their arms were "sons of our sons, and brothers of our husbands, and all in legal marriage" is included:

> *Although this contemplation is not done by a numerical calculation, nevertheless it is set down here because to some the matter may be new and merry.*

This collection of mathematical recreations, though it may be no more original in content than the rest of the *Problems*, is one place where Chuquet may have influenced the development of mathematics, or at least the communication of mathematical culture. Several of the recreations (and the more recreational problems in the body of the work) are included in Bachet's *Problemes plaisans et delectables*, and though neither Chuquet nor de la Roche is named among Bachet's many sources, yet de la Roche at least was sufficiently well-known in Italy that there may have been links in the chain that extends from Alcuin and his sources to Martin Gardner and his successors.

Chapter 7

THE GEOMETRY

Chuquet's *Geometry* begins with two phrases which delineate its character and its interest for the history of mathematics. It is a "small treatise in practical geometry" which illustrates "how the science of numbers can be applied to the measurements of geometry." The first part of the *Geometry* is concerned with basic geometrical ideas. Chuquet also discusses measuring devices such as the quadrant and gauges for wine vessels, and their uses in practical measurement; these topics were part of the tradition of practical geometry. The algebra developed in the *Triparty* is illustrated in the longest section of the *Geometry*, the third section, in which a series of problems applies the rules developed both in the *Triparty* and in the first section of the *Geometry*.

The tradition of practical geometry in the later Middle Ages rests on the distinction between practical and theoretical geometry. An early instance of this distinction was that of Hugh of St. Victor, in the early 12th century, who wrote "practical [geometry] is that which is done by means of certain instruments and which judges by inferring proportionally some distance from some others." (Cited from Stephen Victor, *Practical Geometry in the High Middle Ages*, Memoirs of the American Philosophical Society, Vol. 134, p. 17, Philadelphia, 1979). Another distinction which contrasts practical with theoretical geometry is that of the author of an influential treatise on the quadrant, Johannes Anglicus of Montpellier, in the 12th or 13th century, who wrote "there are two parts of geometry: the theoretical and the practical. The theoretical is that which contemplates proportions, quantities and their measures by the speculation of the mind alone. The practical is when we measure the unknown quantity of some thing by the experience of the senses." (Cited from Stephen Victor, *op. cit.*, pp. 21-22).

Roughly a third of Chuquet's *Geometry* falls within the tradition of practical geometry.

Practical geometry has been linked with questions to do with connections of mathematics with society. There has been some discussion of whether practical geometry had any impact on practical conerns such as architecture or surveying in the high middle ages. Stephen Victor discusses this question at some length, and argues that the most that can be claimed is that there is a likelihood that practical geometry had some such influence; more than that cannot be firmly established for the time he considers - that of the twelfth and thirteenth centuries. Michael Baxandall, in looking at the Renaissance (*Painting and Experience in 15th-century Italy*, Oxford University Press, 1972, pp. 87-102) has argued that the tradition of practical geometry and its educational uses had a considerable influence on artists in the Renaissance, as well as being intertwined with a number of everyday concerns. The education of artists included some mathematics, which meant that they had a good grounding in geometry. Consequently, Baxandall argues that one cannot 'read' a Renaissance painting correctly without taking account of the geometrical knowledge that was, so to speak, ingrained in the painter's way of seeing.

A non-artistic use of practical geometry, which is unfamiliar to later generations living in an age of standardized measurements, is the determination of the volumes of contents of wine containers of varying sizes. Baxandall (p. 86) points out that standard sizes of containers for wine came into use only in the nineteenth century. Chuquet's geometry concludes with a discussion of how to devise and use gauges to measure the volume of wine in a cask.

It is not fair to Chuquet to try to compare the stature of the *Geometry* with his algebraic innovations in the *Triparty*. The *Geometry* is not a mathematical work on the same level as the *Triparty*. It can be viewed in relation to the tradition of practical geometry and to that of the teaching of geometry; almost all the problems which

Chuquet sets recur in other geometries, with which he may have been familiar. It is also of interest as an application of the algebra developed in the *Triparty*. In both these respects it is a work of historical significance, even though it does not rank as a major mathematical work of the same order as the *Triparty*.

There are two manuscripts of the *Geometry*, and this translation has followed the definitive edition of L'Huillier (1979) which is based on a careful study of both manuscripts. The *Geometry* was probably written, in both manuscripts, as a textbook for students. Its content may have been designed so that it could be used for students of varying interests and levels of ability. For the most part it is clearly and concisely written, although at times the style is telegraphic, but this may simply reflect the purpose for which it was written.

The *Geometry* is an early vernacular work using syncopated algebra; a comprehensive terminology for both algebra and geometry in French was not yet well-defined when Chuquet wrote. The difficulty that this caused can be seen in the second passage translated below in the first section of the *Geometry* under the heading "on triangles"; it is paragraph 15 in L'Huillier's edition. In this problem Chuquet uses the word *partie* to refer both to the side of an equation and to a part of the side of a triangle. The difficulty of expressing equations in words with a limited mathematical vocabulary stands out in this particular passage. However, passages such as this are the exception rather than the rule. Chuquet is characteristically concise and expresses himself clearly in the *Geometry*. His reasoning can usually be followed without difficulty with the help of pencil and paper and perhaps a pocket calculator to assist with some of the more involved arithmetical calculations.

The *Geometry* consists of six sections, some of which are more practically oriented than others. It begins with a section that characterizes the scope of geometry, as Chuquet sees it, and defines the line and the circle, and discusses the measurement of standard two- and three-dimensional shapes. The second section is on the construction and use of the quadrant. The quadrant is described and a series of problems is given which illustrate its use. Chuquet also sets some problems on measuring objects without the quadrant. Making do and measuring with what was at hand was part of the tradition of practical geometry.

The third and longest section of the *Geometry* is concerned with applying geometrical principles in conjunction with algebra. Chuquet begins with some fairly simple problems and advances to more complex ones, which make use of the rules developed in the *Triparty*. In this section there are, in total, 43 problems which involve solving equations for an unknown using Chuquet's rule of first terms. The selection of problems translated includes those which involve quadratic and some cubic equations. The problems with cubic equations which come at the end of this section are quite simple and straightforward. Those which make use of quadratic equations sometimes involve complicated arithmetic calculations. Chuquet's arithmetical competance was impressive, and one wonders how far his students progressed in arithmetic as well as in algebraic and geometric skills with Chuquet as their teacher. The problems using quadratic and cubic equations in the latter part of the third section were probably intended for Chuquet's more able students.

The fourth section gives instructions for making geometrical constructions with a compass. The fifth section is on quadrature by lunes, and Chuquet's discussion may be compared with other late medieval discussions. Chuquet here makes one of his extremely rare references to another writer by name: in this case Ramon Lull, the Catalan neo-Platonic theologian who lived from c. 1232 to 1316. It is likely that Lull was one of Chuquet's sources on quadrature. Lull's attempt to construct a symbolic representation of reasoning influenced Leibniz in his quest for *Mathesis universalis*, and may thus be seen as a first step towards mathematical logic and the theory of computation. However, the framework of Christian apologetics within which he located his combinatory art has meant that some writers view his work as philosophy rather than science. The work to which Chuquet refers, the *Ars Magna*, appeared in 1308 and was a later working out of Lull's ideas.

In the sixth and final section, Chuquet returns to the tradition of practical geometry with a discussion of ways and means of measuring the volume contained in a vessel of wine. Since standard sizes of wine vessels did not exist in the fifteenth century, as has been noted above, this was clearly an important practical skill if buyers and sellers were to obtain their money's worth.

The first part of the *Geometry* begins, as stated above, with a characterization of the scope of geometry. Chuquet then discusses various kinds of geometrical figures and how to find their areas and volumes. In this section, Chuquet introduces some of the geometric terms he uses throughout the work. In particular, it is worth noting that, in writing of triangles, Chuquet uses the term "hypotenuse" to refer to any side of a triangle other than its base.

How the Science of numbers can be applied to the measurements of geometry

Here begins a small treatise of the practice of geometry, which contains the manner of measuring all things. Some of these are measured by one dimension, some by two dimensions, and the others by three. Those which are measured by one dimension are all quantities which have length without width, such as a line, of which there are two sorts, that is a straight line and a curved line. Of those that are measured by two dimensions, that is those which have length and width without depth, as do all surfaces, there are endless kinds. Some are circular, some triangular, some quadrangular, some pentagons, some hexagons and so forth. Of those which are measured by three dimensions, that is those which have length, width and depth or height, there are also endless kinds. Some are spherical, some pyramidical, some cubes and so forth.

Of the measurement of linear quantities

The straight line can be measured in two manners. One manually by fathoms or by another measure. The other manner of measuring the same is by the quadrant or by the back of the astrolabe, as appears hereafter in the treatise on altimetry. The circular line is measured, even as it will presently be said, in the measurement of circular surfaces. The other curved lines are reduced as much as possible to the straight line or to the circular line.

Of the measurement of surfaces

Some surfaces are circular as has been said above, and to measure these one should know that there are three terms in a circle, that is, centre, circumference, and diameter. The centre is the point equally distant from all points of the circumference. The circumference is the circular line which encloses in itself the ends of the circular figure. The diameter is a straight line passing by the centre dividing the circle in two equal parts.

To measure and to know how to reduce to the true square all circular figures, according to what the ancients have bestowed on us, there are many manners thereof of which one is as follows: multiply the circumference by itself and take $\frac{7}{88}$ of the product and it will be done. Example: let us propose that the circumference be 22 feet. Therefore multiply 22 by 22. The multiplication amounts to 484, which you should further multiply by 7 and then divide by 88, and one will find 38 square feet and $\frac{1}{2}$. And this same figure contains so much.

Or otherwise: multiply the diameter by itself and take $\frac{11}{14}$ of the product and it will be done. Example: let us propose that the diameter be 7 which multiplied by itself amounts to 49, which one should further multiply by $\frac{11}{14}$ and one will find $38\frac{1}{2}$ square measures.

Or otherwise: multiply the circumference by its diameter and take the quarter of the product. Example: let us propose that the circumference be 22. Thus the diameter will be 7, as will be said presently. Now multiply 22 by 7. The multiplication amounts to 154, of which the fourth part is $38\frac{1}{2}$.

Further, another rule: multiply half of the circumference by half of the diameter, and it will be done.

Whoever multiplies the area of a circular figure by $1\frac{3}{11}$, the second root of the product is the diameter line of this same. And whoever multiplies this area by $12\frac{4}{7}$ and takes the R^2 of the product, he has the circumference.

And one should know that a diameter of 7 gives a circumference of 22. And also a circumference of 22 yields a diameter of 7. And hence when one knows the circumference, one can easily know the diameter by the rule of three. And also when one knows the diameter, one can easily know the circumference by this same rule. Example: a circular figure having a diameter of 4, how much is its circumference? Answer by the rule of three: if a diameter of 7 gives me a circumference of 22, what will 4 give me? Multiply and divide even as the rule of three requires, and you will find $12\frac{4}{7}$, and this same figure has so much of a circumference.

Further: a circular figure having a circumference of 16, how large is its diameter? Answer by the rule of three: if a circumference of 22 gives me a diameter of 7, what will 16 give me? Multiply and divide even as the rule of three enjoins, and you will find $5\frac{1}{11}$. And this same figure has so much for a diameter. And one can understand others likewise.

However, one should understand that all these rules of the circular shape set forth above are conjectural and very close to truth. The ancients have used them and the moderns use them still, in the absence of better rules, since the squaring of the circle is a science which has not yet been found.

Of triangles

All triangles are constituted by three straight lines of which one is taken for the base and the two others for the two hypotenuses. The line descending perpendicularly from the angle of the hypotenuses onto the base is called the cathetuse. Of triangles some are equilateral, for the two hypotenuses and the base are equal. Others are unequilateral, and there are two kinds of these. For some have two equal lines only and the others three unequal lines. Further, of these unequal triangles there are some of which one of the hypotenuses constitutes a right angle with the base. Such triangles are called orthogonal. Whoever understands well the nature of triangles knows a large part of geometry. Hence some of their properties follow...

Another style of investigating the cathetuse is by the rule of first terms. To do this we take this triangle which has 15 and 12 for its hypotenuses and 17 for the base. Now to find its cathetuse, it is necessary to make two parts of 17. One of these multiplied by itself and this product subtracted from 15 times 15 and the remainder put aside and the other part also multiplied by itself and this product taken from 12 times 12, the remainder is equal to the other amount put aside. And for this let us propose that the first part be 1^1 *which multiplied by itself and subtracted from 225, there remains* $225\ \bar{m}\ 1^2$*. Then the other part which is* $17\ \bar{m}\ 1^1$ *multiplied by itself yields* $289\ \bar{m}\ 34^1\ \bar{p}\ 1^2$*, which subtracted from 144 there remains* $144\ \bar{m}\ 289\ \bar{p}\ 34^1\ \bar{m}\ 1^2$ *equal to* $225\ \bar{m}\ 1^2$*. Simplify your sides and you will have* 34^1 *for the divisor and 370 for the number to be divided, and consequently* $10\ \frac{15}{17}$ *for the quotient and for the first of the parts.*

Chuquet next discusses a classical mathematical formula for the base and the cathetuse of scalene triangles, the area of triangles, and the special case of right triangles. He then passes on to four-sided figures.

Of quadrangles

There are some quadrangular figures which have orthogonal angles, that is, their angles are right and drawn squared; and there are others which do not have right angles. And there are still others of which some of their angles are right and others are not. And there are some of the former where the four lines are equal and they are right-angled. Such figures are square because they are as long as they are wide. And there are others which have right angles and are longer than they are wide; they are called tetragons.

The next sections discuss determining the area of quadrangles, starting with right-angled quadrangles; the area of a quadrangle without any right angles is found by dividing it up into triangles. Chuquet then turns to figures with more than four angles and to composite figures.

Pentagons, hexagons and all other figures of many angles

Similarly figures which have 5 angles or 6 or 7 or 8 angles etc. can be divided by lines drawn from one angle to its opposite angle and thus they will represent several triangles which can be measured as was said above.

There are other figures of which one side is curved and the others are straight, such as a quarter or other part of a circular figure. Others which are composed of two such figures are in the shape of an egg. Figures such as the first should be measured thus. First one should draw a triangle against this figure towards the straight line. This line will be able to be the base of the triangle, and the point of the two hypotenuses should end at the lost point, that is, at the centre of the curved line. Consequently one should see what a circular figure of the size that is indicated by the curved line drawn on the centre of this will contain. And then say by the rule of three: if the circumference of the whole circular figure gives me the area of the same, what will give me the curved line of the figure which one wants to measure? Then after, multiply and divide as the rule of three requires, and one will find the number from which it is necessary to subtract the area of this triangle. What will remain will be the area of this region or part of the circle.

Chuquet illustrates this with several examples, and also considers its application to an egg-shaped figure. He then turns to the measurement of the areas of three-dimensional surfaces such as that of the sphere, to which he gives the title "Of the measurement of crooked and hilly surfaces".

The surface of a round pyramid, that is, a body pointed at one of its ends and large at the other like a steeple or a round bell, can be reduced to a square thus. Multiply the circumference of the large end by the distance of this same to the pointed end which is the length of the surface and take half the product and you will thus have what you seek. Example of a pyramid having a circumference of 22 feet and a length of 7 feet, that is, from the circumference to its point. To determine: how many squared feet its surface contains. Answer: multiply 22 by 7, resulting in 154 of which half is 77, and the surface of the same pyramid will contain so many squared feet.

And in order to understand the cause of this rule, one should know that the surface of a round pyramid reduced to the flat represents part of a circular figure, as may appear in this manner: let part of a circular figure such as a half or a third, etc. be drawn upon a paper and let this be abc. And then let this figure be taken up from the paper and let a be joined with b and you will thus have a pyramid. And since pyramidal surfaces descend from circular ones, and circles are such that whoever multiplies the diameter by the circumference and takes the quarter of the total has the area of the circle. Consequently, whoever multiplies the semi-diameter by the circumference and takes half of this, likewise has its area. Thus it is with pyramidal surfaces. Whoever multiplies the length of the surface of this, which is from the point to the large end on the outside, which length represents the semi-diameter, by the circumference of the large end of this same, which represents part of the circumference of a circular figure made on the same semi diameter, and takes half of this product, has the pyramidal surface reduced to an area.

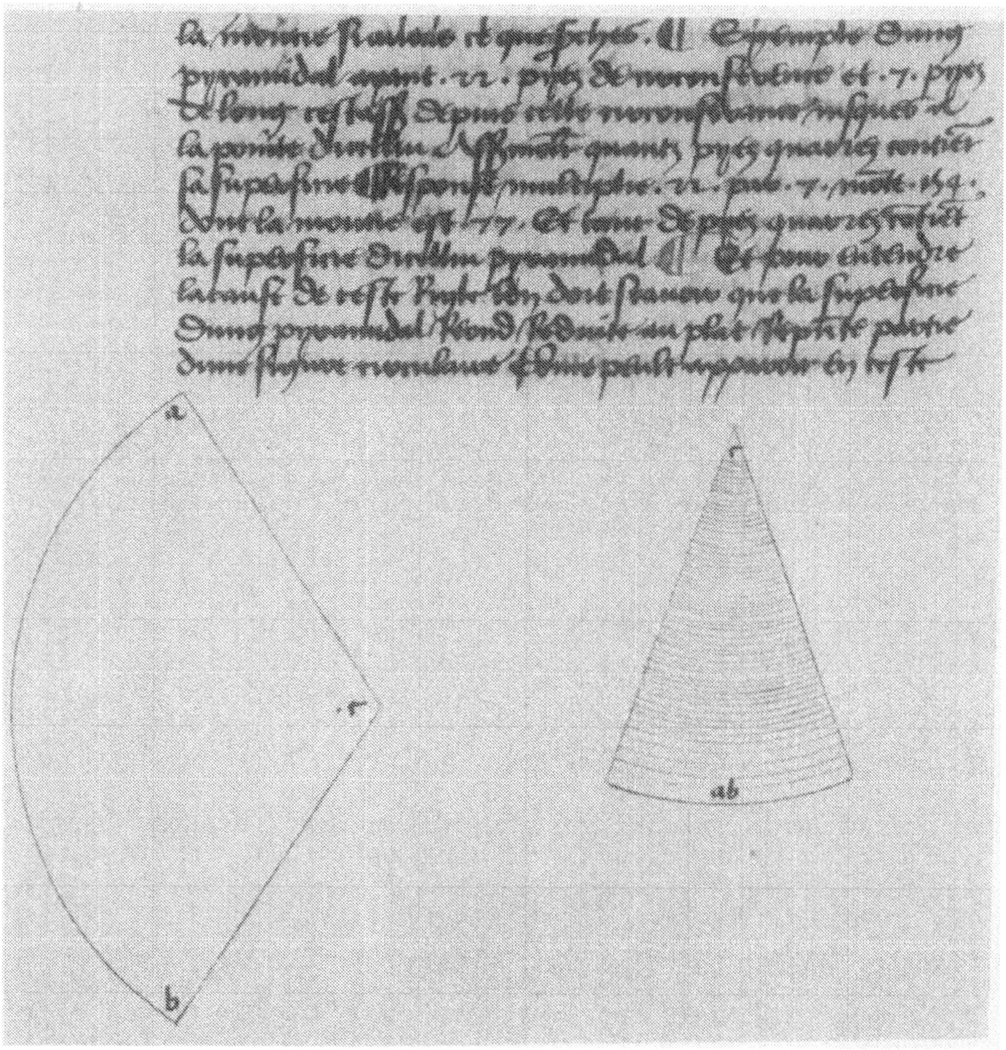

The next sections briefly give rules for finding the diameter of a circle with the same area as a given square or triangle, and other standard conversions of figures. A rule, which Chuquet later uses to determine the dimensions of a wine vessel with certain specifications of its volume, is the following:

Whoever wishes to reduce and to know how to put all plane or surface figures both triangular, quadrangular and others into a circular or round figure, there is the following means. Multiply the area to which you wish to give a curved shape by $1\frac{3}{11}$, for the second root of the product will be the diameter of the figure made round.

Chuquet describes the final section of the first part of the geometry as:

Concerning objects which are measured in three dimensions

Solid objects which have length, breadth and depth or height are reduced and measured by cubic portions, just as surfaces are measured by squared portions. And first, to measure and to know how many cubic portions a spherical body contains there is the following rule. Multiply its diameter cubically, that is, once by itself and what results multiply again by it, and take the $\frac{11}{21}$ of this latter product and that will be done. Example: a spherical body has a circumference of 22 feet. To determine: how many cubic feet this body contains. Answer: since it has a circumference of 22, it follows that it has a diameter of 7 feet, which multiplied by itself yields 49, and again by 7, the multiplication yields 343, of which $\frac{11}{21}$ are $179\frac{2}{3}$. And this body contains $179\frac{2}{3}$ cubic feet. This rule is conjectural and close to the truth just as is the measurement of the circle, for it is a science which has not yet been found.

The following sections on three-dimensional figures discuss finding the volumes of a variety of kinds of pyramids, with circles and different polygons as bases. Truncated pyramids are also considered.

And if it happens that one of the ends is larger or thinner than the other, or larger or thinner in the middle than at the two ends, in order to know the contents of such a body by a conjectural means close to the truth, it is necessary to measure the area of each end by itself and then the two areas added together should be halved, and this half multiplied by the length, and it will be done. And if the body to be measured is larger or thinner in the middle than at the two ends, it is necessary that one carefully determine the area of the middle just as if it were cut. This area is added with the area of one of the ends, and then halved, and multiplied by the length as above. And whoever would more skillfully seek the contents of such a body and to tell the truth as it is yielded by the science of geometry, one should know that a straight body which is larger or thinner at one of the ends than at the other is like a cut pyramid whose pointed part has been removed. And for this cause it is necessary to seek and ask for the entire quantity of this same pyramid, and to calculate its contents according to the rules of pyramids, and to put this aside. Then after, one should seek the contents of the pointed part removed from this pyramid, and to take this from the other put aside, for the remainder is that which is sought.

Chuquet then gives a practical means of determining the volume of oddly shaped bodies that cannot be reduced to combinations of standard geometrical figures. These passages illustrate the practical needs of measurement that contributed to the development of practical geometry in the later middle ages.

Further there are other kinds of bodies to which one can hardly assign length, width nor other dimensions, since they are misshapen and crooked, as are some stones and rocks. Such bodies can be measured by this means. One should put such a body in a vessel and fill it with water. Then one should measure the body of the water by the aforesaid rules and thus one will have the contents of both the water and of the body which one wants to measure. This done, one should remove it from the water and let it drip a little and then measure again the water by itself. It is necessary to subtract this contents from the first contents, for the remainder will be that which is sought.

Or otherwise. If the body to be measured were of metal, one might take one pound or whatever one would like of another metal of a similar kind, and make a cubic body of it, whose volume can easily be known. Then do afterwards as follows. Let us take it that the body to be measured is of iron, and that it weighs 543 pounds. Let the cubic piece of iron weighing 3 pounds have a content of 8 cubic inches. One should now use the rule of three in saying: if three pounds contain 8 cubic inches, what will 543 pounds contain? Multiply and divide as the rule of three requires, and you will find 1448. And this metal body weighing 543 pounds contains so many cubic inches. And whoever would wish to put these same inches into cubic feet must do this. Let us propose that the foot is 10 inches long, and thus the cubic foot will contain 10 times 10 times 10. Now divide 1448 by 1000 and you will find $1\frac{56}{125}$*, and any such body contains one cubic foot and* $\frac{56}{125}$ *of a foot.*

The first part of the *Geometry* concludes with a brief discussion of converting bodies into spheres or cubes.

The second part of the *Geometry* begins with a discussion of the quadrant or astrolabe. The quadrant and the very similar instrument of the astrolabe were standard late-medieval tools for measuring distances and heights that could not be measured directly. The principle involved was that of similar triangles. L'Huillier suggests that Chuquet may not have distinguished between the quadrant and the astrolabe, because he reverses the *umbra recta* and the *umbra versa* in his descriptions, and places them where they should be in the astrolabe. However, it may be that a sharp distinction was not drawn between the two instruments at the time. Although some of the terms which Chuquet uses in his description of the quadrant may be unfamiliar to modern readers, he indicates clearly and concisely what these terms mean.

There follows the construction of the quadrant and its use

The quadrant is a square set in the fourth part of a disc, and each one of the two sides of this same square which are next to the circumference of the disc is divided into twelve equal parts. If anyone should wish, one might divide it into 24 parts or into 60, according to the size of the quadrant. One of these sides thus divided into 12, which is the undermost side of this square, is called the umbra versa; the other divided side, umbra recta. And if anyone should wish, one might make two squares in the two lower quarters of the circle, and make use of whichever one might wish, for one might avail oneself of the one as well as of the other. Then there is the alidade, which is a flat piece of about the width of a small finger, which turns now here, now there on the centre of the circle, and is of the length of its diameter. And at each end of it there is a square plate with a small hole made in it. And also there should be on the upper part of this same disc an anelet by which one can hang the instrument perpendicularly. In order to elucidate and explain the use of the quadrant, four letters are put on its four angles, that is to say a, b, c, d and also f, which is the point where the alidade touches on the aforesaid umbras. However, whoever might wish might, in place of the alidade, use a plumb-line, that is a small lead weight hanging from a thread attached to the centre of the disc. And in doing this two square plates should have holes made and be attached to the one quite close to d and the other to a. And this might be used like the alidade, but the umbra recta should be made the umbra versa, and the versa the recta.

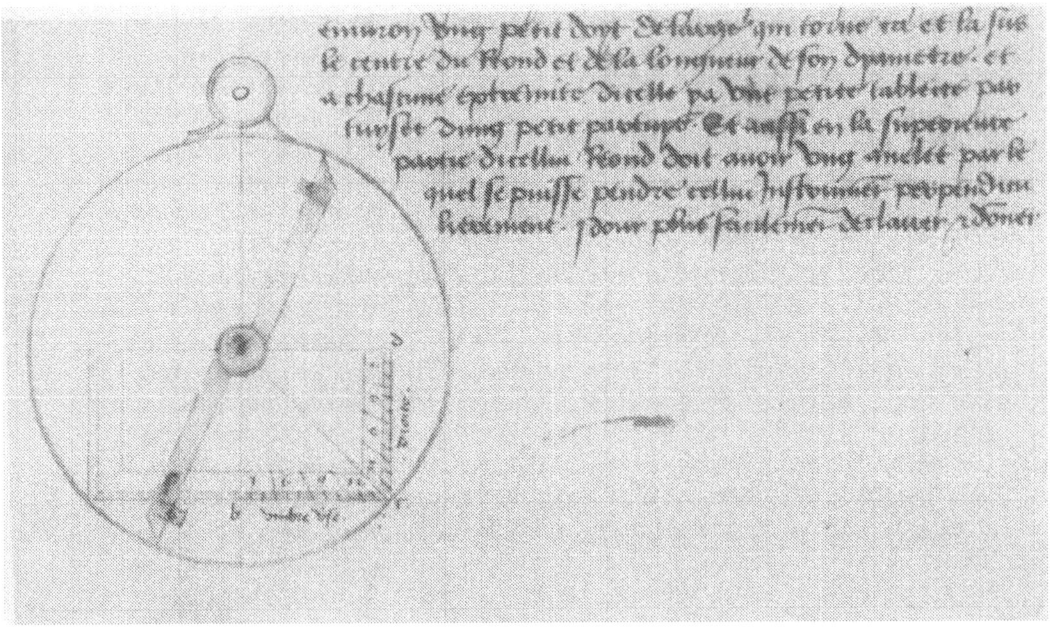

Concerning the use of this same quadrant, four chapters have been made, of which the first teaches the measurement of the length of anything. The second chapter teaches the width, the third the depth, and the fourth the height of elevated things. With the completion of these four chapters, this present treatise will end, of which some have made three parts, of which one was called altimetry, which taught the investigation and seeking of the height of things. The second is called planimetry, and this same one teaches the size of things. And the third, called stereometry, considers the length, the width and the depth of things.

The first chapter, which is about the length of any object.

The length of any object which is equally elevated in a straight line on the horizon, that is to say, as high at one of its ends as at the other, can be known in the following way. You have the aforesaid instrument and you are at one of the ends of the thing to be measured. Look through the two holes of the sights or the square plates, it makes no difference, at the other end of the thing to be measured while hanging the instrument by the anelet. Afterwards note the points of one of the umbras on which the alidade falls, for if it falls on the umbra recta it is a sign that the thing which you are measuring is longer than the distance from your eye to the earth. If on the umbra versa, it is the opposite. And if it is in the middle of these same, that is on 12 points, then your height is equal to the length which you wish to measure. Now let us propose that f, that is the point of the alidade, touches on 2 points of the umbra recta. Then you should know that the proportion which 2 has to 12 is the same as has your height from your eye to the ground to the length which you seek, and for this reason you can say by the rule of three: if 2 of height gives me 12 of length, how much will your height give me? And thus you will have that which you seek. And whoever may not have been instructed in the rule of three should divide 12 by 2 and multiply the result by his height and it will be done...

The third chapter which is about the depth of objects

Whoever would have the depth of anything like a well should put the aforesaid quadrant on the edge of the well, that is to say the line ab, in hanging it by the anelet. Then look at the bottom of the well on the opposite side by the two holes of the sights. For an example: let the diameter of the well be eg and the depth thereof be gh at least up to the water. Then put your instrument, that is to say the line ab, on e. In looking through the two holes of the sights, turn the alidade so that you may see h. Then determine on which points of the umbra versa the alidade falls. For such proportion as they will have to 12 will be the same proportion as eg will have to gh. And therefore, if the alidade falls on 1 point $\frac{1}{2}$ of the umbra versa, divide 12 by $1\frac{1}{2}$ and multiply that which results by the length of eg, and you will have the depth of this same well.

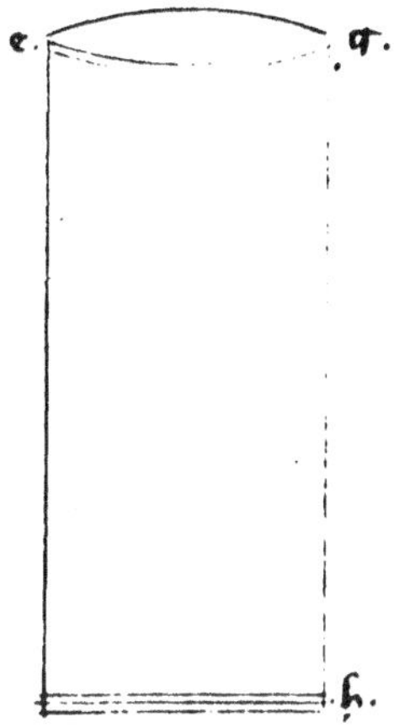

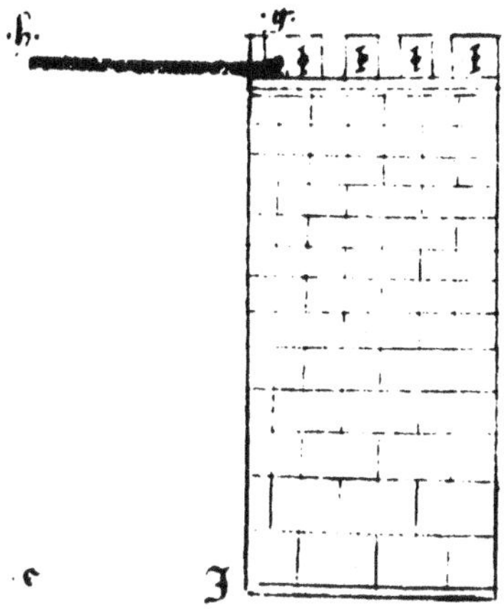

If you are at the top of a tower and wish to know how much it is from the very top to the very bottom, given that this tower is set on flat ground, you may proceed thus. Have a staff which is fairly long and at the end of this same staff put carefully a stone, and put this same staff outside the wall of the tower by a fathom or as far as the staff can extend. Then let the stone fall gently, and let the place where it falls quite straight be e. After hanging your aforesaid instrument, move the alidade now here, now there, so that you may see e by the two holes. And then see on which parts of the umbra versa the alidade falls, for such proportion which they will have to twelve, the length of the staff from the wall of the tower, to the point where the stone was put will have the same proportion to the height of the tower.

Example: let us propose that in looking at e by the two holes of the sights, the alidade falls on the ½ point of the umbra versa, and that the length of the aforesaid staff be a fathom, and let it be called gh and the foot of the tower b j. Now divide 12 by ½ and 24 is the result. Multiply by the length of the staff which is a fathom and you will have 24 fathoms, and the height of this same tower to the earth, that is from g to j, will be so much...

Chuquet also discusses making do with what was at hand, and taking measurements of objects in the absence of the quadrant.

Another way of measuring the height of something without the aforesaid instrument.

Have a staff of whatever length you wish and plant it straight in the earth fairly near to the thing to be measured. And let the place where it is planted be b. Then distance yourself sufficiently from this same staff and put your eye as close as possible to the earth, and in closing the other eye distance yourself or approach the staff in such a fashion that the visual ray which passes through the top of the staff which is c may go to the summit of the thing to be measured which is d. And let the foot of this same be e. Then judge how much there is from a to b and from b to c, for if the distances are equal, know that ae is equal to the height of ed, on the condition that this labour be done on flat ground, and if ab and bc are not equal, one might multiply ae by bc and then divide by ab and one will find ed by the rule of three in saying thus: if ab gives me the height bc, what will ae give me? etc... .

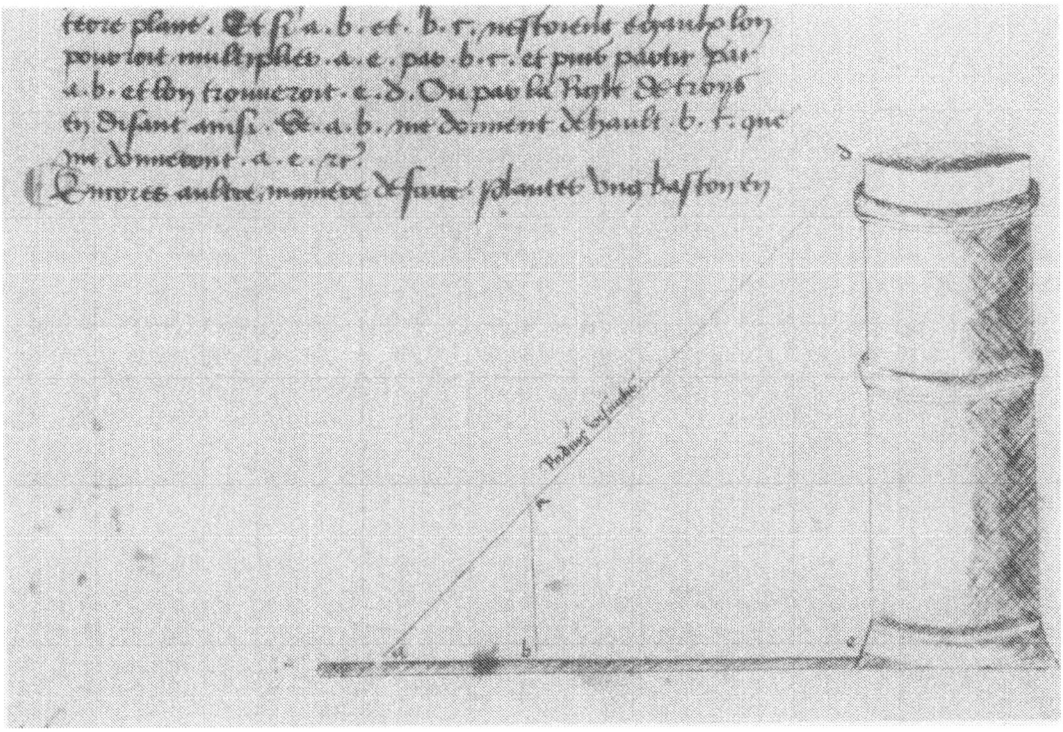

Yet another way of working: put a mirror on the ground, fairly close to the object to be measured. Then distance yourself from the mirror in such a way that it remains between you and the object which you wish to measure and also at such a distance that you may see the top of the thing which you measure. Then multiply the distance which there is from the said mirror up to the thing to be measured by your height, that is from the ground to your eye, and divide the product by the distance of the person doing the measuring from the mirror, and you will have what you want, as it appears in the picture in which ab is the height of the person doing the measuring, bc is the distance from the said mirror up to the foot of the person doing the measuring, and cd is the space from the said mirror to the measured object, and de is the height of the same.

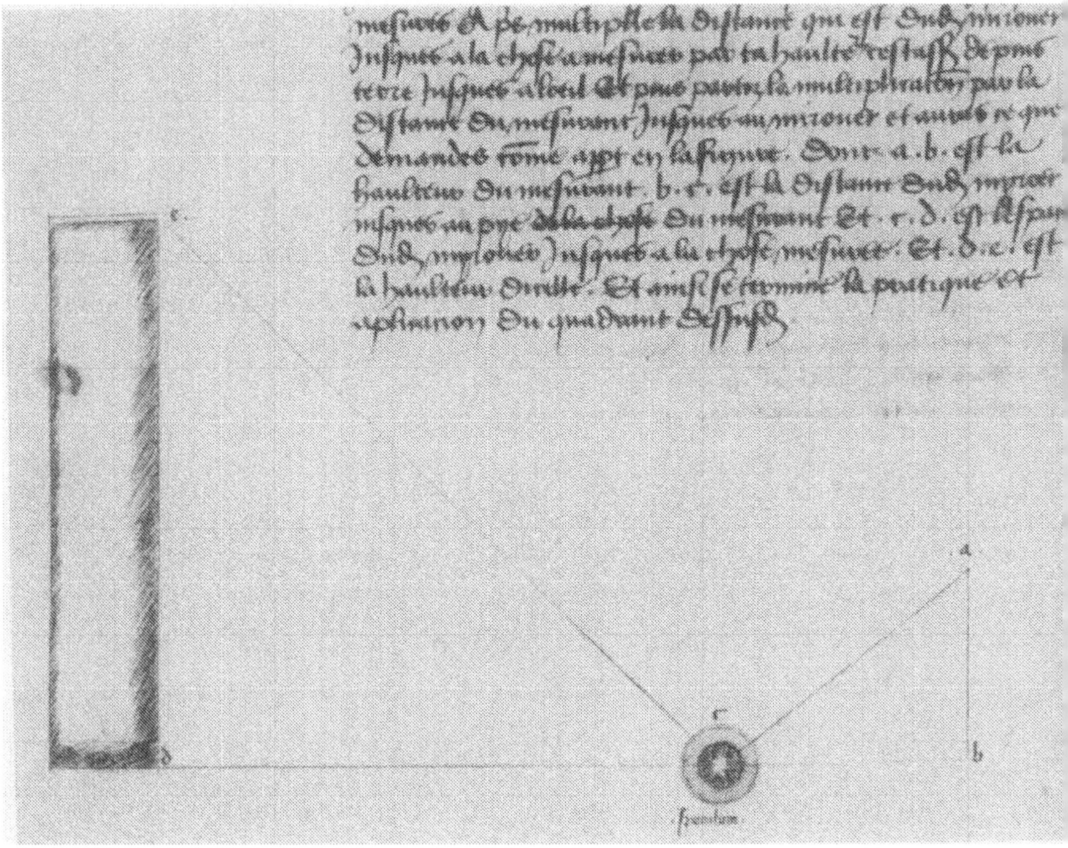

And thus ends the practice and application of the aforesaid quadrant.

The third and longest part of the geometry is devoted to the "application of the aforesaid rules". It begins with simple problems and proceeds to problems that illustrate the algebra of the *Triparty*. It is the algebra involved in the later geometric problems of this section that distinguishes Chuquet's geometry from almost all other 14th- and 15th-century geometries.

Application of the aforesaid rules

In order better to understand and know how to have the help of the aforesaid rules some problems are set hereafter, of which the first is the following.

A spherical stone has a diameter of 4 feet. Determine how many square feet are contained in its surface. Answer: multiply its circumference which is 12 $\frac{4}{7}$ by 4 which is its diameter, yielding the product 50 $\frac{2}{7}$, and the surface of this same stone contains 50 square feet and $\frac{2}{7}$.

A piece of wood 24 feet long has one end two feet wide and 1 foot ½ thick. At the other end it is 2 feet wide and 0 thick. Determine how many cubic feet this piece contains. Answer: determine the area of the large end by multiplying 2 feet by 1½, yielding 3 square feet, which multiplied by the half of 24 which is 12 yields 36 cubic feet, and this piece contains so much.

The next few problems all deal with a kind of measurement of considerable practical utility, namely, that of finding the volume of various sizes of containers of wine. Chuquet returns to this kind of problem in the section on gauging with which the geometry ends. In this section, all but the first problem are phrased in terms of a cooper or a master making a wine vessel to certain specifications. Two of these problems are given here to illustrate this kind of measurement.

A vessel full of wine has a diameter of 2½ feet, that is in the breadth of bottom, and 4 feet of length. One asks how many cubic feet this vessel contains. Answer: it is necessary for the start to know the area of the bottom by the manner by which one measures circular figures, and you will find that it contains 4 feet square and $\frac{51}{56}$. This area must be multiplied by the length of the same vessel, just as the aforesaid rules for the measurement of bodies prescribe. The product 19 $\frac{9}{14}$ results, and the inside of this same vessel contains 19 cubic feet and $\frac{9}{14}$. For the following problems let us take it that it holds 20 cubic feet.

A cooper wants to make another vessel containing once and ½ as much as the aforesaid. This is 30 cubic feet. And the same length of 4 feet. The question is how many feet wide should he make the bottom. Answer: divide 30 cubic feet by the length of the vessel which is 4 feet and there results 7½. Then afterwards it is necessary to use the rule which teaches how all plane figures may be reduced to a circular surface. Multiply 7½ by 1 $\frac{3}{11}$. The product 9 $\frac{6}{11}$ results, of which you take the second root, and you will thus have the length of the bottom.

The next problems deal with the lending of measures of wheat and the volume to be returned in differently shaped containers. Then there are problems which illustrate the commonly used piece-work basis for paying workers. In the first, a ditch-digger is paid by the volume of earth that he has excavated in digging a ditch. In the second, a mason is paid by the volume of the tower that he has constructed. There follow problems to do with the rates of revolution of chariot wheels of different sizes, and in the work of a clockmaker.

In a chariot which has 4 wheels, those in front are 5 feet in diameter and those in back are 7. The question is when the first have made 100 turns, how many turns will the back ones have made. Answer: It is necessary for the start to know how many feet in circumference the first wheels are by the aforesaid rules. They are 15 feet $\frac{5}{7}$, which must be multiplied by 100. The multiplication yields 1571 $\frac{3}{7}$. And then the circumference of the other wheels must be sought, which is 22 feet. Now divide 1571 $\frac{3}{7}$ by 22 and you will find 71 $\frac{3}{7}$, and the rear wheels have made so many turns when the front wheels have made 100.

A clockmaker has made a wheel which has a diameter of 7 inches, and he wants to make another such that, when that which is made will have turned three turns, the other makes 5 turns. The question is: with what diameter in inches should it be made? Answer: for the start it is *necessary to multiply the circumference of the wheel already made, which is 22 inches, by 3 turns, which yields 66, which it is necessary to divide by 5 turns, and there results 13 inches $\frac{1}{5}$. And it should be made with a circumference of that amount, and consequently with a diameter of 4 inches and $\frac{1}{5}$.*

The next problems are concerned with determining the areas of figures formed by the circumferences of two others.

Two circular figures each contain 12 inches square of area. To determine, if the two circumferences are joined together and a circular figure is made of this, what will be its area. Answer: determine the circumference of one of these two figures by multiplying the area which is 12 inches square by 12 $\frac{4}{7}$, yielding the product 150 $\frac{6}{7}$. Thus R^2 150 $\frac{6}{7}$ is the circumference of the one. Doubled it yields R^2 603 $\frac{3}{7}$ for the circumference of the two together. This multiplied by itself yields 603 $\frac{3}{7}$, of which $\frac{7}{88}$ is 48, and the area of this circular figure whose circumference is equal to the circumference of the two aforesaid circles will contain so much. Or otherwise. Since the circumference is doubled, the area should be quadrupled, and to do this multiply 12 by 4 and you will have 48 for the area of this circle.

Another special rule: join the two areas together, that is to say 12 with 12, and you will have 24 which you should put aside. Then multiply one of the areas by the other. That is 12 by 12. It yields 144 which it is necessary to quadruple. It yields in total 576 of which the second root which is 24 added with the 24 put aside make 48 above...

The second of these special rules covers the case of two circles of differing sizes. The problems that follow make use of algebra. L'Huillier points out that what distinguishes Chuquet's presentation of these standard problems in medieval geometry is that he often deliberately chooses to have the solutions of the problems be irrational numbers.

There is a circle whose diameter and area are in equal proportion. To determine: what is its diameter? Answer: I propose that its diameter is 1^1, which multiplied by itself yields 1^2, of which the $\frac{11}{14}$ is $\frac{11}{14}^2$ for the area, equal to 1^1 which is the diameter. Now divide 1 by $\frac{11}{14}$ and you will have $1\ \frac{3}{11}$ for the diameter of this same circle, of which the area is $1\ \frac{3}{11}$ ....

Further, there is a circle which has a diameter of R^2 30. To determine: what is its circumference? Answer: multiply R^2 30 by $3\ \frac{1}{7}$ reduced to an R^2, that is by $R^2\ 9\ \frac{43}{49}$, and take further the second root of the product and you will have $R^4\ 314\ \frac{34}{49}$, and take further the second root of the product and you will have $R^4\ 314\ \frac{34}{49}$ for its circumference....

Moreover there is a triangle whose two hypotenuses are equal and are db and eb. And the base dae which is 20 and fcg 13 and ac 17. To determine: how large is the cathetuse of this triangle which is ab? Answer: subtract fcg which is 13 from dae which is 20. There remains 7. Now say by the rule of three: if I obtain 7 from 17 which is ac, from how much will I obtain 20? Multiply and divide as the rule of three prescribes and you will have $48\ \frac{4}{7}$, and ab is so much...

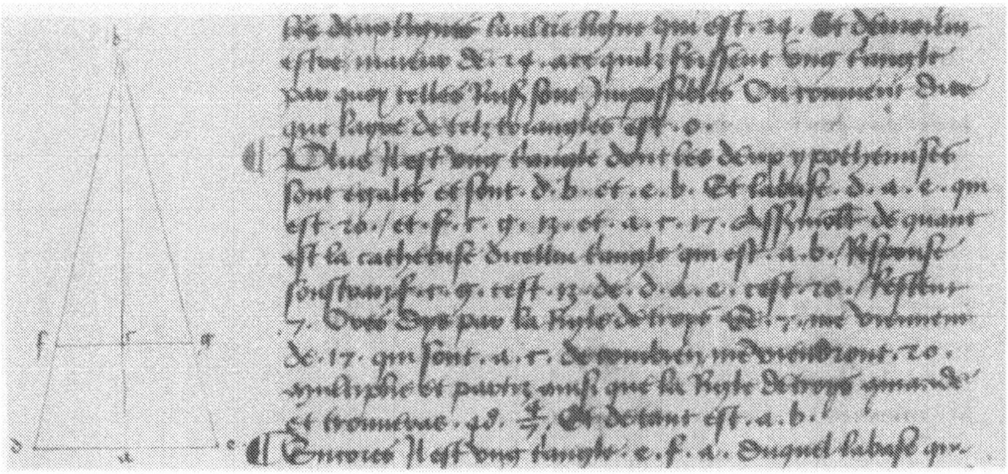

Moreover, there is a right angle quadrangle containing 100 square portions of which the length is the double superbipartiens third of the width. To determine: how much are its length and width? Answer: I propose 3^1 for the width and 8^1 for the length. Whoever multiplies the one by the other obtains 24^2 equal to 100. Now divide 100 by 24^2 and you will have $R^2\ 4\frac{1}{6}$ which you should multiply by 3 and by 8 which are the two positions and you will have $R^2\ 37\frac{1}{2}$ for the width and $R^2\ 266\frac{2}{3}$ for the length....

Further, there is an orthogonal quadrangle containing $R^2\ 200$ in its area whose longitude exceeds its latitude by 3. To determine: how much is its diameter, that is, the main line of the quadrangle coming from one angle to the other and passing through the centre? To do this, first of all I want to seek out the latitude and longitude of this same quadrangle by proposing 1^1 for the latitude and 1^1 plus 3 for the longitude, which multiplied the one by the other make 1^2 plus 3^1 equal to $R^2\ 200$. Now dispose of the rest according to the canons of the rule of first terms and you will find $R^2\ R^2\ \underline{200\ \bar{p}\ 2\frac{1}{4}}\ \bar{m}\ \frac{1}{2}$ for the latitude, to which you add 3, and you will have as a sum $R^2\ R^2\ \underline{200\ \bar{p}\ 2\frac{1}{4}}$ plus $1\frac{1}{2}$ for the longitude. And whoever multiplies the latitude by the longitude, the multiplication will amount to $R^2\ 200$, which is the area of this same quadrangle. Now in order to find its diameter, multiply the latitude by itself. The multiplication amounts to $R^2\ 200$ plus $2\frac{1}{4}\ \bar{m}\ R^2\ R^2\ \underline{1012\frac{1}{2}\ \bar{p}\ 5\frac{1}{16}}$ m $R^2\ R^2\ \underline{1012\frac{1}{2}\ \bar{p}\ 5\frac{1}{16}}$ plus $2\frac{1}{4}$ which it is necessary to simplify by adding plus $2\frac{1}{4}$ to plus $2\frac{1}{4}$ which amounts to $4\frac{1}{2}$. Then add minus $R^2\ R^2\ 1012\frac{1}{2}\ \bar{p}\ 5\frac{1}{16}$ with minus $R^2\ R^2\ \underline{1012\frac{1}{2}\ \bar{p}\ 5\frac{1}{16}}$. This amounts to minus $R^2\ R^2\ \underline{16200\ \bar{m}\ 20\frac{1}{4}}$. Thus this multiplication completely simplified amounts to the sum of $R^2\ 200\ p\ 4\frac{1}{2}\ \bar{m}\ R^2\ \underline{16200\ \bar{p}\ 20\frac{1}{4}}$. And the length multiplied by itself and also reducea amounts in all to $R^2\ 200\ \bar{p}\ 4\frac{1}{2}$ plus $R^2\ R^2\ \underline{16200\ \bar{p}\ 20\frac{1}{4}}$. Now add these two products in this manner. First add $R^2\ R^2\ \underline{16200\ \bar{p}\ 20\frac{1}{4}}$ with $\bar{m}\ R^2\ R^2\ \underline{16200\ \bar{p}\ 20\frac{1}{4}}$. And they amount to 0. Then add $R^2\ 200\ p\ 4\frac{1}{2}$ with $R^2\ 200\ \bar{p}\ 4\frac{1}{2}$. This addition amounts to $R^2\ 800$ plus 9. The second root of this sum is $R^2\ R^2\ \underline{800\ \bar{p}\ 9}$ and is the diameter of this same quadrangle...

After further examples - on pentagons, hexagons, heptagons and octagons - Chuquet discusses the geometric representation of roots.

The properties of lines enclosed in a circle are such that two lines which intersect and divide one another in equal or unequal portions, at right or oblique angles, are of such a nature that if the one portion is multiplied by the other portion of the self-same line, and similarly the two portions of the other line are multiplied the one by the other, then the two products are equal.

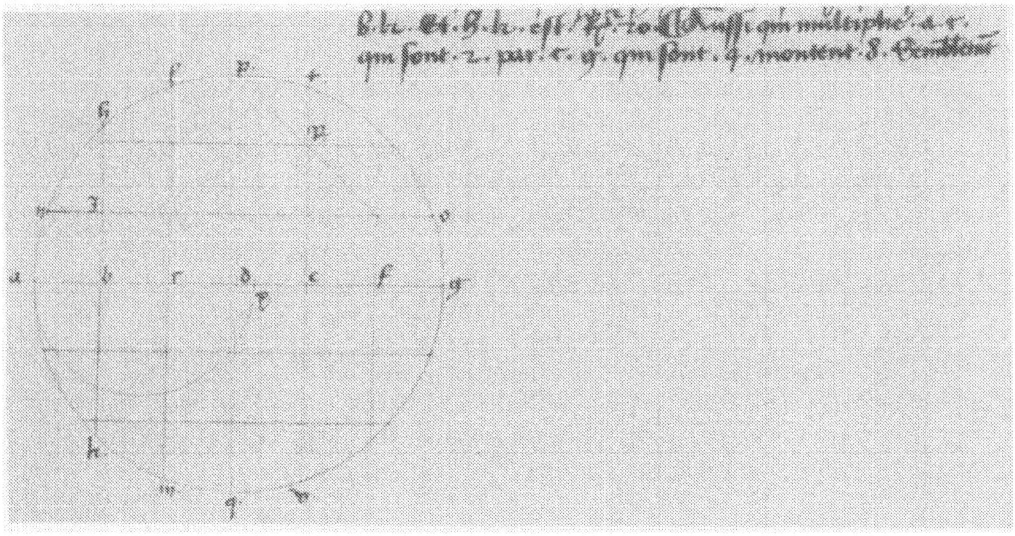

Example: *let a circle be drawn whose diameter is divided into six equal portions which are six units a, b, c, d, e, f, g by means of six lines which cut the diameter at right angles which are h, b, k, l, c, m, and the others follow likewise. Let these six lines be likewise divided into equal portions by six other lines which make*

right angles on the first lines. Now whoever multiplies ab which is 1 by bg which is 5, the multiplication comes to 5. Also whoever multiplies hb by bk obtains 5. From which it follows that hb is R^2 5. And bk likewise. And hk is R^2 20. Also whoever multiplies ac which is 2 by cg which is 4 obtains 8. Similarly, whoever multiplies lc by cm obtains 8. By which it follows that lc and cm each are R^2 8, since these two lines are equal. Further whoever multiplies ad which is 3 by dg which is also 3 obtains 9. Also whoever multiplies pd by dq makes 9. pd must therefore be R^2 9, which is 3, and similarly dq. Again whoever multiplies nj which is R^2 8 minus 2 by jo which is 2 plus R^2 8, obtains 4. Also whoever multiplies hj which is R^2 5 $\bar{m}$ 1 by jk which is R^2 5 plus 1, the multiplication comes to 4. Moreover whoever multiplies pr which is R^2 2 by rg which is R^2 8 obtains R^2 16 which is 4. Likewise whoever multiplies tr which is R^2 8 minus 2 by rv which is 2 plus R^2 8, this multiplication comes to 4. And thus must all other lines and also all numbers be understood.

By this teaching it is apparent that the square roots of numbers can be given by lines. For whoever would represent R^2 5 by a line, let a line of 5 units of length be drawn, such as bcdefg. And at the end of it let 1 be added to it. This is ab. On the point b let there be drawn an orthogonal line of indeterminate length bh. Then on the whole line ag let there be made a semicircle agh. Let it be known that from the point of intersection of the orthogonal line and of the curved line which is h up to b, that is to say the line bh, is R^2 5. And one can do likewise with all other numbers.

Or otherwise. One should take two numbers such that, multiplied the one by the other, they yield precisely the number whose root is sought. If one should wish to find the second root of 8, it is true that 2 multiplied by 4 yields 8. Thus one can draw a line of 2 and of 4 such as ac and cg. And on c one should draw an orthogonal line of an indeterminate length. Then on the line ag let there be drawn a semi-circle, for from c to the intersection of this same orthogonal line and of the semicircle which is 1 is the R^2 8.

The fourth root of a number can also be sought by the same manner. As for example if you would have the fourth root of 5. First let the R^2 5 be found, even as was said above, and let it be bx. To which let 1 be added, which is ab. And on b let there be drawn an orthogonal line of indeterminate length such as bk. And on ax let there be drawn a semi-circle. And from the place where it will cut the orthogonal line to b is R^4 5. And all other numbers can be understood likewise.

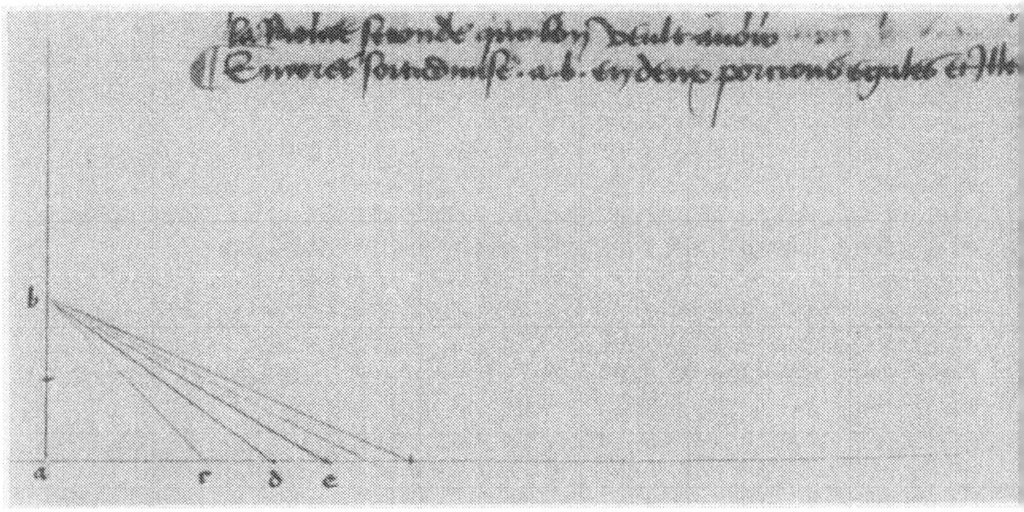

Another way of investigating the roots of numbers by means of continuous quantities of geometry. Let two orthogonal lines be drawn and let their intersection be a. And let c be on one.of these and b on the other. And let ab and ac each be 1. Then let a line be drawn from b to c. Let this line be extended with compasses over the line ac and thus let d be drawn, for both ad and bc are each R^2 2. Next let a line be drawn from b to d, for it is R^2 3, which is extended in its full length on the line acd, and there where it ends is e. Moreover from b to e let there be drawn a line which is R^2 4. And the reason thus is because abe constitute a right triangle of which ab can be the cathetuse and ae the base and be the hypotenuse. Now if the cathetuse which is 1 is multiplied by itself, the result is 1, and the base which is R^2 3, multiplied by itself, makes 3 which added with 1 makes 4 which is the square of the hypotenuse, by which it follows that R^2 4 which is 2 is the line ae. And in this fashion one can continue until one has the second root which one wishes to have.

Further let ab be divided into two equal portions and there let a point be put and on this point let one of the points of the compass be put and the other on c. Then let the compass be lifted and let one of the points be put on a. And with the other let a point be made on the line from which point to a is R^2 1 $\frac{1}{4}$. Then let the compass be extended from this same point which is R^2 1 $\frac{1}{4}$ to the point which is between a and b, for this is R^2 1 $\frac{1}{2}$. And continue thus up to d and you will find the next point which is R^2 1 $\frac{3}{4}$. And by this means it is manifest that whoever pursues this work on the half of ab divides cd into 4 parts and whoever would do this work on $\frac{1}{3}$ of the same would divide cd into 9 unequal parts. And whoever would work on the quarter of the line ab would divide cd into 16 parts. And de will be divided into as many parts, and the others, if there were more of them. And this is for the roots of square numbers.

Or otherwise: let there be sought or found two equal or unequal numbers and such that when each is multiplied by itself and then the two products added together, they make the number of which one wants to have the second root. And then one of these numbers is laid out on one of the perpendicular lines and the other on the other. Let a line be drawn from the end of one of these lines to the end of the other, for this same line is the second root of the number which is sought.

Example: whoever would give the second root of 13 by geometry in a continuous line. The two numbers, which multiplied each by itself and then the two products added together make 13, are 3 and 2. Now let 2 be laid out on one of the perpendicular lines and let it be gk and let gp be 3. Now let a line be drawn from k to p, for that same line is R^2 13. Even as the nature of right *triangles plainly demonstrates. And since it is made of 2 and of 3 in order to have R^2 13, so can other ones be made in their place.*

Chuquet then turns to the famous Delian problem of finding a second cube whose volume is proportional to that of a first cube. It is one of three classic Greek mathematical problems. The historical origins of this problem have not been established; legend has it that the problem arose in connection with the construction of an altar.

How some third or cubic roots may be exhibited and presented by lines. Let two perpendicular lines of indeterminate length be drawn, cj and ck, and let a square abcd be drawn in the angle of these same lines. This square is divided into two equal portions by a line fg which makes of this same square two quadrangles or two tetragons, of which the first is afgc. At the intersection of its diameters, which is the centre of this same one, let there be h. Now let the fixed foot of the compass be put on h, and let the other one be opened so much that it may make two points on the two perpendicular lines, in such a fashion that the straight line coming from one of these same points to the others passes through the angle f and let these two points be jk. Now let us suppose that the aforesaid square abcd be the surface of a cube. Now one can say that a cube made on the line gk is the subdouble of the first cube. And the cube made on aj is its subquadruple. And that made on af or fb is its suboctuple. Now if the cube abcd were 64, the cube gk would be 32, from which the line gk would be R^2 32, and the cube aj would be 16. Thus the line aj would be R^3 16. And the contents of the cube fb would be 8. Thus fb would be R^3 8 which is 2. Thus we have four cubes whose contents are in double proportion, which are 8, 16, 32, and 64. And thus we have four third roots, that is to say R^3 8, R^3 16, R^3 32, R^3 64, which is ab or cd.

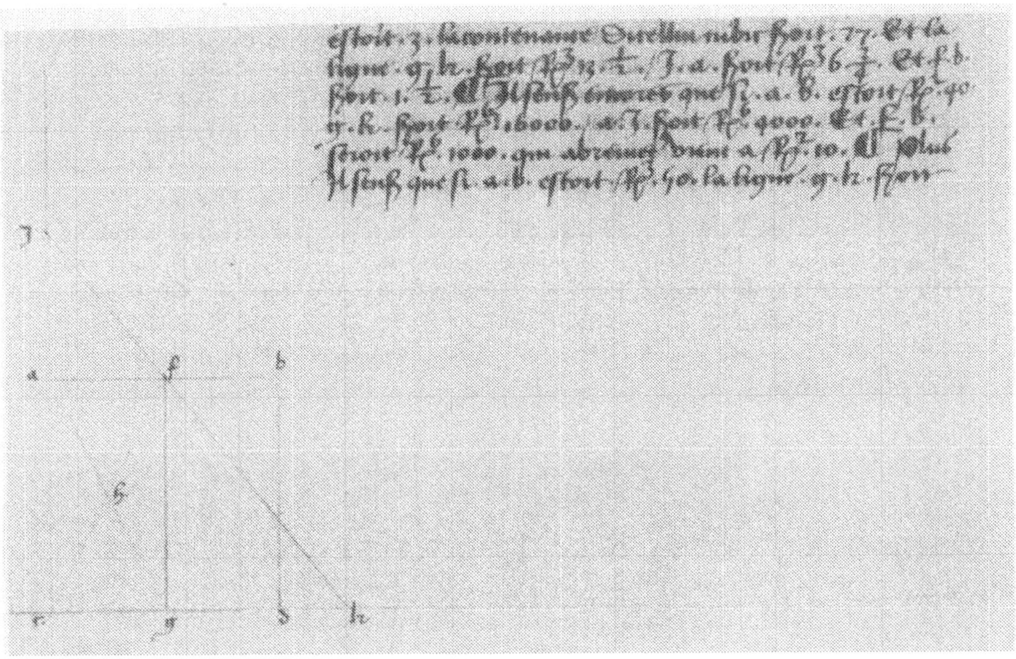

And if the line ab or cd were 3, the contents of that same cube would be 27. And the line gk would be R^3 13 $\frac{1}{2}$. ja would be R^3 6 $\frac{3}{4}$. And fb would be 1 $\frac{1}{2}$.

It follows that if ab were R^2 40, gk would be R^6 16000. aj would be R^6 4000. And fb would be R^6 1000, which simplified comes to R^2 10.

Furthermore it follows that if ab were R^3 50, the line gk would be R^9 15625, the line aj would be R^9 1953 $\frac{1}{8}$ and fb would be R^9 244 $\frac{9}{64}$, which simplified comes to R^3 6 $\frac{1}{4}$. And all other numbers must be understood thus....

Chuquet then sets problems to do with inscribed figures.

Three circles are included and contiguously enclosed in the circumference of another which has the diameter of 12, which is ab. *To determine, what are the diameters of each of these contained circles. To do and seek this, let there be drawn a line from the centre of one of the contained circles to the centre of the other and continue thus from centre to centre. And thus you will have an equilateral triangle* pcd *of which triangle* e *is its centre and also the centre of the containing circle. And this done, it should be understood that from* o *to* e, *that is from the base to the centre of the triangle, is a third of the cathetuse in all equilateral triangles. And also as was said before in the consideration of equilateral triangles, the square of the cathetuse is the triple of the square of half the base.*

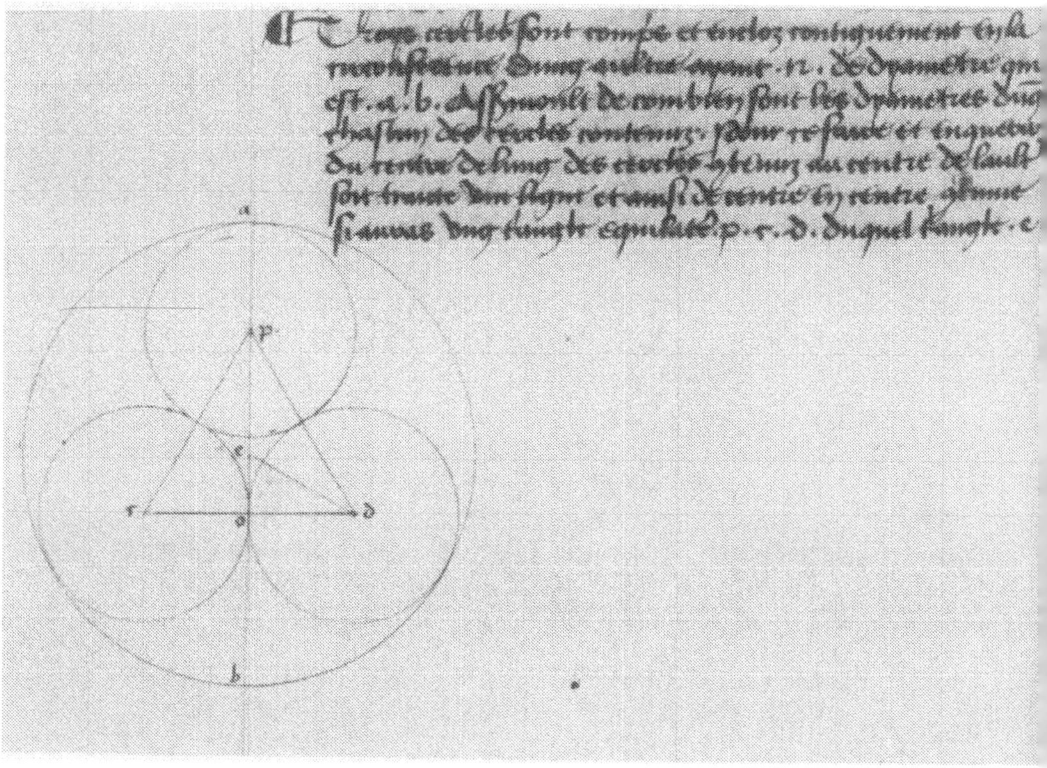

Following upon these considerations I propose 1^1 for half the base which multiplied by itself yields 1^2. Thus the square of the cathetuse po will be 3^2, by which $R^2\ 3^2$ is this same cathetuse, of which the third which is eo will be $R^2\ \frac{1}{3}^2$, which multiplied by itself yields $\frac{1}{3}^2$, which added with 1^2 which is the square of half the base makes $1^2\ \frac{1}{3}$ whose second root, which is $R^2\ 1^2\ \frac{1}{3}$, is the line ed. When this is added with do which is 1^1 it yields in all $R^2\ 1^2\ \frac{1}{3}$, plus 1^1 equal to eb or ea which is half of 12, that is 6. Now simplify your sides and seeing that one of the little parts is R^2, 1^1 should be taken away from this same and also from the other side, and you will have $R^2\ 1^2\ \frac{1}{3}$ on the one hand and 6 minus 1^1 on the other. So then multiply each side by itself and you will have $1^2\ \frac{1}{3}$ on the one hand and 36 minus 12^1 plus 1^2 on the other. Simplify your sides further by giving 12^1 to each of these. And also take from each one 1^2. And thus you will have at the end $\frac{1}{3}^2$ plus 12^1 for one of the sides and 36 for the other. Dispose of the rest of this calculation according to the canons of the rule of first terms and you will thus have R^2 432 minus 18. And od, which is the semi-diameter of one of the three little circles, is so much, and as a consequence the whole diameter will be R^2 1728 minus 36.

Four contiguous circles are enclosed within a circle whose diameter, which is bc, is 12. To determine: how large is the diameter of one of the contained circles? In order to consider this matter, I draw four lines from one centre to another and thus I find the square efd whose side equals two semi-diameters as it is clear from ef. And also I see that the diameter of the square which is de with the two semi-diameters cd and eb is equal to the diameter of the circle containing cb, and therefore the diameter of the square de with one of its sides ef is equal to cb which is 12. And therefore let us propose that ef be 1^1 and df as well. I multiply each one by itself and then add them, which yields 2^2 of which the second root which is $R^2 2^2$ is the diameter de, to which I add ef, which yields in all $R^2 2^2$ plus 1^1 equal to 12 which is cb. Simplify your sides by removing 1^1 from each of them and you will have $R^2 2^2$ on the one side and 12 minus 1^1 on the other. Now multiply each side by itself and thus

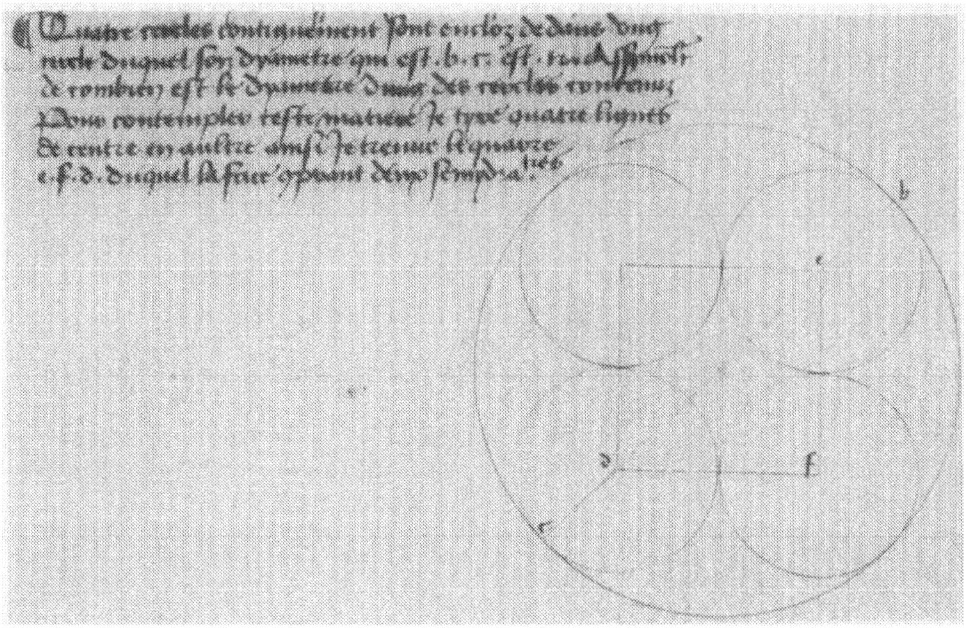

you will have 2^2 on one side and $144\ \bar{m}\ 24^1\ \bar{p}\ 1^2$. Lend 24^1 to the one and the other sides and take from each of these 1^2 and you will have $1^2\ \bar{p}\ 24^1$ on one side and 144 on the other. Dispose of the rest of the calculation according to the canons of the rule of first terms and thus you will find R^2 288 $\bar{m}$ 12 and ef amounts to that much. And as a consequence, so does each diameter of the contained circles. And to check this work I multiply ef by itself, which is R^2 288 $\bar{m}$ 12, and also df by itself, and each multiplication yields 432 minus R^2 165888, which I add together, which yields in all 864 minus R^2 663552. The second root of this number is R^2 864 $\bar{m}$ R^2 663552 and is the diameter de. This root simplified by extraction comes to R^2 576 $\bar{m}$ R^2 288, which simplified again by extraction of R^2 comes to 24 $\bar{m}$ R^2 288, to which I add the two semi-diameters cd and eg which are themselves both together R^2 288 $\bar{m}$ 12. This addition yields 12 which is the perfection of this examination.

There is an equilateral triangle contiguously enclosed in the circumference of a circle having a diameter of 12.

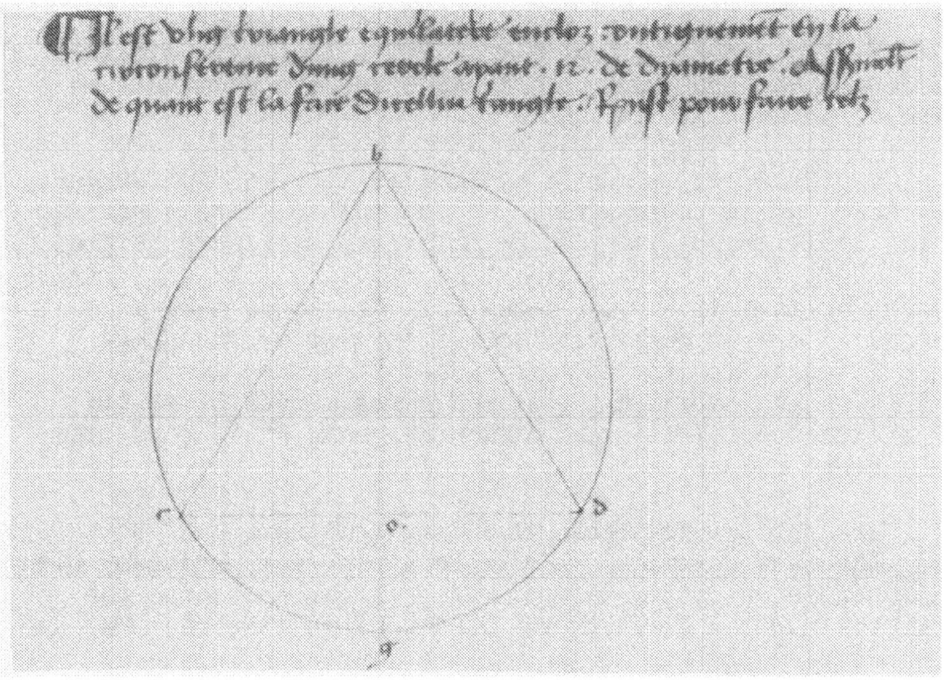

To determine: how large is the side of this triangle? Answer: in order to do such calculations one should know that the cathetuse of equilateral triangles is always $\frac{3}{4}$ of the diameter bg and therefore bo is 9 for the cathetuse and og is 3. Now whoever multiplies 3 by 9 obtains 27, and so much should result from co which is the semibase multiplied by cd which is the other semibase. By which it follows that the semibase is thus R^2 27. And therefore the whole base will be R^2 108. And each of the three faces of this triangle is so much.

Or otherwise. Since the cathetuse is 9, its square is 81, and the square of the semibase is a third of 81 which is 27. These added with 81 make 108 which is the square of one of the hypotenuses. By which R^2 108 is the hypotenuse line.

And if the side of the triangle bc were 8, how large would the diameter be? 8 multiplied by itself yields 64. The root of $\frac{3}{4}$ of this which is R^2 48 is the cathetuse of which the sesquitierce is the diameter of the circle bg. This sesquitierce can be found by the rule of three in saying: if 3 requires 4 what will R^2 48 require? Multiply and divide and you will find R^2 85 $\frac{1}{3}$ and the diameter bg would be so much if bc were 8.

Or otherwise: whoever multiplies co by od obtains 16. Also go multiplied by ob yields 16. And since go is the subtriple of ob, it is therefore necessary to find two numbers in a triple proportion and such that multiplied the one by the other they yield 16. And to do this I propose 1^1 and 3^1 which multiplied the one by the other yield 3^2 equal to 16. Divide 16 by 3^2 and you will have R^2 5 $\frac{1}{3}$ for the subtriple go and consequently R^2 48 for the triple ob. These added together make R^2 85 $\frac{1}{3}$ for the whole line bg ...

Three of the following problems deal with quadrilaterals inscribed in circles. The next problem, with a quadrangle inscribed in a circle, illustrates both the strengths and the limitations of Chuquet's algebra. The use of negative exponents is an isolated instance in the *Geometry*. L'Huillier comments that he knows of no similar problem in medieval geometry. The working is long-winded and cumbersome.

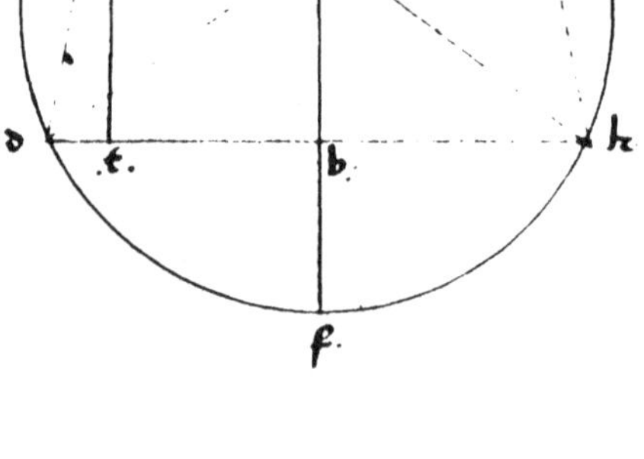

There is a quadrangle pgkd contiguously contained in the circumference of a circle. In this quadrangle one of the two equidistant lines dk is 12 and the other pg is 8. The two other sides pd and gk are each 6.
To determine: how large is the diameter of the circle mf? In order to do this calculation, we should first seek out the distances of the two equidistant lines which is cb and as a consequence pt which is the cathetuse of the triangle pdk. And since pc is 4 and db is 6, it follows that dt is 2 which multiplied by itself makes 4, which should be taken away from the square of pd which is 36, leaving 32 for the square of pt. By which it follows that this line pt is R^2 32 and cb as well. So then multiply pt by itself which yields 32, and tk which is 10 by itself makes 100, which added with 32 makes 132 which is the square of the diameter of the quadrangle pk, from which R^2 132 is thus the line pk.

Or otherwise: in order to seek out the diameter of the quadrangle, one should know that all squares or quadrangles, whether right-angled or oblique, contiguously enclosed in a circle have the property that one of the four sides multiplied by its opposite and the other side by the other, and then the two products added together are equal to the one of the diameters multiplied by the other. The example of the aforesaid quadrangle: let pg be multiplied by dk yielding 36 which added to 96 makes 132 and pk multiplied by gd should amount to as much, by which it follows that pk is R^2 132 and similarly for gd. So then, since we know that bc is R^2 32, it is necessary to seek the whole diameter fm. And to do this let us propose that fb be 1^1. Then I multiply db by bk which amounts to 36 which I divide by 1^1. The division comes to $36^{1\bar{m}}$ for bm from which I subtract R^2 32 which is cb. There remains $36^{1\bar{m}}$ minus R^2 32 for mc. Now I multiply mc which is $36^{1\bar{m}}$ minus R^2 32 by cf which is R^2 32 $\bar{p}$ 1^1, yielding the product R^2 $41472^{2\,m}$ minus 32 plus 32 minus R^2 32^2 equal to pc multiplied by cg which make 16. Now simplify your sides and you will have 12 plus R^2 32^2 on the one hand and R^2 $41472^{2\bar{m}}$ on the other. And seeing that one of the sides is a second root of a minus second term, multiply each of these same by 1^1, and you will have R^2 41472 on the one side and 12^1 $\bar{p}$ R^2 32^4 on the other. And seeing that R^2 of a fourth term is equivalent to a second term, thus we have here first and second terms equal to a number or to R^2 of a number which is all the same. And hence divide the preceding terms, that is 12 and R^2 41472 by the sequent, that is by R^2 32^4. And you will have R^2 1296 which is 36 and R^2 4 $\frac{1}{2}$ for an intermediate of which half, which is R^2 1 $\frac{1}{8}$ multiplied by itself and added with 36 makes 37 $\frac{1}{8}$ whose second root is R^2 37 $\frac{1}{8}$, from which it is necessary to take away the half of the intermediate which is R^2 1 $\frac{1}{8}$. There remains R^2 37 $\frac{1}{8}$ $\bar{m}$ R^2 1 $\frac{1}{8}$ and fb amounts to that much. So then divide 36, which is db multiplied by bk, by R^2 37 $\frac{1}{8}$ $\bar{m}$ R^2 and by the division you will find for bm R^2 37$\frac{1}{8}$ plus R^2 1 $\frac{1}{8}$. Now add fb with bm. Thus you will have R^2 148 $\frac{1}{2}$ which is the amount of the diameter of the circle mf. mc is R^2 37 $\frac{1}{8}$ minus R^2 21 $\frac{1}{8}$. cf is R^2 37 $\frac{1}{8}$ plus R^2 21 $\frac{1}{8}$. Now multiply mc by cf because the multiplication amounts to as much as pc multiplied by cg, which is the probation of this work.

Or otherwise: multiply dt by tk yielding 20 which divided by pt which is R^2 32, the division comes to R^2 12 $\frac{1}{2}$, which is tr. Now add pt with tr amounting to in all R^2 84 $\frac{1}{2}$ for the whole line pr which multiplied by itself makes 84 $\frac{1}{2}$ which should be added to 64 which is the product of pg multiplied by itself, yielding the sum 148 $\frac{1}{2}$ of which sum the second root which is R^2 148 $\frac{1}{2}$ is rg which is the diameter of the circle.

Further problems:

There is a triangle acb of which bc which is its base is 8 and its cathetuse ag is as much. In it there is an equilateral triangle, the largest which can be included there, whose base is de which is upwards, and the point of the two hypotenuses is g. To determine: how large is the side of the contained triangle? To do this one should understand that the lines af and de are equal. And however I propose that one of these same be 1^1. Thus df will be $\frac{1}{2}^1$ which multiplied by itself yields $\frac{1}{4}^2$ which subtracted from the square of dg which is 1^2 there remains $\frac{3}{4}^2$ for the square of gf. Thus R^2 $\frac{3}{4}^2$ is for the line gf

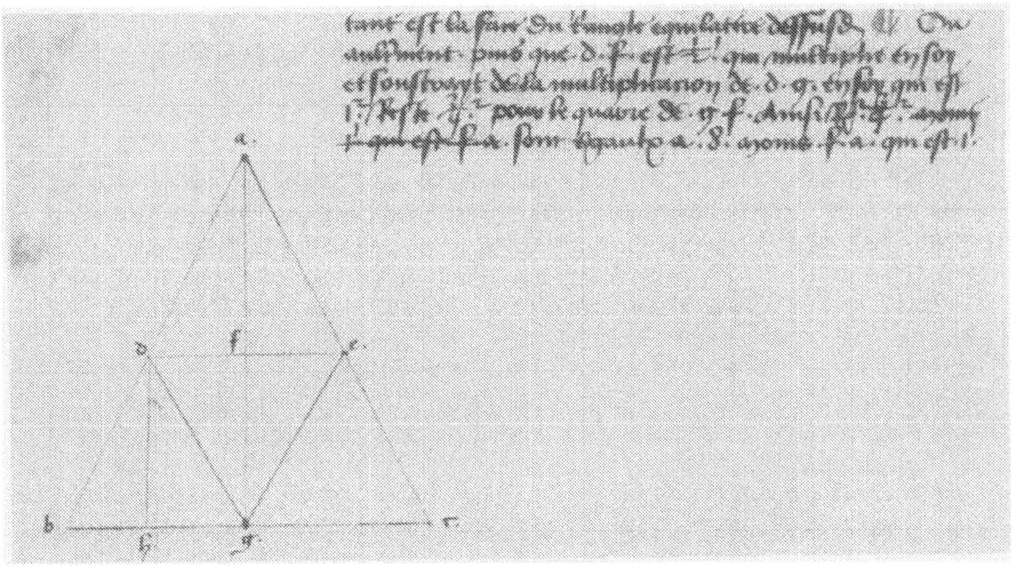

which added with *fa* which is 1^1, the addition yields $R^2\ \frac{3}{4}^2\ \bar{p}\ 1$ equal to 8 which is *ag*. Now simplify your sides and reduce them to non-roots. You will thus have $64\ \bar{p}\ \frac{1}{4}^2$ for one side and 16^1 for the other. Then complete the rest of this calculation according to the canons of the rule of first terms. You will thus have 32 minus R^2 768 and so much is the side of the aforesaid equilateral triangle.

Or otherwise: since *df* is $\frac{1}{2}^1$ which multiplied by itself and subtracted from the product of *dg* multiplied by itself which is 1^2, there remains $\frac{3}{4}$ for the square of *gf*. Thus $R^2\ \frac{3}{4}^2$ are equal to 8 minus *fa* which is 1^1. Now complete the rest according to the canons of the rule of first terms, and you will find as above.

Note that the line fg is thus $R^2\ \overline{1344\ \bar{m}\ R^2\ 1769472}$ which simplified by extraction of the second root comes to R^2 768 minus 24 which subtracted from 8 which is ag, there remains 32 $\bar{m}$ R^2 768 for af.

And therefore the line ad will be $R^2\ \overline{2240\ \bar{m}\ R^2\ 4915200}$ which simplified by extraction of the second root comes to R^2 1280 $\bar{m}$ R^2 960.

The line bd can be found thus. Multiply dh which is equal to fg by itself and also multiply bh by itself which is R^2 192 minus 12 and then add the two products together and from this sum take the second root which is $R^2\ \overline{1680\ \bar{m}\ R^2\ 2764800}$ which simplified by extraction of the second root comes to R^2 960 $\bar{m}$ R^2 720. Now whoever adds bd which is R^2 960 $\bar{m}$ R^2 720 with ad which is R^2 1280 $\bar{m}$ R^2 960, he finds that everything yields R^2 80 which is ab. Also whoever multiplies ag by itself and bg by itself and then adds everything together, he finds R^2 80 for ab.

Moreover, the area of the triangle bdg is then R^2 3072 minus 48. The area of the equilateral triangle gec is then R^2 3072 minus 48. The area of the equilateral triangle deg is then R^2 602112 minus 768. The area of the triangle ade is then 896 $\bar{m}$ R^2 786432. The first three areas added together make R^2 786432 minus 864, which sum if it is joined with the fourth area, one will find in total 32 which is the area of the 4 aforesaid triangles, and equal to the area of the triangle abc.

There is an equilateral square acdf contiguously contained within an equilateral triangle hgp of which the cathetuse is he. Each side of the square is 4. To determine how large is the side of the containing triangle. In order to do such problems one must understand that the little triangle hac is equilateral and its side is equal to that of the square which is 4. Thus hc will be 4 and consequently hb will be R^2 12 which added with be make 4 plus R^2 12 for he which multiplied by itself make 28 plus R^2 768 for the square of the cathetuse he. Now it is true that in equilateral triangles the square of the cathetuse and the square of the hypotenuse are always in sesquitierce proportion as 3 and 4 and hence if you want to find the one by the other, multiply the square of the cathetuse by $1\frac{1}{3}$. Now follow the rule of three in saying: if 3 are worth 4 what would $28\ \bar{p}\ R^2\ 768$ be worth? Multiply and divide as the rule of three requires and you will find $37\frac{1}{3}$ plus $R^2\ 1365\frac{1}{3}$ for the square of the hypotenuse hg or hp. And therefore $R^2\ \underline{37\frac{1}{3}\ \bar{p}\ R^2\ 1365\frac{1}{3}}$ is the hypotenuse line of the triangle hgp.

Whoever would make inquiries into the area of this same triangle containing the square, it is necessary first to multiply $4\ \bar{p}\ R^2\ 12$ which is he by itself, yielding $28\ \bar{p}\ R^2\ 768$ for the square of the cathetuse. Afterwards take the semibase eg or pe which is $R^2\ \underline{9\frac{1}{3}\ \bar{p}\ R^2\ 85\frac{1}{3}}$ which multiplied by itself yields $9\frac{1}{3}$ plus $R^2\ 85\frac{1}{3}$ for the square of the semibase. Now multiply this square of the semibase by the square of the cathetuse. The multiplication yields $261\frac{1}{3}\ \bar{p}\ R^2\ 66901\frac{1}{3}$ plus $R^2\ 66901\frac{1}{3}$ plus $R^2\ 65536$ which simplified by extracting the R^2 of 65536 which is 256 which one should join with $261\frac{1}{3}$. This amounts to $517\frac{1}{3}$. Then it is necessary to add $R^2\ 66901\frac{1}{3}$ to $R^2\ 66901\frac{1}{3}$. This amounts to $R^2\ 267605\frac{1}{3}$. Then this multiplication completely simplified results in $517\frac{1}{3}\ \bar{p}\ R^2\ 267605\frac{1}{3}$ from which the second root which is $R^2\ \underline{517\frac{1}{3}\ \bar{p}\ R^2\ 267605\frac{1}{3}}$ is the area of this same triangle, which root simplified or extracted results in $R^2\ 256\ \bar{p}\ R^2\ 261\frac{1}{3}$ which again simplified comes to $16\ \bar{p}\ R^2\ 261\frac{1}{3}$ and the area of the triangle is so much.

And if each side of this same triangle hpg were 8, how large would be the side of the contained square? To do this calculation I propose that it be 1^1. Thus ab will be $\frac{1}{2}^2$ and ha 1^1. Now from the square of ha which is 1^2 subtract the square of ab which is $\frac{1}{4}^2$. There remains $\frac{3}{4}^2$ which is the square of hb. By which it follows that $R^2\ \frac{3}{4}^2$ is the line hb which added with be which is 1^1 yields $1^1\ \bar{p}\ R^2\ \frac{3}{4}^2$ equal to $R^2\ 48$ which is the cathetuse he when hp is 8. And hence if $R^2\ \frac{3}{4}^2$ are equivalent to first terms, thus 1^1 plus $R^2\ \frac{3}{4}^2$ are all first terms and are the divisor. And $R^2\ 48$ which is equivalent to a number is the number to divide. And because the divisor is compound, one must simplify it in multiplying by 1 m̄ R^2 $\frac{3}{4}$ and this multiplication yields $\frac{1}{4}$ for the divisor. Also R^2 48 must be multiplied by 1 minus R^2 $\frac{3}{4}$ yielding R^2 48 minus 6 which it is necessary to divide by $\frac{1}{4}$. There results from the division R^2 768 minus 24. And the side of the square ac or ad is so much. And whoever would seek further he would find R^2 1008 m̄ R^2 995328 which simplified comes to R^2 576 m̄ R^2 432, which further simplified comes to 24 minus R^2 432 for the line hb which added with be which is R^2 768 minus 24, in adding minus 24 with plus 24 it makes 0. Then whoever adds m̄ R^2 432 with plus R^2 768 obtains plus R^2 48 for the line he and also for the checking of this work... .

Moreover there is a tetragon which has a breadth of 7 and a length of 9, inside of which there is another equidistant, whose area is 60. To determine, how much is there between the two. Answer: the circumference of the containing tetragon is 32. So then to have the circumference of the contained tetragon let us propose that the distance from the one to the other be 1^1. Thus its breadth would be 7 m̄ 2^1 on the one hand and 9 minus 2^1 on the other. So thus its circumference is 32 m̄ 8^1 which added with the circumference of the containing tetragon which is 32 yields 64 minus 8^1 of which half is thus 32 m̄ 4^1, which half, if it is multiplied by 1^1 which is the distance from the one to the other, yields the product 32^1 minus 4^2 equal to 3. For if the area of one tetragon which is 60 is taken from the area of the other which is 63, there remains 3 for the area which is between the two. Now give 4^2 to the one and to the other sides, and you will have 32^1 on the one hand and 3 p̄ 4^2 on the other. Then dispose of the rest of this calculation according to the general canons of the rule of first terms and you will find 4 m̄ R^2 15 $\frac{1}{4}$. And the space between one tetragon and the other is so much. The breadth of the contained tetragon is found thus. From 7 which is the breadth of the containing one take away the double of 4 m̄ R^2 15 $\frac{1}{4}$ which is 8 minus R^2 61. There remains R^2 61 m̄ 1 for its breadth. And from 9, which is the length, 8 m̄ R^2 61 must be taken away and there remains R^2 61 plus 1 for its length. By which it follows that its circumference is thus R^2 976. And *one should know that the length and breadth of the contained tetragon are not in the same proportion as the length and breadth of the containing tetragon, for if they were of one same proportion, they could not be equally distant from the length and from the breadth as they have been presupposed to be.*

This problem can be done in another way when the length and the breadth of the large quadrangle go in a progression of 2. Also it is necessary that the length and the breadth of the little quadrangle go in a progression of 2. For this find a number which multiplied by itself plus 2 amounts to 60. Answer. I propose 1^1. Multiplied by itself plus 2 it amounts to 1^2 plus 2^1 equal to 60. Now divide 60 and 2^1 by 1^2 and the result of it is 60 and 2^1. Then halve 2^1 which is 1^1 multiplied by itself amounts to 1, added to 60, amounts to 61, from which it is necessary to take the R^2 which is R^2 61 from it is necessary to take the half of the first term which is 1, and you will have R^2 61 $\bar{m}$ 1 for the breadth of this little quadrangle. Add 2 to it and you will have R^2 61 plus 1 for the length....

Further problems concern volumes:

Moreover there is an orthogonal body whose dimensions are in double proportion and its contents are 40 cubic units. To determine: how large are its dimensions. Answer: let us propose that the first and smallest dimension be 1^1. Thus the second will be 2^1 and the third 4^1, which multiplied the one by the other make 8^3 which is equal to 40. Now divide 40 by 8^3 and thus you will have R^3 5 for the first dimension, which should be doubled and one will have R^3 40 for the second one, which should be doubled and one will have R^3 320 for the third.

In the next two problems, Chuquet's language does not make clear the operation to be performed. When he says "divide the number by the third", what he means is "take the third root of both sides of the equation".

Furthermore there is an orthogonal body containing 40 cubic portions, of which its three dimensions are in a proportion of quadruple superbipartiens third. To determine: how large are its dimensions. Answer: let us propose that the first and least be 9^1. The second will be 42^1 and the third 196^1 which multiplied the one by the other up to the third yields the product 74088^3 equal to 40. Divide 40 which is the number by the third and you will have $R^3\ \frac{5}{9261}$ which multiplied by 9 yields $R^3\ \frac{135}{343}$ for the first dimension. Then whoever multiplies $R^3\ \frac{5}{9261}$ by 42 and also by 196, he finds R^3 40 for the second dimension and $R^3\ 4064\frac{5}{27}$ for the third. Moreover, there is an orthogonal body containing 40 cubic portions of which the dimensions are in the same proportions as 2, 5 and 9. To determine: how large are its dimensions? Answer: let us propose that they be 2^1, 5^1 and 9^1, which multiplied the one by the other yield 90^3 equal to 40. Divide the number by the third and you will have $R^3\ \frac{4}{9}$. This multiplied by 2, by 5 and by 9, which are the three positions, one will have $R^3\ 3\frac{5}{9}$ for the first dimension, $R^3\ 55\frac{5}{9}$ for the second and R^3 324 for the third....

Furthermore there is an orthogonal body containing 40 cubic units of which the length and the breadth are in double proportion and the breadth and the depth are in subtriple proportion. To determine: how large are its three dimensions? Answer: let us propose 3^1 for the length. Thus the breadth will be 6^1 and the depth will be 2^1, which multiplied the one by the other, the multiplication amounts to 36^3 equal to 40. Now divide 40 by 36^3 and you will thus have $R^3\ 1\frac{1}{9}$ which should be multiplied by 3, by 6 and by 2. And one will find R^3 30, R^3 240, $R^3\ 8\frac{8}{9}$ for its three dimensions....

The fourth section of the book is on geometric constructions, of which a small sample is given below.

How some geometric lineations and figurations are drawn

At the very beginning in order to investigate and seek the lost centre of a curved line such as acb of which a and b are the two ends of the line. Let c be equally or unequally distant from a and b. Then let the fixed foot of the compass be put on a and let the other movable foot be opened to c, or more or less, as you wish, and with this movable foot let there be drawn the line fcg

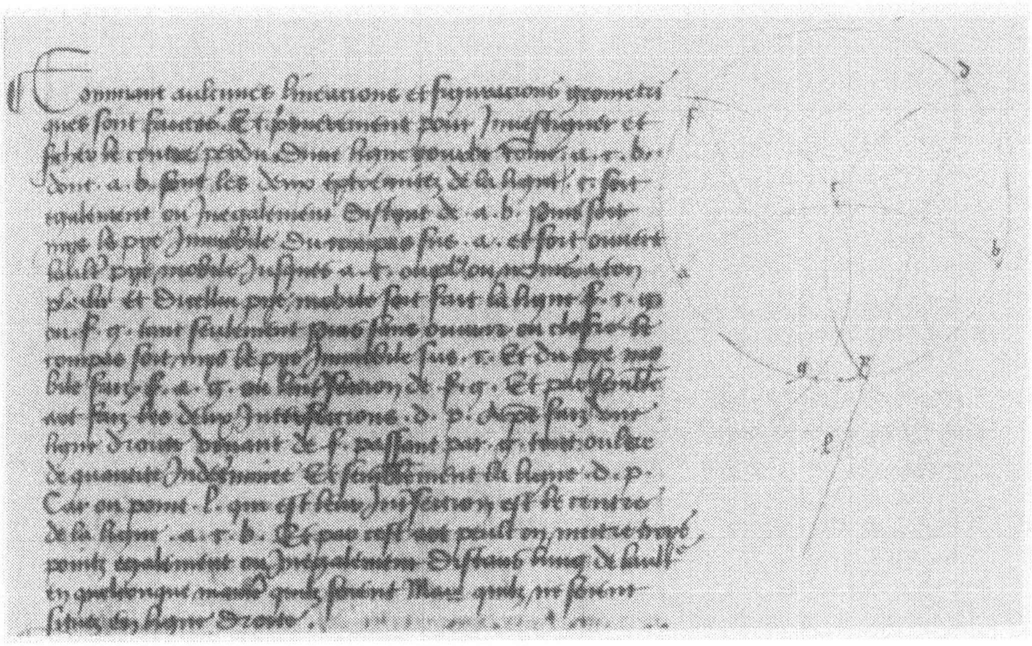

or fg only. Then without opening or closing the compass, let the fixed foot be put on c and with the movable foot draw fag or the intersection of f and g, and draw the two intersections d and p by a similar art. Then draw a straight line coming from f passing by g and well beyond, of an indeterminate length, and similarly the line dp, because at the point l which is their intersection is the centre of the line acb. And by this art one can put three points equally or unequally distant the one from the other, in whatever manner they may be, but let them not be placed in a straight line.

On the line bc and at the end of the same I wish to draw a perpendicular line making a right angle with it. To do this, with a compass opened at will, I put one of its feet on b and the other one now here, now there at will, as on d, and with the foot that is on b I make the point g and further a light mark through k. Then I put a line on g and on the point d, and make an intersection on the light mark k. And from this intersection I draw a straight line up to b. Thus I have achieved and accomplished that which I wished to do.

Furthermore on the middle of ad I wish to make a perpendicular line making 4 right angles. First, the compass opened at will, I put the fixed foot on a and make two light marks, one through b and the other through c. And then, without moving the legs of the compass, I put its fixed foot on d and make two other light marks as above through b and through c. And from the intersection of the light marks at b through the intersection of those at c, I draw a line which is of the aforesaid condition....

I wish to make two circles whose areas are truly in double proportion. First let a circle be made at will and divided into 4 equal quarters. Then let the compass be opened from d through c, or of the size of another quarter, and with the compass opened thus, let another circle be drawn, for its area will be the double of that of the first....

On an orthogonal triangle cbd I wish to form several triangles which are equal to it, despite their angles being dissimilar or unequal. And in order to do this, I draw the line bc of an indeterminate length and divide it into several equal portions, as many as the number of triangles that I want to make. And let each division be equal to bc and let the divisions of this line be f, g, k. Then from d to each one of them I draw a line, and thus I have constructed three other non-orthogonal triangles: that is dcf for one, dfg for the second, and dgk for the third. The area of each of them is equal to the orthogonal triangle dbc....

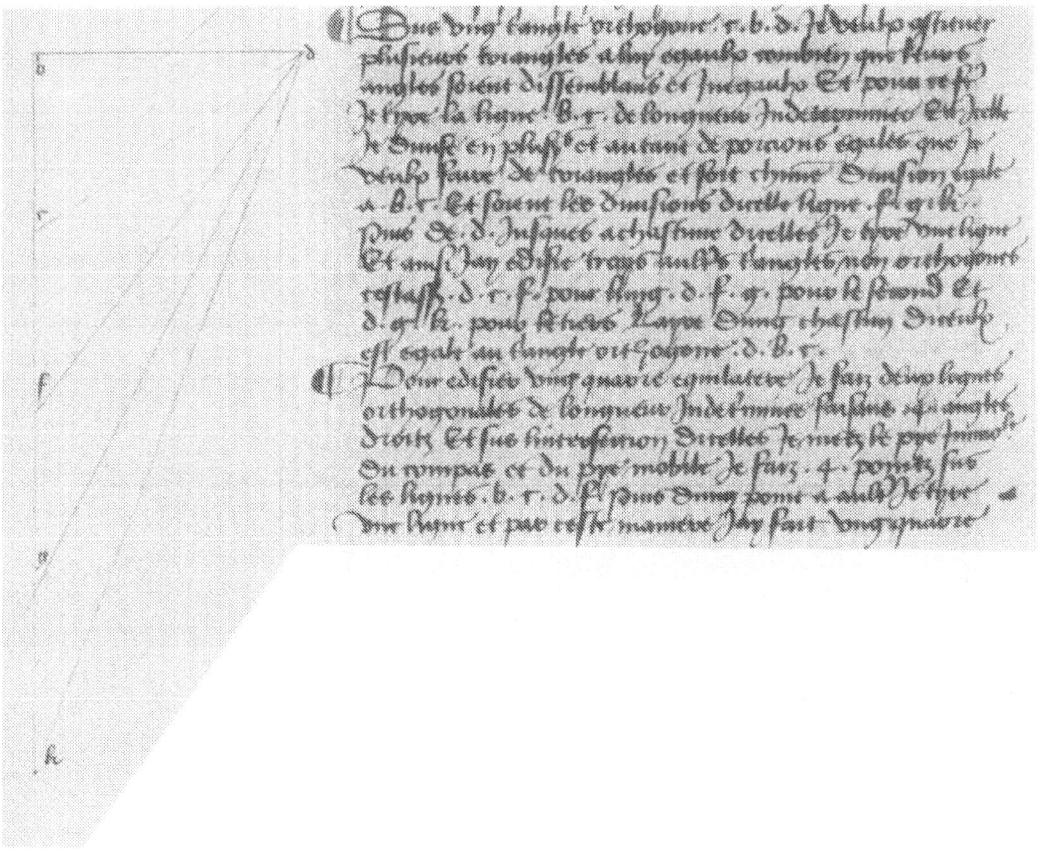

Sur ung triangle orthogone .c.b.d. je veulx construire
plusieurs triangles a luy equaulx combien que leurs
angles soient dissemblans et inequaulx Et pour ce faire
je tire la ligne .b.c. de longueur indeterminee Et icelle
je divise en plusieurs et autant de porcions egales que je
veulx faire de triangles et soit chascune division egale
a .b.c. Et soient les divisions dicelle ligne .f.g.k.
puis de .d. jusques a chascune dicelles je tire une ligne
Et ainsi jay edifie trois aultres triangles non orthogones
cestassavoir .d.c.f. pour lung .d.f.g. pour le second Et
.d.g.k. pour le tiers Lequel chascun diceulx
est egale au triangle orthogone .d.b.c.

Pour edifier ung quarre equilatere je faiz deux lignes
orthogonales de longueur indeterminee faisans .4. angles
droitz Et sur lintersection dicelles je metz le pie ferme
du compas et du pie mobile je faiz .4. pointz sur
les lignes .b.c.d.g. puis dung point a laultre je tire
une ligne et par ceste maniere jay fait ung quarre

The fifth section discusses quadrature by lunes.

Arrangement and argument about the inquiry into the quadrature of the circle

To do this I draw a circle divided into four equal fourths, divided by two orthogonal lines at the points b, c, d and g whose centre is p. Then from one quarter to another I draw a line bc on which I draw a semi-circle bkc. This

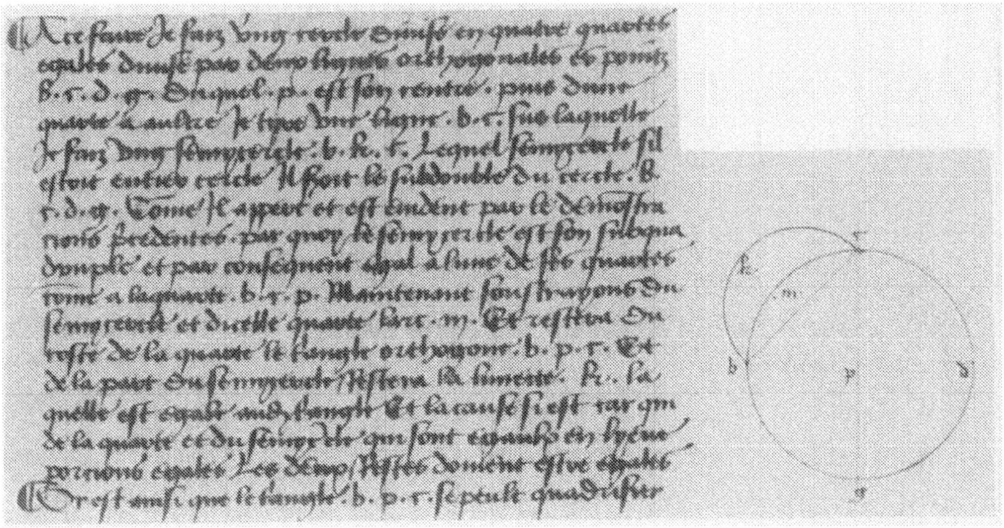

semicircle, if it were the whole circle, would be the subdouble of the circle bcdg, as it is apparent and evident from the preceding demonstrations. From which the semicircle is its subquadruple and as a consequence equal to one of its quarters, such as to the quarter bcp. Now we subtract from the semi-circle and from this same quarter the arc m and there will remain on the side of the quarter the orthogonal triangle bpc and on the side of the semi-circle will remain the lune k, which is equal to the said triangle, and the reason is because if equal portions are taken from the quarter and from the semi-circle, which are equal, the two remainders should be equal. Now the triangle bpc can also be squared, by which it follows that the lune k which is part of the circle can also be reduced to a square.

Moreover, within the circle bogl I build a hexagon of which each of its sides is equal to the semi-diameter of the said circle and on each of its three sides I make a semi-circle, that is to say bc for the one, cd for the second, and dg for the other. Now it is certain as is apparent from the preceding conclusions that the area of each of these semi-circles is equal to the fourth part of the semi-circle bck. By which it follows that the said

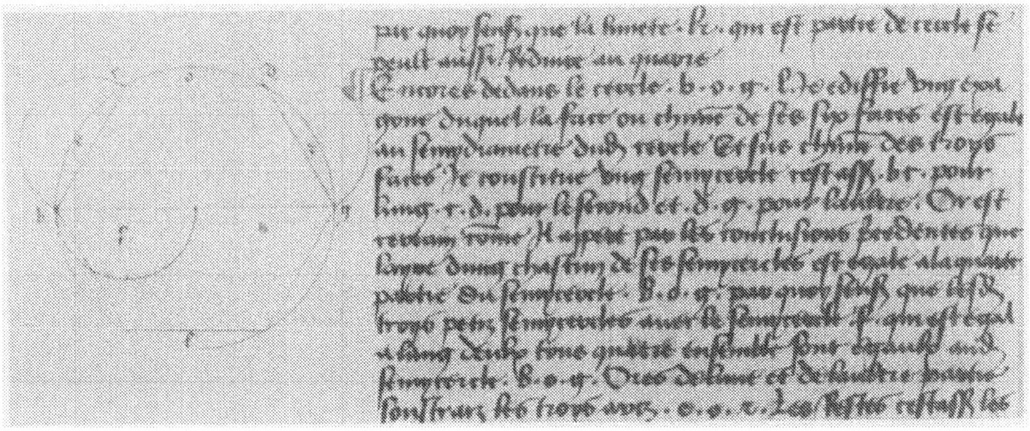

three little semi-circles with the semi-circle f which is equal to one of them, all four together are equal to the said semi-circle bog. Now subtract the three arcs eor from the one and the other side. The remainders, that is to say the three lunes with the semi-circle f are equal to the semihexagon bcdg. Now it is thus that the semihexagon as well can be squared, because it can be resolved into three equilateral triangles. By which the three lunes with the semi-circle can also be converted into a square.

Another consideration on the quadrature of the circle. If the circumference of a circle is extended in a straight line, and on any part of this same, let it be at one of the ends or in the middle or elsewhere, this semi-diameter is put perpendicularly and from each extremity of this same extended line a straight line is drawn to the upper part of the semidiametrical line, one will have a triangle whose area is equal to the area of the circle. Example: let the extended line of the circumference be bd, the semi-diameter cp be raised perpendicularly on bd. Then let the line bp and pd be drawn. You will have the triangle bdp of which the area is equal to the area of the circle. And seeing that the semi-base of the triangle multiplied by the cathetuse produces the area of the triangle, it results from this that the semi-diameter multiplied by the semi-circumference produces the area of the circle.

Common usage holds that the proportion of the diameter to the circumference is as 7 to 22, but this is something which cannot be proved by any demonstration. Master Ramon Lull in his Ars Magna holds that the circular

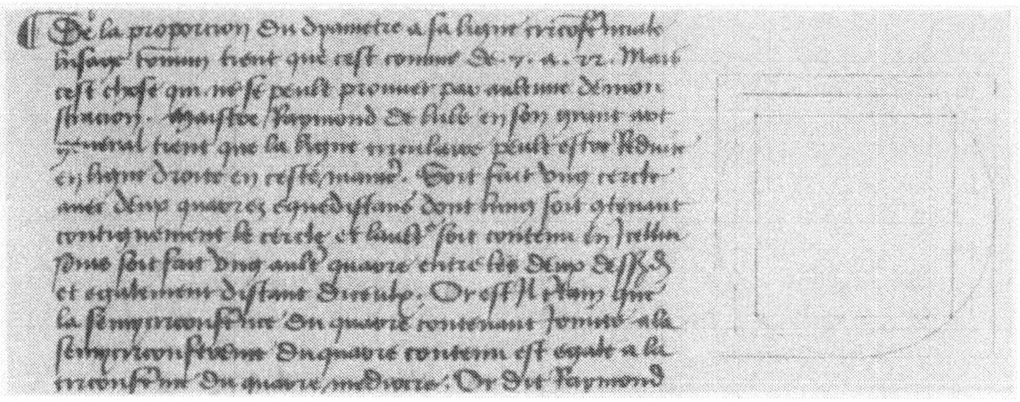

De la proportion du dyametre a sa ligne circonferencialle lusage commun tient que cest comme de .7. a .22. Mais cest chose qui ne se peult prouver par aucune demonstracion. Maistre Raymond de lulle en son grant art general tient que la ligne circulaire peult estre reduicte en ligne droite en ceste maniere. Soit fait ung cercle avec deux quarrez equidistans dont lung soit contenant contiguement le cercle et laultre soit contenu en icellui Puis soit fait ung aultre quarre entre les deux dessusd. et egalement distant diceulx. Or est il certain que la semicirconference du quarre contenant joincte a la semicirconference du quarre contenu est egale a la circonference du quarre moien. Or dit Raymond

line can be reduced to a straight line in this manner. Let a circle be drawn with two equidistant squares of which one contiguously contains the circle and the other is contained in this same. Then let there be drawn another circle between the two aforesaid ones and equally distant from these same. Now it is certain that the semi-circumference of the containing square joined to the semi-circumference of the contained square is equal to the circumference of the intermediate square. Now Ramon says that the circumference of this same intermediate square is equal to the circumference of the circle and therefore the diameter *would be contained in the circumference by two times plus R^2 2. And although this opinion may have considerable plausibility, nevertheless it is something which cannot be ascertained. At another time I drew three equilateral triangles of which the largest contained the circle, the smallest was contained in this same, and the third was equally between the two, from which it appeared that its circumference was equal to the circular line. Also similarly I attempted the task with three hexagons and one circle, and it appeared to look as if the intermediate hexagon was equal to the circular line. But when I examined everything by considering the numbers, I find great variety and I believe that the intermediate square is closer to equalling the circular line than the intermediate triangle and moreover the middling hexagon is closer than the intermediate square, but in none of them is there anything certain....*

The final section concerns constructing and using measures of the volume of wine vessels. Some of the words used in the translation have fallen out of common usage in English, since there is no longer occasion to determine the volume of wine vessels of a variety of shapes and sizes. Although Chuquet makes clear the meaning of each of these terms when they are introduced, a brief glossary might be of use. Three words are names for measuring devices: diapason, gauge and sampling cup or sample. The gauge is a rod on which measurements are marked. The sampling cup is a cup whose volume is known.

How one can make the diapason or the gauge to measure circular vessels

It is necessary for the start to have a well-made little round vessel of 16 or 20 inches in diameter or thereabouts or more, for the larger the better. Let it be as deep as its diameter or as you will, just as you would like it. Then let there be put in this vessel a measure of water of the quantity that one wants. Then prudently establish the depth of the water and let the depth of this same be the line bd. And the diameter of this vessel called the sampling cup is the line cg. Then we have two measures. The one is for the diameter of the sampling cup, which is one measure, and the other is for the length or the depth of this same.

Another style for obtaining quite precisely the length and breadth of the sampling cup. One should have a fairly large round vessel which holds 40 measures or thereabouts. And in this same let a large number of measures be put, such as 36, which can be found by the multiplication of 4 and 9. Now take the second root of 4 which is 2. Now divide the diameter of this same large vessel by 2. It is in two parts, for the one of these will be the true diameter of the sampling cup. So then let the depth of the said 36 measures of water be divided by 9, for one of these same will be for the length of your sampling cup. And whoever would put only 24 wine cups or measures in this large vessel, which number can be obtained from the multiplication of 4 by 6. Of the lesser number, as of 4, the second root must be taken, which is 2, and do as above, and one will have the diameter of the sampling cup. Or in this manner, and it all comes to one. Let the diameter of this same large vessel holding 24 measures be divided by 4 or in 4 equal parts and of one of these same 4 parts take two thereof which are the R^2 of 4 and one will have the diameter of the sampling cup. And the sixth part of the depth of the water will be the length or the depth of the sampling cup.

This manner of taking the length and the breadth of the sampling cup by a large number of measurements is more precise and more certain than that taken by one measurement, for the error that one can make with one measurement causes a larger error than that which is made with a large number.

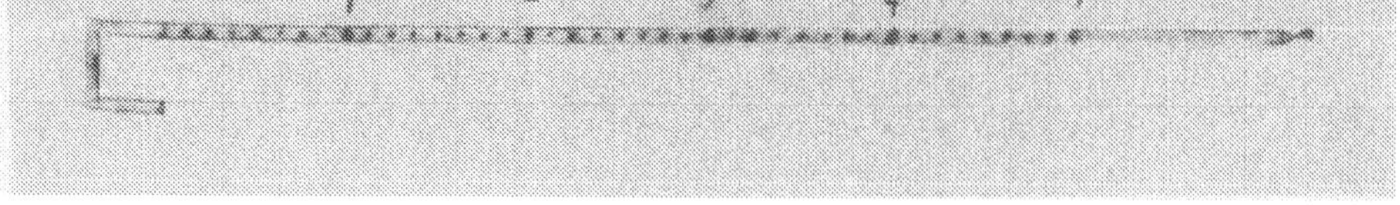

So then one has the diameter and the length of one measuring vessel, that is of the sampling cup. One should have a rod which is quite straight on which there is a hook at the end. It should be about half a foot long and with as much of an opening. It should be divided into several parts equal to the line bd which is the length or the depth of the sampling cup. And each of these same lengths should be divided further into 8 or 12 smaller parts. And this rod divided thus is to measure the length of vessels.

The rod for measuring the diameters of these same is divided in this manner. One of the sides of the aforesaid square rod should be divided into several portions equal to the diameter of the sampling cup. And each of these can be further subdivided in 8 or 12 parts just as the aforesaid one.

The usage of this rod, otherwise called a gauge, is as follows. Let it be laid out on the length of the vessel to be measured in such a fashion that its hook touches the outside of the end of the vessel which one wishes to measure. And judge how many lengths of the sampling cup are contained in the length of the vessel to be measured. That is to say from the one of the ends to the other, the estimated thickness of the two ends deducted. And let us suppose that this depth, the estimated thickness of the two ends deducted, is $8\frac{1}{2}$, which one should keep in mind. Then by the other part of the rod which is applied to the diameter of the end, let the end be measured. And also let the depth of the vessel be similarly measured from the bung. And the mean measure between that of the end and that of the bung be taken. If the measure of the bung were 7 and that of the end were 6, the mean measure for the true diameter of this same vessel would be $6\frac{1}{2}$, which one should always multiply by itself, yielding $42\frac{1}{4}$, which one should further multiply by $8\frac{1}{2}$ which is the length of the vessel on which you work. And the multiplication yields $359\frac{1}{8}$ and this vessel contains so many measures.

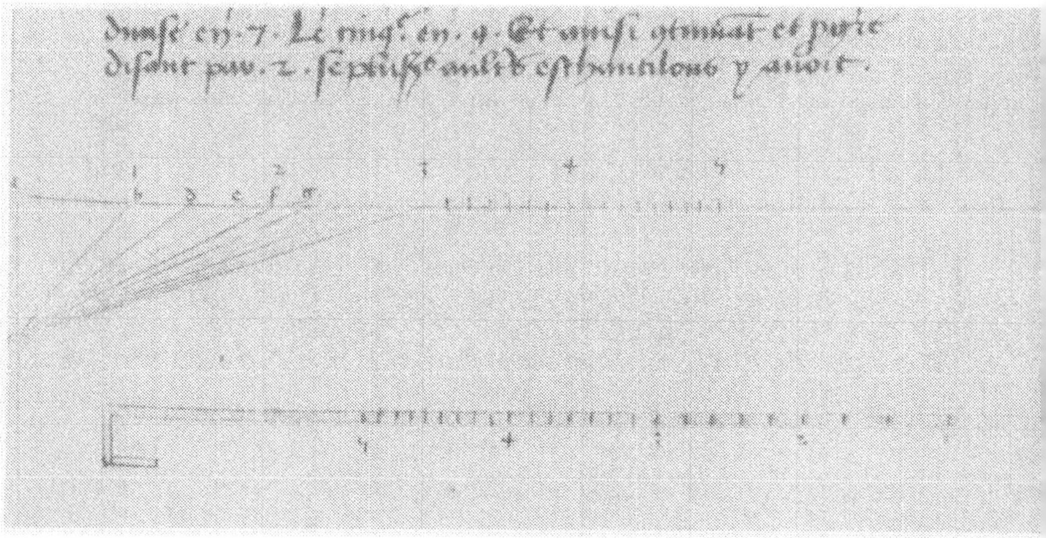

Another manner of subdividing the rod appropriately for the diameters or ends of vessels. The aforesaid rod is already divided into equal portions according to the line cg which is the diameter of the sampling cup. Each of these parts can be subdivided into several other unequal parts. And in order to do this, on a well-smoothed wooden table, it is necessary to make a straight line along the length of this table, and at one of the ends of this line, let another orthogonal line be made across this table, and let each of these two lines be of an indeterminate length. And on the intersection of these two lines let a be noted. So then, let the compass be opened to the size of the line cg which is the diameter of the sampling cup, and the fixed foot put on a. Let two points be made on the two lines of which one is c and the other b. And let the line which is on the length of the table be divided into several equal portions, as many as suffice, and in portions such as ab. And let these equal divisions be noted by numbers in setting 1 on b and 2 on the second point, and then 3 and 4 etc. This done, let the compass be opened to b and with the fixed foot put on a, let d be drawn. Then the compass opened from d to c and the fixed foot put on a, let e be drawn. Further let the compass be opened from e to c and then the fixed foot of the compass put on a. The other movable foot should fall on f which is the second measure of the diameter of the sampling cup. Furthermore let the compass be opened from f to c, and the fixed foot put on a, and let g be made. And continue thus as much as suffices. And by this manner, the first sample ab remains undivided, the second bf is divided into three unequal parts, the third is divided into 5 unequal parts, the fourth is divided in 7, the fifth in 9. And thus continuing and progressing by 2, there are thus several other samples.

And since the line on the length of the table is divided and subdivided by similar and equal divisions and subdivisions, one should divide and punctuate one of the sides of its rod. One should know that the first space ab is worth one measure, bd another, de another, and consequently each of the others is worth a measure, that of a pint or of a wine-cup etc. And this rod is thus divided in such a manner that each division or each point carries with it the properties of a circle in the manner that it appears in this following figure.

The usage of this gauge is that the length of the vessel taken by the aforesaid manner with the practice and usage of the other rod, and similarly the intermediate diameter taken by this rod in the manner set forth above, this diameter multiplied by the length of the vessel demonstrates its contents.

Example: let the lengths of the vessel to be measured open from within, that is between the two ends, be 8 and the intermediate diameter 30. Multiply then 8 by 30. The multiplication yields 240, and the vessel measured contains so many measures.

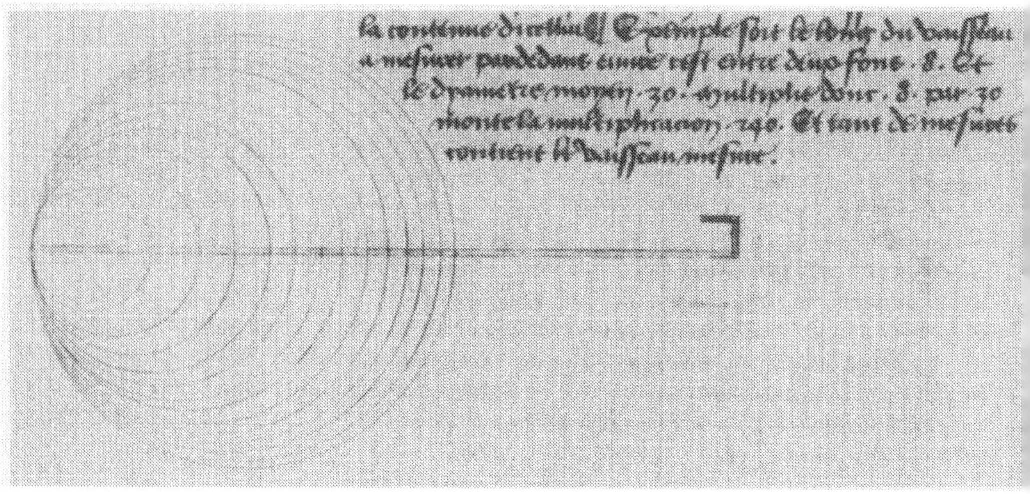

With this detailed account of skills using practical geometry which have fallen into desuetude, Chuquet's *Geometry* ends. It has been possible here to present only representative extracts from the work. For the full French text, see l'Huillier (1979).

Chapter 8

THE COMMERCIAL ARITHMETIC

In an age where mathematics was praised and commended for its utility in all trades and professions, there was a considerable gulf between the theoretical arithmetic and algebra of the *Triparty* and the needs of the market-place. A teacher might add to his prestige by writing on algebra, but his pupils' motives (or in the case of children, the motives of their parents) in learning arithmetic were almost entirely practical, and the lives of merchants and their apprentices were not devoted to solving quadratic equations or to inscribing one regular polygon inside another. Evidence about Renaissance Florence collected by van Egmond (1976) suggests that the pupils at arithmetic schools were usually the children of merchants and craftsmen, destined to work in the family business, who would study mathematics for two years at about the age of eleven after a basic education in reading and writing their native language. The minority who wished to take their studies further might do so while acting as assistants to the master.

The character of Chuquet's *Commercial Arithmetic* may suggest that his students were rather older than the boys who attended the abacus schools of Florence. The book is set out, like the *Geometry*, as a series of problems and their solutions. Occasionally a rule is stated, and alternative methods are offered for those who do not wish to use the rule of three. It is quite possible that the work reflects the content of his teaching. Some knowledge of arithmetic is assumed, but much less than is covered in the first part of the *Triparty*, and there are indications that this knowledge may have grown rusty. On the other hand, Chuquet makes no concessions in the direction of giving simple examples of arithmetical working before moving on to more complicated figures. His discussions of the rule of companies, which governed the division of money between partners in a business or between a bailiff and his students, and of the practice of exchanging goods by barter show a concern for justice, and, in the former case, go into more detail than would be expected in a first course. The last two chapters are devoted to problems associated with the minting of gold and silver, and would be particularly suitable for workers and apprentices at the mint.

Advanced mathematics is avoided almost entirely. Some questions on compound interest occur in the *Problems*, but it is not treated directly in this part of the manuscript, perhaps because Chuquet did not wish to discuss the subject without considering its more advanced aspects. There is one question, concerning how much a certain sum of money 'earns' in a given time, to which Chuquet offers an algebraic solution based on the assumption of compound interest, but the reference is probably to money invested in a company rather than to money lent by an individual, and most writers of the day would have been satisfied with Chuquet's first solution, assuming something more like simple interest (see Example 5.1.1, p. 315).

Lyon and its fairs

In the sixteenth century, Lyon had a flourishing printing industry, inferior only to Venice and Paris in number of titles published, and particularly renowned for the publication of scientific works. Silk-weaving was established at the southern end of the town. The spice market was the most active in Europe, and Lyon was also an important commercial centre on account of the banks established by the Florentine community. This prospertiy was due in no small measure to the quarterly fairs, established by Louis XI in 1463 after a false start earlier in the century, and continuing (apart from a break from 1484 to 1494) until the late 18th century. For a detailed history of the Lyon fairs, the reader is referred to Brésard (1914).

The privilege of holding fairs was much sought after in the Middle Ages. Lyon was so honoured largely because of Louis XI's links with the town. Its location was also propitious, with navigable rivers enabling not only goods from the south but even from the Loire Valley and from Paris (via the Yonne) to be brought into the town by water, with only a short journey overland from one river valley to another. As a frontier town, it offered great possibilities for international trade. Merchants came from Spain, Portugal Italy, Germany, the Netherlands and all parts of France. Only the English, the old enemy, were excluded by decree.

Each fair lasted a fortnight, during which the streets and open spaces of the town were taken over by the multitude of visiting merchants and a number of regulations were relaxed. For example, charging interest on money lent during the fairs was explicitly authorized on the grounds that the money was borrowed as an instrument of production, and not as the last resort of the poor (Brésard 1914, p 110). This encouraged the establishment of banks, though the Medici had had an agent in Lyon since 1455. Foreign visitors enjoyed special privileges of safe conduct to and from the fairs, and exemption from taxes if they owned property in Lyon on account of the fairs. They were also allowed to import and export money without the usual restrictions during the fairs, which led to the complaint that the French currency was being debased, one of the main reasons given for the suspension of the fairs in 1484.

In Chuquet's day, about half the total trade was in various kinds of cloth, not produced in Lyon. Chuquet mentions *blanchet*, a white woollen material for making linings and petticoats, and *brunette*, a fine black cloth, as well as the more exotic *camlet*, which was a mixture of hair with silk or other materials. The market in silk was controlled by the Italians, who were accused of bringing in unnecessary luxuries and taking out the best of the coinage in exchange. Charles VIII on his accession in 1483 promoted the development of a silk industry in Lyon, and by the middle of the sixteenth century the French were able to compete with the Italians.

Just as Charles VIII encouraged the silk industry, so Louis XI had opened up the market for spices, offering an exemption from the usual 10% import duty. Merchants from central and eastern France and from Germany came to Lyon to buy spices from the French and Italian ports.

Although the first printers in Lyon had already set up their presses before Chuquet wrote down his manuscript, the industry was still in its infancy. By 1538, on the other hand, it was so well established that Gilles Huguetan included two tables especially for printers and booksellers in his Appendix to the second edition of de la Roche's *Larismethique*. At least one elementary book on arithmetic was printed in Lyon around the turn of the century; more significant works followed in 1515, including Ortega's *Oeuvre tressubtile...*, written by a Spaniard and previously published in Latin, preparing the way for the first edition of de la Roche in 1520. De la Roche's *Larismethique* (1520) describes a wide variety of goods, especially fabrics, and their places of origin. His chapter on barter is particularly rich in this respect. Although this is not direct evidence, it probably gives some indication of the merchandise bought and sold in Lyon. Jean Trenchant (1557) devotes a section to the commercial practices of the fairs, particularly the quarterly settlement days when letters of credit were exchanged or traded until all debts had been settled.

The modern reader may regret that Chuquet is sparing with such details, but it was no part of his purpose to leave a record for social historians. His *Commerical Arithmetic* outlines the state of the subject, as observed by a practising teacher of arithmetic in a fifteenth-century town full of vitality but with no tradition of academic learning or commercial sophistication. Brésard (1914, p 285) points out that the Florentines, who controlled not only the banks in Lyon but also the settlement days, were far ahead of their French comtemporaries in their financial practices.

Units of measurement

Although a number of different coins were in circulation in France during the fifteenth century, their values were expressed in terms of *livres*, *sols* and *deniers*, descended from the Roman *libri*, *solidi* and *denarii*, and related to the English £.s.d (pounds, shillings and pence) which remained in use until 1970. One sol was worth 12 deniers, and one livre worth 20 sols. However, the livre and the sol would seem to have been used almost entirely as accounting money rather than as actual coins. The high-value coin most frequently mentioned in Chuquet is the *ecu (écu)*, quoted at various values between 27 sols 6 deniers and 35 sols 9 deniers. Florins, ducats and reals are also mentioned, but at least some of these were probably foreign coins. At the lower end, a number of small coins worth a few deniers are also mentioned. An account of life in Paris in the first half of the fifteenth century (*Anon*, 1963) describes numerous revaluations and withdrawals of certain coins, sometimes because they had become worn away, or because of forgeries, or because they bore the head of the wrong king, but more frequently in response to severe shortages of food and fuel under siege conditions. The denier itself was only a local standard; as one of Chuquet's examples (5.3.1, p. 317) shows, the Paris denier, the Lyon denier and the Geneva denier could all have different values.

A further complication is that *denier* and *sol* also had particular meanings in the context of gold and silver. A denier of pure metal was the twentyfourth part of an ounce; but *denier* was also used to measure purity, so that pure silver was described as 12 denier or a sol of pure silver. 11-denier silver would contain 11 parts of silver to one part of base metal, and so on. Thus a coin could be described as having a face-value of a certain number of deniers, silver content a different number of deniers, and alloy yet a third number of deniers.

The only units of length in the *Commercial Arithmetic* are the *aulne* (ell) and the *palm*. The latter can mean either a handsbreadth or a span (also *grand palme)*, but since Chuquet quotes $5\frac{1}{3}$ palms to the aulne, and the aulne in various places measured between $2\frac{1}{2}$ and $4\frac{1}{2}$ feet, the meaning here must be 'span'.

With the exceptions of silk and precious metals, most weights are expressed in some combination of *ounces*, *pounds (livres)*, *quintals* and *charges*. However, the number of ounces in a pound appears to vary. Chuquet does not make it clear whether, when he writes $15\frac{1}{2}$ ounces to the pound, he is talking about a single system of weights, or whether he is claiming that one pound in some other region's system of weights was equivalent to $15\frac{1}{2}$ lyonnais ounces. The latter appears more probable. A quintal is usually (but not always) 100 pounds, and a charge is three quintals, or the load an ass can carry. The measure *sester*, used for corn, was also based on an ass's load, but as a measure of capacity rather than weight; a sester of corn weighed about 220 pounds.

On the whole, Chuquet's names for the units have been preserved in the translation, with minor changes in spelling to conform to equivalent English units (*sester* for *sestier*) or to modern French spelling (*grands blancs* for *grans blans*). *Livre* has been preserved for money, but translated as *pound* for weight. Chuquet normally abbreviates *livre*, *sol* and *denier* to l.s. and d., and an *obol* (half a denier) to ob, while *livre* in the sense of weight becomes lb; in the interests of readability, these abbreviations have been used only in marginal calculations.

How the science of numbers may be applied in matters of merchandise, and firstly of addition

1.1 *Someone owes a man all the sums which follow, that is to say, 12 livres, 15 sols, 11 deniers for one sum. Item, 35 livres 6 sols 9 deniers. Item, 9 livres 19 sols 3 deniers. Item, 7 sols. Item, 7 deniers. Item, 17 livres. Where a livre is 20 sols and a sol is 12 deniers. To determine how much all these sums amount to. To do this, it is first necessary to add together all the deniers, which are 30. These should be divided by 12 to make sols from them, and one will find 2 sols 6 deniers, of which the 6 deniers should be placed below and aligned with the other deniers. And the 2 sols should be added to the other sols, which are 49 in all, which one should divide by 20 to make livres from them. One will have 2 livres 9 sols. The 9 sols should be placed below the sols, and the 2 livres should be counted with the other livres, which are 75 in all. Thus all these sums amount to the total of 75 livres 9 sols 6 deniers.*

12	15	11
35	6	9
9	19	3
0	7	0
0	0	7
17	0	0
75	9	6

1.2 *Item. A man is owed all the sums which follow. That is to say, 45 ecus, 17 sols, 9 deniers. Item, 54 ecus, 11 sols, 10 deniers, an obol. Item, 9 ecus, 28 sols. Item, 11 deniers. Item, 540 ecus, 12 sols. Item, 81 ecus, 7 deniers, an obol. Item, 13 ecus, 4 sols, an obol. In which an ecu is at 30 sols 3 deniers. To determine how much they all amount to...*

1.3 *Item. A man is owed 25 florins, 7 gros, 5 deniers. Item, 63 florins, 9 gros, 10 deniers. Item, 52 florins, 16 gros, 3 deniers. Item, 36 florins, 9 deniers. Item, 8 florins, 6 gros, 4 deniers. In which the florin is 21 gros and the gross is $13\frac{1}{3}$ deniers.*

1.4 *Item. Again, a man is owed 12 ecus $\frac{3}{4}$. Item, 27 ecus $\frac{3}{8}$. Item, 15 ecus $\frac{1}{2}$. Item, 93 ecus $\frac{2}{3}$. Item, 40 ecus $\frac{5}{6}$. To determine what it all amounts to. To to this, reduce all the fractions to 24ths...*

1.5 *Item. A man has bought 14 aulnes $\frac{1}{2}$ of cloth. Item, 19 aulnes $\frac{3}{4}$. Item, 16 aulnes $\frac{2}{3}$. Item, 20 aulnes. Item, 13 aulnes $\frac{1}{4}$. Item, 18 aulnes, 2 quarters $\frac{1}{2}$. Item, 9 aulnes, 2 thirds and $\frac{1}{2}$. Item, 11 aulnes and $\frac{1}{2}$ third. Item, 10 aulnes, 1 quarter and $\frac{1}{2}$. Item, 15 aulnes and $\frac{1}{2}$ quarter. To determine how much it all comes to...*

The wording of this question shows that simple fractions have been assimilated into the vocabulary of the marketplace, while more complicated ones have not; thus $\frac{5}{8}$ is expressed as 2 quarters $\frac{1}{2}$.

1.6 *Item. A man has sold 5 quintals 10 pounds of spices. Item, 3 quintals, 97 pounds, 10 ounces. Item, 7 quintals, 28 pounds, 13 ounces. Item, 2 quintals, 83 pounds. Item, 1 quintal, 9 ounces. Item, 4 quintals. Item, 20 quintals, 36 pounds, 11 ounces. Item, 19 pounds, 7 ounces. To determine how much they all amount to, counting a quintal as 100 pounds and a pound as 16 ounces...*

Two methods are given: adding the ounces first and reducing to pounds, then adding the pounds and reducing to quintals; or changing the quintals to pounds before proceeding to add the pounds. This shows that even the relationship between pounds and ounces could not be regarded as fixed.

1.7 *Item. Someone owes a merchant the sum of 5 quintals 16 pounds, 12 ounces of wool. Item, 3 quintals, 91 pounds, 7 ounces. Item, 1 quintal, 10 pounds, 9 ounces. Item, 4 quintals, 81 pounds, 8 ounces. Item, 2 quintals, 106 pounds. In which a quintal is 117 pounds* *and a pound is 13 ounces $\frac{1}{2}$...*

1.8 *Item. Someone has sold 25 marks, 4 ounces of silver. Item, 18m, 7 ounces, 13d, 15 grains, 12 primes. Item, 46m, 3 ounces, 19 grains. Item, 67m, 21d, 22 primes. Item, 30m. Item, 3m, 2 ounces, 16 deniers, 12 grains...*

The relationship between the units in the question is given in the working; it is the standard system of weight for silver presented below in Chuquet's chapter 6. Each of these problems is accompanied by the calculation set out in numbers, either in the margin or below the description of method.

There are 16 problems on subtraction. None involves more than one subtrahend, and the scope is similar to that of the problems on addition, except for the last two questions, which relate to dates. Here is the second of these:

1.2.16 *A decree was made in the year 1391, on the 13th day of October. To determine how many years it is since it was made, now that it is 1484 and the 2nd day of May. Reckoning 30 days for a month, and 12 months for a year, and the beginning of January as the beginning of the year. Subtract 1390 years, 9 months, 13 days, from 1483 years, 4 months and 2 days, and you will find 92 years, 6 months and 19 days. That this is how long it is since it was made.*

Marre (1880) suggests that the date 2 May 1484 tells us when this part of the manuscript was written. This fits well enough with the date at the end of the *Triparty*. The details of the calculation, with 30 days to the month, would not in general lead to an exact answer. However, it is interesting that Chuquet regards 1 January as the beginning of the year; this was not the official New Year in France until the following century.

1.3.3 *Item. Again, a merchant has laid out in goods 546 livres 17 sols 9 deniers $\frac{1}{3}$ in such a way that each livre made a profit of $\frac{3}{4}$ of an ecu. To determine how much all this profit comes to. To do this, it* *is necessary to consider 17 sols 9 deniers $\frac{1}{3}$. What part is it of one livre? And it is $\frac{8}{9}$ of a livre. Now multiply $\frac{3}{4}$ of an ecu by 146 livres $\frac{8}{9}$ of a livre, and you will find 410 ecus and $\frac{1}{6}$ of an ecu. And so much does all the profit amount to.*

There are 9 questions in the section on multiplication, and the same 9 questions, posed in reverse, in the section on division; Chuquet was familiar with this aspect of the schoolmaster's art. These are not, of course, the simplest possible examples on the four rules as applied to money, weights and measures.

How the rule of three may be applied in matters of merchandise. And firstly, of money

2.1.1 *Someone owes a merchant the sum of 200 ecus at 30 sols 3 deniers each. To determine how many livres at 20 sols should be paid to him. Answer. Take 20 sols for the first number in the rule of three. Then take 30 sols $\frac{1}{4}$ for the second number, and 200 ecus will be the third number. That is to say, if 20 ecus are worth 30 livres $\frac{1}{4}$, what will 200 ecus be worth? Then multiply and divide as the rule of three requires, and one will find 302 livres $\frac{1}{2}$. And so many livres are the 200 ecus worth.*

And one should know that if the value of one of the coins which are owed is given in sols, then the value of one of the coins with which the payment is to be made should also be expressed in sols. And if the one is expressed in gros, the other should likewise be expressed in gros, so that they may be set down appropriately in the rule of three. As for example, whoever should say, from 18 florins at 18 gros each, how many reals are they worth at 30 sols? In such a case, one should put the 30 sols in gros, or the 18 gros in sols, so that they are alike, and then set them down in the rule of three in the manner aforesaid.

Or otherwise, reduce the 20 florins into petits

blancs or liars or deniers, or obols if necessary, and then divide them by as many petits blancs or liars or deniers etc. as a real of 30 sols is worth.

2.1.9 *Or otherwise, one may thus propose such calculations. 46 florins at 10 gros $\frac{2}{3}$ each, how many florins are they worth at 12 gros, with the gros at 15 deniers in the one and the other sum. Then proceed according to the aforesaid manner, taking 12 for the first number in the rule of three, then 10 $\frac{2}{3}$ for the second number, next 46 florins for the third number. And then multiply and divide as the rule of three requires, and it will be done. Or otherwise, if 9 are worth 8, what will 46 be worth, etc.*

Notable facts concerning money

60 groups of 4 coins of any coinage, counted 4 by 4, or 80 groups of 3 groups counted 3 by 3, are worth as many livres as one of the coins counted is worth in deniers. And by this rule, 60 groups of four lyarts or 80 groups of three are worth 3 livres, for one lyart is worth 3 deniers. And 60 groups of four or 80 groups of 3 grands blancs at 10 deniers are worth 10 livres, and so for others.

36 groups of four, or 48 groups of three, of any coin are worth as many livres as one of the coins is worth in forts, counting the forts as 1 denier $\frac{2}{3}$. Thus 36 groups of four or 48 groups of three grands blancs at 6 forts each are worth 6 livres. And so should other coins be understood, according to this rule.

This note, which seems out of place in a section on the rule of three, conjures up a picture of a money changer counting out large numbers of small coins - something that his pupils must have seen many times in the streets of the city during the fairs. In the margin at the bottom of the page, Chuquet has added a problem which belongs more to recreational than to commercial mathematics, about a man who wished to change 4 gold ecus at 35 sols each into equal numbers of coins valued at 3, 5 and 12 deniers.

Other calculations

2.3.1 *Someone owes a pension of 5 ecus, from a time when a gold marc was worth 64 ecus. One asks what this pension is worth now that that a gold marc is worth 77 ecus...*

2.3.2 *Another pension. Someone owes a pension of 5 ecus from a time when an ecu weighed 2 deniers and was 21 carat alloy. One asks what this pension is worth, now that an ecu weighs 3 deniers and is 22 carat...*

2.3.3 *Item. Someone owes a man 15 sols 8 deniers from a time when an ecu was worth 27 sols 6 deniers. To determine how much he owes him now that an ecu is worth 30 sols 3 deniers...*

2.3.4 *Item. Someone owes a merchant 240 livres from a time when a silver marc was worth 15 livres. To determine how much he owes him how that a silver marc is worth 8 livres 12 sols...*

2.3.5 *Item. Someone owes a man 12 grands blancs from a time when a silver marc was worth 6 ecus. To determine what he owes him now that a silver marc is worth 7 ecus $\frac{3}{4}$...*

These five problems suggest considerable uncertainty about what monetary standard, if any, should be adopted. In the third problem, the ecu is taken as standard; in the others, it is seen as variable with respect to a silver or gold standard.

How the rule of three may be applied in the case of merchandise which is sold by length

2.4.1 *A piece of linen measuring 11 aulnes costs 4 livres. One asks how much one aulne comes to...*

Solutions are given both using the rule of three, and by division alone, which is the obvious way of tackling this question. The two questions which follow are also solved both with and without the rule.

2.4.3 *Item. A piece of linen measuring 10 aulnes, 3 quarters and ½ costs 54 livres 15 sols. To determine what one aulne costs...*

2.4.4 *And whoever should resell it at 8 sols 9 deniers the aulne, to know what profit he would make on the whole...*

The next 7 questions, all straightforward examples on length and price, are omitted.

2.4.12 *Item. A merchant bought a piece of cloth measuring 13 aulnes which cost him 27 livres. One asks for how much he should resell an aulne to gain 5 livres on the whole piece.*

2.4.13 *Item. A merchant bought a piece of cloth measuring 18 aulnes $\frac{3}{4}$ which cost him 56 livres 10 sols. To determine for how much per aulne he should resell it to profit at the rate of 15 percent...*

2.4.14 *Item. If a piece of cloth 12 aulnes long and 5 quarters wide is sold at 21 livres, for how much should 5 aulnes of such cloth 3 quarters wide be sold?*

2.4.15 *Item. If an aulne of cloth $\frac{2}{3}$ and $\frac{1}{2}$ wide sells for 26 sols, for how much should 16 aulnes $\frac{5}{8}$ at 5 quarters $\frac{1}{2}$ sell?...* ($\frac{2}{3}$ and $\frac{1}{2}$ is interpreted as $\frac{2}{3} + \frac{1}{2}$.)

2.4.16 *Item. A merchant bought a piece of cloth measuring 16 aulnes, 5 quarters wide, which cost him 28 livres. To determine how many square aulnes this piece contains, and also how much it cost him per square aulne...*

2.4.17 *Item. A man bought 3 square aulnes $\frac{2}{3}$ of this cloth. To determine how many aulnes of the aforesaid cloth, at 5 quarters wide, should be delivered to him...*

2.4.18 *Item. One asks for a piece of cloth measuring 19 aulnes $\frac{1}{2}$ at 5 palms $\frac{1}{3}$ the aulne. How many measures of 8 palms does this piece contain?...*

2.4.19 *Item. A man bought 9 aulnes of cloth at 6 quarters wide to make a garment. One asks how much blanchet he needs at 4 quarters wide to line his garment...*

2.4.20 *Item. Again, another man bought 4 aulnes $\frac{2}{3}$ of brunette 5 quarters $\frac{1}{2}$ in width to make a coat. One asks how much linen he needs at 3 quarters $\frac{3}{4}$ in width to line his coat...*

2.4.21 *Item. A piece of cloth measuring 16 aulnes $\frac{1}{2}$ is sold for 29 livres 10 sols. To determine how much cloth one would have for 6 livres. Answer. If $29\frac{1}{2}/16\frac{1}{2}/6$. Multiply and divide in the way you know, and you will find 3 aulnes and $\frac{21}{59}$ of an aulne which is a little more than $\frac{1}{3}$. And so much of this cloth should one have for the 6 livres.*

The last of these problems is one of the few in which Chuquet offers any simple approximation to the answer.

How the rule of three may be applied in the case of merchandise sold by weight

2.5.1 *If 1 pound of silk is worth 6 livres 10 sols, what would 10 pounds 5 ounces be worth, counting 12 ounces to the pound...*

2.5.2 *Item. At 5 livres $\frac{3}{4}$ the pound of silk, what would 15 pounds 3 ounces, 5 treseaux be worth, reckoning 12 ounces to the pound and 8 treseaux to the ounce?...*

The next five problems probably relate to silk; the following group refer to much cheaper items, and the units of weight, apart from the pound, are the *quintal* (usually, but not always, 100 pounds) and the *charge* (300 pounds, supposed to be the load of one ass). Questions 2.5.12 to 2.5.15 ask about the profit on resale.

2.5.16 *One asks: 7 quintals at 100 pounds per quintal and 16 ounces per pound. How many quintals are they worth at 15 ounces per pound?*

The assumption in the working is that the ounce remain fixed.

2.5.17 Item. One asks: 10 quintals 78 pounds at 13 ounces $\frac{1}{2}$ per pound. How many quintals are they worth at 16 ounces per pound...

2.5.18 Item. One asks: 4 quintals, 65 pounds, 10 ounces at 119 pounds per quintal and 15 ounces $\frac{1}{2}$ per pound. How many quintals are they worth at 100 pounds per quintal and 12 ounces $\frac{2}{3}$ per pound?...

2.5.19 Item. If a charge of spices costs 121 livres, to know how much of that spice one should have for 40 sols 4 deniers...

Other special brief rules

Whoever multiplies the deniers that a pound is worth by $\frac{5}{12}$, he finds how many livres in money a quintal is worth.

Whoever multiplies the deniers that a pound is worth by $\frac{5}{4}$, he finds how many livres of money a charge is worth...

The opposite of these two rules

Whoever multiplies the livres of money that a quintal is worth by $\frac{12}{5}$, he finds how many deniers a pound is worth.

Whoever multiplies the livres of money that a charge is worth by $\frac{4}{5}$, he finds how many deniers a pound is worth...

Other calculations with various goods

2.6.1 A merchant bought 7 charges and 5 bichets of wheat at 27 sols 6 deniers per charge, reckoning 6 bichets to the charge, plus 9 charges, 4 bichets at 25 sols the charge, plus 8 charges, 3 bichets at 32 sols the charge. To know, the one bearing the other, how much it cost him the charge...

2.6.2 Again, One asks for how much the charge of that wheat should be resold to make a profit of 5 sols per charge, and the twentieth part is paid for duty.

2.6.3 Again. One asks how much the duty comes to on a charge sold at 34 sols $\frac{89}{114}$...

2.6.4 *A merchant sold wine for 370 livres, of which he had to pay [duty of] an eight on one side and a tenth on the other. To know how much the duty came to on the 370 livres...*

2.6.5 *A merchant bought 280 barrels of herrings, which cost him when he first bought them 4 livres, 12 sols and 6 deniers the barrel. Then they cost him, in carriage, tolls and other expenses, the sum of 136 livres, 16 sols and 8 deniers. To know how much a barrel of herring cost him when everything is reckoned in...*

The next two chapters are devoted to particular applications of the rule of three: the rule of companies or the rule of fellowship, about contracts between partners or between master and steward, and the practice of barter - the exchange of goods, sometimes at a special notional barter price.

. *How one may be helped by the rule of three concerning companies*

The rule concerning companies is such. Multiply by each and divide by all together.

3.1.1 *Example. Three merchants formed a company together concerned with merchandise. One of them put in 10 ecus, the other 6 ecus, and the third put in 8 ecus. And all together in a certain time found 15 livres profit. To determine how much of that profit comes to each one of them, considering what each has put in.*

Answer. For the first, add 10 ecus, 6 ecus and 8 ecus, making 24 ecus for the common divisor. Next, multiply 15 livres which is the profit by 10 which is the stake of the first, and then divide by 24, and you will find 6 livres $\frac{1}{4}$ for the share of the first... And thus the share of each is found. And likewise one may do all such calculations, whether with plus or with minus. And these calculations by the rule of three may be laid down in this way.

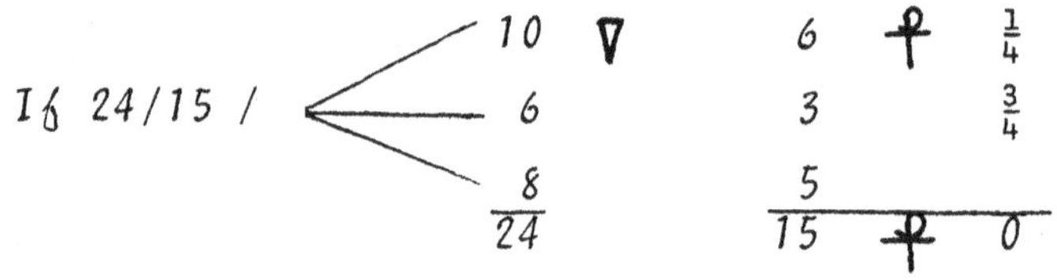

Another good way of doing such calculations. Consider how much profit comes to each ecu by dividing 15 livres by 24 ecus and you will find that $\frac{5}{8}$ of a livre comes to each. And therefore multiply $\frac{5}{8}$ of a livre by 10 ecus, which is the stake of one of the three merchants, and you will find 6 livres $\frac{1}{4}$... And thus one may do other similar ones.

The inclusion of this second method suggests that some of Chuquet's pupils were not at ease with the rule of three, and preferred to avoid it.

3.1.2 *Item. Two merchants made a company together, one of whom put in 5 ecus $\frac{2}{3}$, and the other put in 8 ecus $\frac{1}{4}$. And in a certain time they made a profit of 10 livres $\frac{1}{2}$. To know how much of this profit comes to the one and to the other, considering what each has put in...*

And whoever should wish to avoid the subtleties of broken numbers could arrange such calculations so that they are done by whole numbers in this manner. Reduce 5 ecus $\frac{2}{3}$ and 8 ecus $\frac{1}{4}$ and you will have $\frac{68}{12}$ for 5 ecus $\frac{2}{3}$ and $\frac{99}{12}$ for 8 ecus $\frac{1}{4}$. Now, take 68 for the stake of the first and 99 for the stake of the second, and then add them and you will have 167 which should be put in $\frac{1}{2}$ by multiplying it by 2 which is the denominator of $10\frac{1}{2}$, and thus you will have 334 for common denominator. Also put the 10 livres $\frac{1}{2}$ into its fraction and you will have 21 for the common multiplier...

3.1.3 *Another calculation. Three merchants formed a company, one of whom put in 10 ecus which remained there for the space of three years. The second put in 6 ecus which remained there for 7 years, and the third put in 8 ecus which remained there for 4 years. And at the end of a period, 20 livres of profit was* found. One asks how much comes to each, considering the money and the time that each has used it. Answer. Multiply each stake by the time it stayed there ... And the calculation is done, according to the style and opinion of some. And in order for such reckoning to be of value, it is necessary to presuppose that the principal or the capital alone has made a profit, and not the profit (itself). And inasmuch as it is not thus, for the profit and the profit on the profit made in merchandise can earn profit and profit on profit in proportion to the principal, from day to day, from month to month, and from year to year, whereby a larger profit may ensue. Thus such calculations are null, and I believe that among merchants no such companies are formed.

<u>How one may settle a tax</u>

3.1.4 There are six householders who have to pay a tax of 33 livres 15 sols. To determine how much each should pay, the strong supporting the weak. To make such calculations, it is necessary to know on each side the means and fortune of each. Well, suppose that the first has means of 50 ecus, the second 29, the third 73, the fourth 35, the fifth 20 and the sixth 63 ecus. Now you can multiply each one's means by 33 livres $\frac{3}{4}$. And then divide by 270 ecus, which is everyone's means added together, and thus you will find what each should pay...

3.1.5 There are two merchants who form a company together, one of whom puts in 1800 livres and his person. The other puts in 1200 livres and also his person. That is, each of them should serve in and handle the business, to govern their merchandise and make it valuable. And their contract is such that he who put more into the company will take proportionally more. After this contract has been made, the one who put less in goes back to his partner and says: Let a value be put on our persons, that is, our services, and let the service of each be set at the same value, and let these values be put in the accounts of our company, and let the service of each of us be valued at a large sum like 3000 or 4000 *livres. This is something that cannot harm either of us, since the values are equal, and so we may do it. And the other consents and says, Let each of us be valued at 4000 livres. To know whether neither of them is cheated in doing this, with regard to the first contract...*

The change in the contract clearly favours the second man, at least for as long as the company makes a profit.

3.1.6 *Again, there are two merchants who form a company together in such a way that one of them should put in 500 livres and his house and shop, without his person, and the other should put in 400 livres without his person. And in doing this the first should take $\frac{2}{3}$ of the profit and the other a third. Now it happened that the one who should have paid 500 livres and his house only paid 300 livres. And the one who should have paid 400 livres only put in 300 livres. To determine, if the first contract is not to be broken, what portion of the profit each of them should take, considering what he put in and what he ought to have put in.*

Chuquet's solution is based on the deduction that, under the terms of the original contract, the house must have been valued at 300 livres, so that the stakes of the two partners were in the same proportion as the proposed share of the profits. Thus their actual contribution should be valued at 600 and 300 livres, and the original arrangement should stand. However, he claims, some people consider that the original contract should be interpreted to mean that the house was to be regarded as one third of the cash assets of the company, that is, as 300 livres if the original contract had been fulfilled, but in fact as 200 livres, since the actual money put in to the company is only 600 livres. He rejects this argument as unjust: after all, if the first man had fulfilled his side of the contract and the other had put in only 100 livres, the cash value of the stakes would again have been 600 livres. Why should the first man's contribution of business premises be devalued through no fault of his own?

3.1.7 *Further, there are two merchants who have contracted to form a company together in such a way that one of*

> *them should put in the sum of 1200 livres without his person, and the other should put in 800 livres and his person, that is, his service. And in doing this they should divide the profit in half. Now, it happened that the one who should have put in 1200 livres only put in 900 livres, and the one who should have put in 800 livres only put in 400 livres and his person. To determine what portion of the profit each of them should take in this case, if the original contract is not to be broken but observed and kept.*

By the method of the previous question, Chuquet finds that the services of the second man were valued at 400 livres in the original contract, hence the profit should be divided in the ratio of 900 : (400 + 400).

> *... And this is a manner of judging such questions. Another consideration and manner of proceeding. It is true that the first put in $\frac{3}{4}$ of what he should have put in, and the other put in $\frac{2}{3}$ of what he should have put in. Thus, the first says that he has been wronged inasmuch as the second did not put in $\frac{3}{4}$ of what he should have put in, as he did. The second replies and says, Let it be considered how much profit one would have made if I had put in $\frac{3}{4}$ of what I should have put in, and I am content that you should have as much as would have fallen to your share...*
> *... in this manner of proceeding, the first takes $\frac{7}{13}$ of the whole profit, and the second $\frac{6}{13}$.*
>
> *Yet another consideration. The second merchant, according to the original contract, should take the profit from 400 livres for his service on 2000 livres, that is, on 1200 livres from the first man and 800 livres from himself. Now in fact he only served on 1300 livres, whence he has not so much merit, from which he should take a lower wage.*
>
> *Next, the first merchant put in 300 livres less than he ought, and this fault should bring no damage to the second, any more than the fault of the second should harm the first. And because it is not held*

against the second that the first did not put in the whole of his 1200 livres, so it is not held against him that he had not served on what, with his 400 livres, would be 1600 livres. Thus, the service of the second should be reckoned to him on 1600 livres. Now, let us say by the rule of three, if his service on 2000 livres gives the second [man] half the profit, what should his service on 1600 livres give? Multiply and divide, and you will find $\frac{2}{5}$. And the second merchant should take $\frac{2}{5}$ of the profit, and $\frac{3}{5}$ will be for the first. Of these three ways of considering such calculations, each one should take that which seems to him the most just, for according as they are considered in divers manners, divers proportions of number emerge.

3.1.8 Another calculation. Let us suppose that, in the aforesaid calculation, $\frac{3}{5}$ of the profit, if profit there be, should belong and appertain to the first merchant, and $\frac{2}{5}$ to the other. And let us put the case that there was no profit, but, by some chance, that the principal was lost, that is, the 900 livres of the one and the 400 livres of the other. To know how this loss should be shared.

Answer. Such portion as one takes of the profit when it comes, the same portion should one bear of the loss also when it occurs. Whereby it follows that the first should bear $\frac{3}{5}$ of the loss, which is 780 livres, and the other, $\frac{2}{5}$, which amounts to 520 livres. And since the second had only 400 livres of his own, subtract them from 520. There remains 120 livres, which he should hand over to the first in recompense.

Another judgment. In time of profit, the first merchant takes his 900 livres and $\frac{3}{5}$ of the profit, and the second draws his 400 livres plus $\frac{2}{5}$ of the profit. And conversely, in time of loss each should take what he takes in time of profit. Thus the second merchant should lose only his 400 livres, and not 520 as decreed above. And the first should lose his 900 livres.

But in doing this it seems that this judgment is false. For he who takes $\frac{3}{5}$ of the profit should bear $\frac{3}{5}$ of the loss... This proposition is true. But when the first merchant loses his 900 livres, he loses $\frac{9}{13}$ of the loss, which is more than $\frac{3}{5}$... by which it appears that the first merchant is wronged by this judgment. To understand and solve this difficulty, one should know that the second merchant in time of profit takes $\frac{4}{13}$ of the profit on account of his 400 livres, and on account of his service he takes the surplus from $\frac{4}{13}$ up to $\frac{2}{5}$, which is $\frac{6}{65}$. Likewise in time of loss he should lose $\frac{4}{13}$ and his service which is valued at $\frac{6}{65}$ of all the substance of the company. Thus in losing $\frac{4}{13}$ and $\frac{6}{65}$ which is his service, he bears the loss which he should support, so that the first merchant is not wronged or disadvantaged by this decree.

Chuquet considers three further cases in which the loss of capital or debts have to be divided between the two partners. In the main body of the text, he appears to support the view that he attacked above, *viz.* that loss of all or part of the capital should be divided according to the original cash stakes (in the ratio 9:4), while maintaining that nay loss over and above the initial capital should be divided in the ratio 3:2. The second merchant, he states, loses his service as well as his capital. However, a marginal note reverts to the opinion quoted above, that all losses should be divided in the ratio 3:2.

3.1.11 *Another problem. The master of a household has handed over to his shepherd 400 sheep to look after and guard them for the space of 6 years, at the end of which they are to share everything by halves, that is to say, half the aforesaid 400 animals and their offspring. It happens that with the consent of both parties they wish to divide [the flock] at the end of 4 years, according to their contract and convenants. They find 1000 beasts in all. To determine how many of these 1000 beasts the shepherd should have in respect of the time that he has served and the time that he ought to serve. Answer. The 4 years that he has served are $\frac{2}{3}$ of his term, whereby he should take $\frac{2}{3}$ of the half, which is $\frac{1}{3}$*

of the whole. So the sheperd should have a third of the whole...

This is a familiar problem in the commercial arithmetics of the time, and acts as a bridge between questions about companies formed between equals and arrangements between a merchant and an agent or factor who is employed to trade with the merchant's money. The agent might invest some of his own money in the business, either as part of the original contract or subsequently, by mutual consent. Although this arrangement sounds little different from the kind of partnership discussed in the preceding section, where one of the partners contributed his service as well as part of the initial stake, the rules would seem to have been quite different; if an agent invested more money than allowed for in the original contract, any profit earned on this money is to be his absolutely "since the agent is charged with, and responsible for, everything, and the merchant concerns himself with nothing". If the merchant, on the other hand, increases his original stake, the agent is still entitled to the same fraction of the profit on the new stake as he was on the original sum. In the case of failure, on the other hand, an agent would appear to be in a less favourable position than a working partner, according to the same principle that losses are to be divided in the same manner as profits.

Chuquet's interest in these problems of contracts appears to transcend the purely mathematical and involve notions of justice, and it may be that, as an educated and literate man, he was called on to advise his neighbours or the parents of his pupils in such matters. Van Egmond [1976] mentions a case where an arithmetic master in Florence was involved in settling disputes about the pay of builders at construction sites.

How the rule of three may be used in barter or exchange of goods

4.1.1 *Two merchants wish to exchange their goods with each other. All the merchandise of the first [merchant] is worth 60 livres. We ask how many things at 9 livres each the other merchant should hand over to him. Answer, and rule for such problems.*

Divide 60 livres, which is the value of all the merchandise of the first, by 9 livres, which is the value of one item of merchandise of the second, and you will find 6 items $\frac{2}{3}$, and so many should be handed over. Or otherwise, by the rule of three, if 9 are worth 1, what are 60 worth?

4.1.2 *Another manner of exchange. Two merchants wish to exchange their goods with each other. All the merchandise of one of them is worth 60 livres, and all the merchandise of the other is worth 93 livres. One asks how much more one is worth than the other. Answer and rule for such problems. Subtract the lesser number from the greater, that is to say, 60 from 93, and you will find 33 livres remaining, and so much more is one stock of goods worth than the other.*

4.1.3 *Another manner of exchange. Two merchants wish to exchange their goods with each other. All the merchandise of one of them is worth 60 livres in ready money, and he wishes to mark them up to 80 livres. One asks how much the other merchant should mark up to him an item of his merchandise which is worth 9 livres to sell for ready money.*
Answer by the rule of three. If 60 come to 80, how many will 9 come to? Multiply and divide, and you will find 12 livres, and that is how much he should mark up his goods.

4.1.4 *Another exchange. Two merchants wish to barter their goods with each other. All the merchandise of the first of them is worth 60 livres when sold for ready money, and he wishes to mark it up to 80 livres. One asks how much the other merchant should mark up to him all his goods, which are worth 73 livres when sold for ready money.*

This last question is not worked by the rule of three but by subtraction; the second man should add 20 livres to the value of his goods if neither is to be cheated. It is hard to tell to what extent barter was current in Renaissance Lyon, and whether it would really have been accompanied by these elevated prices. One variant of the barter problem does explain the high prices; if payment for the bartered goods is not required until several months later, the elevated barter price acts as a disguise for the charge of interest. This situation does not occur in Chuquet's *Commercial Arithmetic*, but de la Roche (1520) devotes a chapter to it, and concludes by naming no fewer than seven different types of problem associated with barter. The main variant in Chuquet is a demand for payment partly in cash and partly in goods, as in the following:

4.1.5 *Another exchange. There are two merchants who wish to exchange goods with each other. All the merchandise of the first of them is worth 60 livres when sold for ready money, and he wishes to oversell it for 80 livres in barter, and he wishes to have one half in ready money and the rest in goods. To determine how the other merchant should oversell a part of his merchandise which is worth 9 livres sold for ready money.*
Answer and rule for such problems. It is necessary to take half of 80 livres, which is 40, from 60. There remains 20 for the first number in the rule of three. Than afterwards 40 must be taken from 80, leaving 40 for the second number. And then one must take 9 livres which is the just price of the other man's merchandise for the third number. And then multiply and divide as the rule of three requires and one will find 18 livres. And he should oversell his merchandise at this price, and none will be cheated.

In some of his questions on barter, Chuquet is rather more specific about the goods exchanged: pepper for cloth, camlet (a fine fabric, of Eastern origin) normally sold at 7½ livres the length, against ginger which cost 5 sols the pound, lead changed for bonnets, a horse for a load of tin plus cash balance, saffron for linen and paper for pins.

The chapter ends with instructions for checking calculations involving barter by working out the cash equivalents of the goods proferred in exchange on both sides, "so that nobody is cheated".

How the three numbers of the rule of three may be joined and composed in many and divers manners

This compound rule of three has 4 parts. The first and second parts always require 5 numbers, of which the second and the fifth are always alike in the first part, whose rule is thus:
Multiply the first number by the second, to be the divisor, and the three other numbers by each other, to be divided.

5.1.1 Example. If 100 gain 15 livres in 12 months, one asks what 60 ecus will gain in 8 months.
Answer. Multiply 100 which is the first number by 12 which is the second, and 1200 comes for the divisor. Then multiply the three other numbers one by the other, that is to say, 15 times 60 times 8, which comes to 7200. Now divide 7200 by 1200 and you will find 6 livres and that is how much the 60 ecus will gain in 8 months. The five numbers may be set down in this manner.

Ecus	Months	Livres	Ecus	Months
100	12	15	60	8

This calculation done immediately above is according to what several masters in this art have ordained. And in order that such calculations should be true and that they should remain in their entirety, one should understand and suppose that the principal alone gains, and not the profit [itself], for otherwise such accounts would be false. However, this supposition is false, for in commerce the profit, and even the profit on the profit, may yield profit from day to day or from month to month like the principal, whence the profit may be greater than what was found in the above problem.

Let us set down another supposition closer to reality, and suppose that the profit and also the profit on the profit are ordinary things governed by rules and that they gain like the principal. To determine, if 100 ecus in 12 months gain 15 livres, what will 60 gain in 8 months?

Answer. Let us see what 60 ecus would gain in 12 months, that is to say, in a year, by saying, if 100 ecus gain 15 livres, what should 60 ecus gain? Multiply and divide and you will find 9 livres, and that is how much the 60 ecus would gain in a year.

Now let us suppose that the ecu be worth 35 sols and the livre 20 sols. Thus the 9 livres would be worth 5 ecus $\frac{1}{7}$, which added to 60 makes 65 ecus $\frac{1}{7}$. Now, inasmuch as 8 months is $\frac{2}{3}$ of a year, for this reason it is necessary to seek two proportional numbers between 60 ecus and 65 $\frac{1}{7}$, in the manner which was shown above in the special application of the rule of first terms in finding proportional numbers. And of these two numbers take the greater which is R^3 254615 $\frac{25}{49}$. From this, take away 60 ecus, and there remains R^3 254615 $\frac{25}{49}$ of an ecu minus 60 ecus. This remainder is less than 6 livres, and that is how much the 60 ecus will gain in 8 months.

This is the only place in the *Commercial Arithmetic* where the rule of first terms is used - indeed, the only place in this part of the manuscript where compound interest is calculated, though Chuquet's comment on the rule of companies with time (problem 3.1.3 above) shows that he was concerned to point out to his students the limitations of the rule of three in such questions. Surprisingly, he does not say by how much the answer differs from the 6 livres obtained by the ordinary rule; the error is roughly 20 deniers.

5.1.2 *When a measure of wheat costs 20 sols, a loaf of bread weighing 15 ounces is sold for 2 deniers. One asks, if this measure cost 30 sols, for what price should one sell a loaf of bread weighing 24 ounces...*
(5 deniers)

A measure was about $1\frac{1}{2}$ bushels.

In the second part of the compound rule of three, the third number should be like the fifth. And for this there is such a rule: Multiply the third number by the fourth to be the divisor, and the three other numbers by each other, to be divided.

5.2.1 *Example. If 100 ecus gain 15 livres in 12 months, in how long 60 ecus will gain 20 livres...*

The nature of this problem, and of those that follow, are analogous to the questions on the first case of the compound rule of three. Once again, Chuquet issued a caveat about the first problem because no allowance is made for compound interest, but this time he does not give an alternative solution. For the next example, he states the special rule that if the third and fifth numbers are equal, it is sufficient to divide the product of the first two numbers by the fourth. The last example on this case of the rule illustrates another context in which it could be applied:

5.2.5 *If a quintal of merchandise costs 36 sols for carriage over 84 leagues, one asks how many leagues one should carry 64 pounds for 20 sols.*
Answer. Multiply 20 sols by 100 times 84 and then divide by 36 times 64, and you will find 72 leagues $\frac{11}{12}$ of a league. And that is how far 64 pounds should be carried for 20 sols.

In the third part of the compound rule of three, the first number and the last should always be unlike. The question should be from the last number to [find the] first. There many be 5 numbers, or as many as one wishes. For this case there is a rule which is thus.

Multiply what you wish to know by all the values, and then divide by the product of the things valued.

5.3.1 *Example. If 4 Paris deniers are worth 5 Lyon deniers, and 10 Lyon deniers are worth 12 Geneva deniers, one asks how many Paris deniers are worth 8 Geneva deniers.*
Answer. Multiply 8 Geneva deniers which you want to know by 4 Paris deniers and 10 Lyon deniers which are the values, and they come to 320. Divide them by the product of 5 Lyon deniers with 8 Geneva deniers, which are the things valued, and you will find 5 Paris deniers $\frac{1}{3}$, and that is how much the 8 Geneva deniers are worth.

Paris	*Lyon*	*Lyon*	*Geneva*	*Geneva*
4	5	10	12	8

The three remaining problems in this section involve seven given numbers, with three identities. One, like the first, relates to the values of small coins from different regions, one to the more valuable livres, ecus, florins and ounces of silver, and one, written in the margin, to apples, pears, nuts and deniers. Finally, the fourth case of the compound rule of three is the converse of the third, just as the first is converse to the second, a fact which Chuquet exploits in giving his rules for checking a calculation in one form by means of the converse form.

The remaining pages of the manuscript, some 50 folios, are devoted to gold and silver coins. This is considerably more space than Chuquet's contemporaries gave to this topic. Pacioli does have a section on gold and silver, but it is mainly concerned with the problem of mixtures (determining the proportion of different metals in an alloy of known value) which Chuquet assigns to the more algebraic parts of the work. Chuquet's treatment is not only purely arithmetical (like most of the *Commercial Arithmetic*) but also includes tables for reckoning the exact content of precious metals in a given weight of some of the alloys distinguished by special names - king's silver, court silver, etc. These are the only tables in the manuscript apart from the multiplication table - evidence that Chuquet was particularly concerned with the needs of workers in the mint at Lyon.

The French currency was very unstable in the later Middle Ages. In part, this was due to irregular supplies of silver, in part to the wearing down of old coins, which left the authorities with a choice between devaluation - issuing new coins at a reduced weight to match the average of the old money in circulation - or demonetarizing the old coins and recalling them for melting down. Numerous royal edicts for the withdrawal of coins were issued during the fifteenth century, but devaluation was also a common solution.

Lyon had a particular problem because of the greater circulation and easier export of money during the fairs. Chuquet gives a purely arithmetical method for deciding whether old coins should be melted down: is the silver content higher than that obtained from silver bullion of the same price? In fact, there is evidence to suggest that once a coin was withdrawn, it became worthless or almost worthless [*Anon*, 1963]. Chuquet also describes the practice of making an assay to test the silver content of an alloy - one of the few references in the manuscript which suggests any familiarity with the work of his fellow-citizens, indicating that he numbered workers at the mint among his acquaintances, perhaps even among his pupils.

The two chapters on gold and silver follow the same pattern: an introduction to the units and special terms used, how to find the content of precious metal in an alloy in both theory and practice, finding the value of foreign coins, and converting coins to bullion. The tables appear together in a final chapter, with full instructions for their use. Because of this duplication of material, the chapter on gold is not presented here.

Of the nature of gold and of silver

This part is subdivided into two. In the first of these, the nature of silver is treated, and in the second the nature of gold. Each of these two parts is again divided into several chapters, of which the first on the nature of silver is:

Of the division and different kinds of mark. And how one may make a sol of fine [silver]

The mark is divided into 8 ounces, the ounce into 24 deniers, the denier into 24 grains, the grain into 24 primes, the prime into 24 seconds, the second into 24 thirds etc.

Another division

The mark is divided into 8 ounces, the ounce into 20 esterlins, the esterlin into 2 mailles, the maille into 2 serlins, the serlin into 2 demiserlins, the $\frac{1}{2}$ *serlin into* $\frac{2}{4}$ *of a serlin.*

And note that a serlin weighs	*28 gn* $\frac{4}{5}$
A maille weighs	*14 gn* $\frac{2}{5}$

A serlin weighs	7 gn $\frac{1}{5}$
A $\frac{1}{2}$ serlin weighs	3 gn $\frac{3}{5}$
A $\frac{1}{4}$ serlin weighs	1 gn $\frac{4}{5}$
And $\frac{1}{8}$ of a serlin weighs	0 gn $\frac{9}{10}$

One should know that there are court marks and Paris marks which are both divided in the aforesaid manners, while the Paris mark is heavier, for 7 ounces $\frac{5}{18}$ of Paris weight make 8 ounces of court, and 8 ounces 19 deniers 1 grain $\frac{37}{131}$ of a grain of court weight make 8 ounces of Paris. Thus 131 marks of Paris are worth 144 of court.

Item. One should know that a sol of fine [silver] is 12 deniers of fine [silver] which is worth 8 ounces in weight.

Item. Ash silver (*argent de cendre*) is 11 deniers 18 grains of fine [silver] which is worth or weighs 7 ounces $\frac{5}{6}$.

Item. King's silver is 11 deniers 12 grains of fine [silver] which weighs 7 ounces $\frac{2}{3}$.

Item. Court silver is 10 deniers 22 grains of fine [silver] which weighs 7 ounces $\frac{5}{18}$.

Item. One should know that there are deniers and grains of alloy, and deniers and grains of weight. The difference between them is that one denier of alloy is worth 16 deniers weight, and 1 grain of alloy is worth 16 grains weight. Also, 12 deniers of alloy are worth one sol, which is a mark. And 12 deniers of weight are worth $\frac{1}{2}$ ounce, for 24 of them to make an ounce. And the denier of alloy is divided into 24 grains of alloy, and the grain into 24 primes of alloy etc. Thus a mark is worth 8 ounces, and 8 ounces [are worth] 192 deniers weight or 12 deniers alloy, and 192 deniers weight are worth 2608 grains weight or 288 grains alloy.

Item. Further, one should know that all fine silver is of 12 deniers alloy, for otherwise it would contain impurities or copper. Therefore, when some bullion is not of 12 deniers alloy, but of less than 12 deniers, that is to say, that it contains 7 parts of fine [silver] and 5 of impurities or copper, yet from bullion which is not fine one makes a reckoning of fine silver (*solt de fin*) to know how much fine silver and how much of the impurities this bullion contains. Several ways of doing this are set down hereafter, of which the first rule is this:

Whoever multiplies the deniers and grains of alloy by the marks, from this will come deniers and grains of alloy.

And if the deniers and grains of alloy are tripled and then multiplied by the ounces, from this will come grains and primes of alloy. And multiplied by deniers weight, from this will come primes and seconds of alloy. And from grains weight come seconds and thirds of alloy.

Or otherwise, all comes back to one:

For each d. of alloy the mark contains	the ounce holds	3 grains
	the denier	3 primes
	the grain	3 seconds
		of alloy

For each gn. of alloy the mark contains	the ounce holds	3 primes
	the denier	3 seconds
	the grain	3 thirds
		of alloy

And for the better understanding of these facts are set down after this several examples, the first of which is this.

8.1.1 One asks how much fine silver does 10 marks of bullion at 9 deniers alloy contain. Answer. Multiply 9 deniers by 10 marks, and it amounts to 90 deniers. Divide by 12 to make sols from this, and you will find 7 sols 6 deniers of fine silver, and so much do the aforesaid 10 marks contain.

There are 14 examples altogether in this section, and six different methods. The rule of three, which is presented throughout the rest of the *Commercial Arithmetic* as fundamental, appears only in fourth position, and the example below shows that even the most primitive methods were appropriate - only halves and thirds are used, as if division by larger numbers might be regarded as too complicated.

Yet another method of making [a reckoning of] fine silver

8.1.5 *One asks how much fine silver does 24 marks 7 ounces contain at 10 deniers 20 grains of alloy. Answer. For the first of the 10 deniers of alloy, we take 6 of them which are ½ sol, for which we take also the half of 24 marks, which are 12 sols of fine silver. Next, of these 10 deniers of alloy we shall take 3, which is half of the 6 deniers, for which we take half of the 12 sols which is 6 sols of fine silver. Again, of these 10 deniers, we take 1 which is a third of the aforesaid 3 deniers, for which we shall take a third of 6 sols which is 2 sols. And thus we have the fine silver for the 10 deniers. Further, of the 20 grains we shall take 12 of them, which is ½ denier, for which we should take half of 2 sols which is 1 sol of fine silver... Again, of the 7 ounces, we shall take 4 which is ½ mark, for which we shall take half of 10 deniers 20 grains of alloy, which is 5 deniers 10 grains of fine silver... Now whoever adds all together, he will reach 22 sols 5 deniers 11 grains ½ of fine silver, and so much does the 24 marks 7 ounces aforesaid contain, as it appears in the working below.*

At 10 d 20 gn	*6 d*	*12 gn*	*4 ounces*
	3	*6*	*2*
24 m 7 ounces	*1*	*2*	*1*

12 s		
6		
2		
1	0 d	
	6	
	2	
	5	10 gn
	2	17
	1	8½

Final sum:

22 s 5 d 11 gn ½

Another style of making a [reckoning of] fine silver in pounds at 12 ounces per pound

8.1.13 *Just as in some regions silver is weighed in marks at 8 ounces per mark, so in other places it is weighed in pounds. The pound may be at 12 ounces, or 16 ounces, or otherwise according to the diverse local custom...*

Chuquet gives examples using both the rule of three and the kind of rule of thumb quoted at the beginning of the section but adapted to pounds instead of marks.

This first section has covered problems about finding the content of pure silver in a given weight of a given alloy. Chuquet proceeds to discuss other problems of conversion and calculation - how much of one of the named alloys (e.g. king's silver, court silver) can be made from a given quantity of one of the others, or would have the equivalent value; finding the value of a given weight of the alloy given the value of 1 sol of that alloy and vice versa; and finally the combination of silver of two different alloys in various proportions.

Of combinations

For the combination of all kinds of bullion, and setting it to whatever alloy one desires, there are two general rules. The first of these teaches the combination of two species of bullion, a fixed quantity of the one with an unknown quantity of the other. The second rule teaches how much the alloy will be when two or more kinds of bullion are melted and mixed together in fixed and limited quantities.

The first rule

From the deficit or surplus in the alloy of the unknown quantity of bullion, make the first number in the rule of three. From the deficit or surplus in the alloy of the fixed quantity of bullion, make the second number in the rule of three, and from the total weight of the fixed quantity of bullion, make the third number.

8.4.1 Example. Whoever should wish to combine 10 marks of fine silver to make 7 deniers alloy, how much copper should be added?
Answer. From 0 which is the alloy of the unknown quantity of bullion, to 7, is a deficit of 7 for the first number in the rule of three. Item, from 12 deniers of alloy in the 10 marcs of fine silver which is the fixed quantity of bullion to 7 is a surplus of 5 for the second number. Item, it is necessary to take the whole weight which is 10 marks for the third number. Now multiply the third number by the second and then divide by the first, and you will find 7 marcs $\frac{1}{7}$ of a mark, and that is how much copper must be combined with the 10 marks of fine silver for an alloy of 7 deniers.

Or thus. Multiply 12 deniers by 10 marcs and it will amount to 120 deniers. Divide this by 7 and you will find 17 marcs $\frac{1}{7}$ of a mark. That is how much the bullion will weigh in all after the combination. Therefore subtract the 10 marks from this, and there will remain 7 marks $\frac{1}{7}$ of a mark which is the copper which must be mixed there...

8.4.4 Item. Whoever should wish to combine into an alloy at 8 deniers 12 grains 14 marks 7 ounces at an alloy of 11 deniers 21 grains, one asks how much bullion he would need at 5 deniers 8 grains alloy...

8.4.5 *Another calculation. A merchant has 10 marcs of bullion at 7 deniers alloy which he wishes to refine to 8 deniers. One asks how much impurity he should remove, since this merchant has no high bullion to mix with it.*
Answer. First determine the fine silver and the impurities in the 10 marks. there is 70 deniers of fine silver and 50 of impurities ... If 8 deniers of fine silver require 4 deniers of impurities, what will 70 deniers of fine silver require? Multiply and divide, and you will find 35 deniers, which one should take away from 50, and there will remain 15. And the merchant should remove 15 deniers of copper from his 10 marks to make them come to 8 deniers.

8.4.6 *Item. One asks how much of the aforesaid 10 marks must be taken to be purified in order to remove neither more nor less than 15 deniers of copper?...*

The second rule

Divide the pure silver by the weight and you will find what the alloy is.

8.4.8 *Example. A merchant has 4 marks of bullion which are at 5 deniers of alloy. Item, 6 marks at 3 deniers alloy, and 5 marks at 7 deniers. One asks, if all the bullion aforesaid were melted and mixed together, what would be the alloy of the whole mass...*

The next section describes the *assay* or sampling of a small quantity of silver bullion in order to determine the alloy or the silver content of the whole. It was convenient to use a sample of ½ ounce in weight (12 deniers weight), since the number of deniers weight in the sample then corresponded to the number of deniers alloy in the mixture. Chuquet sets one problem in which two samples are taken and the values averaged. He also describes the custom of making a set of weights (*semelles*) for use in the assay to reduce the complexity of calculation.

8.5.5 *Example. An assayer made an assay of 12 deniers by his semelle, and the assay yielded the result that his fine silver weighed 5 deniers 15 grains ½ by his*

semelle. To determine how much is the alloy of the bullion from which he made his assay.
Answer. It is 5 deniers 15 grains ½. And so should all the others be understood. However, if the lead which was put into the assay contains silver, as it often happens, one should make allowances for this as skilful practice and experience may show.

Of patterns, and of the making of coins

To make a pattern (*deneral*), it is first necessary to know how the prince values a silver mark or a sol of pure silver. It is also necessary to know the alloy and face value of the coin for which one wishes to make a pattern. And according to how the lord values a sol of fine silver, one should note what a mark of that money is worth. This value should be divided by the face value for the quotient will be the number of coins in one mark. For each coin there is in a mark, one should count 1 denier, and 1 sol for 12 deniers, and 1 livre for 20 sols. Thus one will have a tally of this coinage in livres, sols and deniers. Further, if the number of coins which make up the mark is 192 or below, one should divide 192 deniers, which is one mark, by the number of those coins. And if it comes to more than 192 coins, then one he should divide 4608 grains, which is one mark, by that number of coins. And one will find how many deniers or grains weight each of the coins should weigh.

8.6.1 Example. Let us suppose that the prince holds a sol of find silver at 9 livres, and that a coin is 4 deniers alloy with a face value of 6 deniers. One asks what the tally of this money is, and also what the deneral or pattern of these coins should weigh.
Answer. For the first, a mark of this coinage at 4 deniers alloy is worth 3 livres at the rate of 9 livres per sol of fine silver. These 3 livres must be reduced and put into deniers, and then divided by their face value which is 6 deniers, and one will find 120 coins. That is how many coins make up a mark. Now, count each coin for 1 denier and 12 for a sol, and you will have 10 sols. The

tally of this money is 10 sols. Now, to know what the deneral or pattern of these coins weighs, divide 192 deniers, which is a mark, by 120 coins, and you will find 1 denier $\frac{3}{5}$, and that is how much the deneral or pattern of these coins should weigh.

8.6.5 *A master made coins with a face value of 3 deniers from 10 deniers alloy, and at a tally of 19 sols. He gave 60 sols for a silver mark. One asks how much that master took in profit for each working mark. Answer. First let us see how many coins there are to the mark which is at a tally of 19 sols, by multiplying 19 by 12. There are 228 coins, each of which is worth 3 deniers. Thus there are in all 684 deniers which is worth 57 sols. Thus the master deals with a working mark at 10 deniers alloy. Now see how much the 10 deniers alloy should be worth at the rate of 60 sols per sol of pure silver, and you will find 50 sols. The master draws 57 from this. Thus he takes 7 sols profit on the working mark.*

There are further problems on the same theme, including some which start with the master's profit and then work out the number of coins to be struck from a mark, or the face value which should be given to the coins if a certain number of them must be struck from a mark of the bullion.

There are two further sections on silver, one giving a short example of working out the relative values of certain silver coins on the basis of their silver content and weight, and one on the economics of buying in coins for melting down. These old coins became worn with use, so there could be a critical weight below which the silver content was worth less than the face value.

8.8.1 *How one may make a deneral or pattern for weighing and melting down coins*

To make a pattern for melting down coins with the intention of making a profit, it is necessary to make the pattern in the aforesaid manner, except that it is necessary to make the deneral on the basis of what the lord gives for pure silver, and not what he takes for it. However, one should know *that bullion which has been melted down and shaped into ingots commands a higher price than bullion to be melted down. Now, suppose that one gives for melted bullion at the rate of 9 livres per sol of pure silver, and for bullion to be melted down at the rate of 8 livres 17 sols. From this must be deducted what one would be content to make in profit on each sol of fine silver. Well, suppose that we should be content to make 5 sols on fine silver. Item. suppose that a coinage was of 3 deniers 20 grains alloy and 7 deniers face value. To make a pattern for melting down this money, it is first necessary to subtract 5 sols, which is the profit one wishes to make, from 8 livres 17 sols, because this is bullion which has not been melted down. There remains 8 livres 12 sols. After this, it is necessary to see what 3 deniers 20 grains are worth at the rate of 8 livres 12 sols for fine silver, and you will find 2 livres 14 sols 11 deniers $\frac{1}{3}$, which is 659 deniers $\frac{1}{3}$.*

This must be divided by 7 deniers which is the face value of the money, and one will find 94 coins $\frac{4}{21}$. Well, to know what the pattern for balancing this money should weigh, divide 192 deniers by 94 $\frac{4}{21}$, and you will find 2 deniers $\frac{38}{989}$ of a denier which is about 1 grain, and 2 deniers 1 grain is what the pattern for balancing these coins should weigh. For all the coins which weigh as much as or more than the pattern may be set aside for melting down and converting to bullion, for there is profit in it.

The chapter on gold repeats almost step by step the chapter on silver, the only difference being that whereas pure silver is equivalent to 12 deniers, pure gold is equivalent to 24 carats. Finally, two tables are provided, together with detailed instructions for their use; the first is a multiplication table to base 24 (or at least, the numbers are expressed in terms of two units, the larger being the equivalent of 24 of the smaller) which finds many applications in the table of units for measuring the gold and silver content, and extends over 4 pages to cover all combinations up to 24 x 72. The second table is for converting fine silver to king's silver and extends to 72 sols 23 deniers of the former. (One page of the multiplication table is shown, together with the conversion table).

	13		14		15		16		17		18		19		20		21		22		23		24	
37	20	1	21	14	23	3	24	16	26	5	27	18	29	7	30	20	32	9	33	22	35	11	37	0
38	20	14	22	4	23	18	25	8	26	22	28	12	30	2	31	16	33	6	34	20	36	10	38	0
39	21	3	22	18	24	9	26	0	27	15	29	6	30	21	32	12	34	3	35	18	37	9	39	0
40	21	16	23	8	25	0	26	16	28	8	30	0	31	16	33	8	35	0	36	16	38	8	40	0
41	22	5	23	22	25	15	27	8	29	1	30	18	32	11	34	4	35	21	37	14	39	7	41	0
42	22	18	24	12	26	6	28	0	29	18	31	12	33	6	35	0	36	18	38	12	40	6	42	0
43	23	7	25	2	26	21	28	16	30	11	32	6	34	1	35	20	37	15	39	10	41	5	43	0
44	23	20	25	16	27	12	29	8	31	4	33	0	34	20	36	16	38	12	40	8	42	4	44	0
45	24	9	26	6	28	3	30	0	31	21	33	18	35	15	37	12	39	9	41	6	43	3	45	0
46	24	22	26	20	28	18	30	16	32	14	34	12	36	10	38	8	40	6	42	4	44	2	46	0
47	25	11	27	10	29	9	31	8	33	7	35	6	37	5	39	4	41	3	43	2	45	1	47	0
48	26	0	28	0	30	0	32	0	34	0	36	0	38	0	40	0	42	0	44	0	46	0	48	0
49	26	13	28	14	30	15	32	16	34	17	36	18	38	19	40	20	42	21	44	22	46	23	49	0
50	27	2	29	4	31	6	33	8	35	10	37	12	39	14	41	16	43	18	45	20	47	22	50	0
51	27	15	29	18	31	21	34	0	36	3	38	6	40	9	42	12	44	15	46	18	48	21	51	0
52	28	4	30	8	32	12	34	16	36	20	39	0	41	4	43	8	45	12	47	16	49	20	52	0
53	28	17	30	22	33	3	35	8	37	13	39	18	41	23	44	4	46	9	48	14	50	19	53	0
54	29	6	31	12	33	18	36	0	38	6	40	12	42	18	45	0	47	6	49	12	51	18	54	0
55	29	19	32	2	34	9	36	16	38	23	41	6	43	13	45	20	48	3	50	10	52	17	55	0
56	30	8	32	16	35	0	37	8	39	16	42	0	44	8	46	16	49	0	51	8	53	16	56	0
57	30	21	33	6	35	15	38	0	40	9	42	18	45	3	47	12	49	21	52	6	54	15	57	0
58	31	10	33	20	36	6	38	16	41	2	43	12	45	22	48	8	50	18	53	4	55	14	58	0
59	31	23	34	10	36	21	39	8	41	19	44	6	46	17	49	4	51	15	54	2	56	13	59	0
60	32	12	35	0	37	12	40	0	42	12	45	0	47	12	50	0	52	12	55	0	57	12	60	0
61	33	1	35	14	38	3	40	16	43	5	45	18	48	7	50	20	53	9	55	22	58	11	61	0
62	33	14	36	4	38	18	41	8	43	22	46	12	49	2	51	16	54	6	56	20	59	10	62	0
63	34	3	36	18	39	9	42	0	44	15	47	6	49	21	52	12	55	3	57	18	60	9	63	0
64	34	16	37	8	40	0	42	16	45	8	48	0	50	16	53	8	56	0	58	16	61	8	64	0
65	35	5	37	22	40	15	43	8	46	1	48	18	51	11	54	4	56	21	59	14	62	7	65	0
66	35	18	38	12	41	6	44	0	46	18	49	12	52	6	55	0	57	18	60	12	63	6	66	0
67	36	7	39	2	41	21	44	16	47	11	50	6	53	1	55	20	58	15	61	10	64	5	67	0
68	36	20	39	16	42	12	45	8	48	4	51	0	53	20	56	16	59	12	62	8	65	4	68	0
69	37	9	40	6	43	3	46	0	48	21	51	18	54	15	57	12	60	9	63	6	66	3	69	0
70	37	22	40	20	43	18	46	16	49	14	52	12	55	10	58	8	61	6	64	4	67	2	70	0
71	38	11	41	10	44	9	47	8	50	7	53	6	56	5	59	4	62	3	65	2	68	1	71	0
72	39	0	42	0	45	0	48	0	51	0	54	0	57	0	60	0	63	0	66	0	69	0	72	0

[illegible]	Argēt le Roy			
	solz	d	gr	24es
1	1	0	12	12
2	2	1	1	1
3	3	1	13	13
4	4	2	2	2
5	5	2	14	14
6	6	3	3	3
7	7	3	15	15
8	8	4	4	4
9	9	4	16	16
10	10	5	5	5
11	11	5	17	17
12	12	6	6	6
13	13	6	18	18
14	14	7	7	7
15	15	7	19	19
16	16	8	8	8
17	17	8	20	20
18	18	9	9	9
19	19	9	21	21
20	20	10	10	10
21	21	10	22	22
22	22	11	11	11
23	24	0	0	0
24	25	0	12	12
25	26	1	1	1
26	27	1	13	13
27	28	2	2	2
28	29	2	14	14
29	30	3	3	3
30	31	3	15	15
31	32	4	4	4
32	33	4	16	16
33	34	5	5	5
34	35	5	17	17
35	36	6	6	6
36	37	6	18	18

[illegible]	Argēt le Roy			
	solz	d	gr	24es
37	38	7	7	7
38	39	7	19	19
39	40	8	8	8
40	41	8	20	20
41	42	9	9	9
42	43	9	21	21
43	44	10	10	10
44	45	10	22	22
45	46	11	11	11
46	48	0	0	0
47	49	0	12	12
48	50	1	1	1
49	51	1	13	13
50	52	2	2	2
51	53	2	14	14
52	54	3	3	3
53	55	3	15	15
54	56	4	4	4
55	57	4	16	16
56	58	5	5	5
57	59	5	17	17
58	60	6	6	6
59	61	6	18	18
60	62	7	7	7
61	63	7	19	19
62	64	8	8	8
63	65	8	20	20
64	66	9	9	9
65	67	9	21	21
66	68	10	10	10
67	69	10	22	22
68	70	11	11	11
69	72	0	0	0
70	73	0	12	12
71	74	1	1	1
72	75	1	13	13

[illegible]	Argēt le Roy		
[illegible]	[illegible]	gr	24es
1	1	1	1
2	2	2	2
3	3	3	3
4	4	4	4
5	5	5	5
6	6	6	6
7	7	7	7
8	8	8	8
9	9	9	9
10	10	10	10
11	11	11	11
12	12	12	12
13	13	13	13
14	14	14	14
15	15	15	15
16	16	16	16
17	17	17	17
18	18	18	18
19	19	19	19
20	20	20	20
21	21	21	21
22	22	22	22
23	24	0	0

[illegible] gr 24es

Chuquet's *Commercial Arithmetic* does not attempt to reach the same level of abstraction and generality as other parts of the manuscript; it was written to meet the needs of students with a certain rather limited arithmetical background. It was probably compiled as a collection of suitable examples for teaching, or as a reference work, rather than for independent study, as there is little explanation; there is also a shortage of very simple examples on the topics covered.

A study of the *Triparty* and *Problems* alone has given some commentators (see Chapter 9) the impression that Chuquet was so far ahead of his age as to be out of touch with his pupils, and that his work was much more limited in scope than that of Pacioli. The *Commercial Arithmetic* shows that Chuquet did not confine himself to algebra, and that he was not simply a student of mathematics driven into elementary teaching for financial reasons. He was aware of the rules of thumb used at the moneychangers' table and in the mint. He offers alternative methods for those who are not happy about using the rule of three. His discussion of contracts drawn up according to the rule of companies indicates an area where a literate, numerate, articulate man could render services to his uneducated neighbours.

Although a number of the early printed arithmetics show more awareness of current business practices and a better grasp of pedagogy than the *Triparty* manuscript, Chuquet's work helps to fill out the picture of the emergence of the teacher of arithmetic in France - some 150 years later than in Italy.

Chapter 9

THE PLACE OF NICOLAS CHUQUET IN THE HISTORY OF MATHEMATICS

In order to assess the importance of Nicolas Chuquet's writings for the history of mathematics, it is first necessary to measure Chuquet's *Triparty* manuscript against its most significant precursor, Fibonacci's *Liber abaci*, and against the most influential exposition of his day, Pacioli's *Summa*. The comparison between Chuquet and Pacioli has already been explored by Cantor (1892) and by Juschkewitsch (1961), but these authors had access only to the material published by Marre in 1880/1, and so were unaware of the contents of Chuquet's *Geometry* and *Commercial Arithmetic*. Lack of space precludes detailed comparison with other mathematicians of the Middle Ages and early Renaissance, apart from a note on Chuquet's sources. A more detailed study, so far as the *Geometry* is concerned, is to be found in L'Huillier's edition (1979).

Next, the work of Estienne de la Roche in transmitting some of Chuquet's ideas will be examined. One commentator has dubbed him a "vile plagiarist", and others have placed on him the blame for the loss of his teacher's major contributions to mathematics. While these criticisms cannot be dismissed entirely, they should be moderated and placed in perspective by considering the course of the development of algebra as well as the practices of other writers of textbooks at that period.

De la Roche's readers included several minor mathematicians of the sixteenth century, but there is only the slightest of evidence that any of Chuquet's ideas could have been brought to the attention of any writer capable of making use of them. The only possibility - and that a remote one - is that Bombelli knew of Chuquet's index notation when he developed his own. But even if index notation was discovered independently by a small number of mathematicians working in different countries, this particular discovery was indeed a major step towards a powerful symbolical algebra, and therefore belongs to the history of mathematics. In the same way, a consideration of Chuquet's other innovations shows what was possible in the late 15th century.

Some of the commentators on Chuquet in the late 19th century were grossly over-enthusiastic in their praise, and a number of general works on the history of mathematics propagate the view that Chuquet's work was neglected because he was so far ahead of his time. The last part of this chapter attempts to construct a more balanced picture of his achievements and his limitations, based for the first time on the whole of the manuscript.

Fibonacci's *Liber Abaci*

Fibonacci almost certainly had access to the major works of Arab mathematics and in some cases went beyond them. It is generally acknowledged that it took two or three hundred years for his ideas to be absorbed, and to a great extent his work was transmitted through the simpler writings of intermediaries.

Among the subjects found in Fibonacci but not in Chuquet are the formula for the sum of squares of natural numbers, a series described by a recurrence relation (the famous Fibonacci series), and methods of iterative approximation to square roots and cube roots. The *Liber abaci* also contains some ideas which were still not generally accepted when Chuquet included them in the *Triparty*: the treatment of negative solutions as debts is an example. The problem of the broken eggs (of Arab origin) and the weights problem are also found here (cf *Problems* 143 and 142).

Fibonacci's other writings on algebra and algebraic geometry establish his stature as a mathematician not merely a communicator. He presents some general methods of solving indeterminate quadratic equations (e.g. find two rational numbers whose squares differ by 5), and attempts to prove that cubic equations cannot be solved in terms of quadratic surds. In computational geometry, he follows Archimedes in finding approximations to π by considering polygons with 96 sides. When his answers involve square roots, he works to the fourth order of sexagesimal fractions, giving a much greater precision than Chuquet who often neglects to evaluate square roots when they are not exact.

It seems unlikely that Chuquet had direct access to the *Liber abaci* (let alone the *Practica geometriae* and the *Liber quadratorum*), or he would surely have incorporated some of the topics mentioned above. Pacioli, on the other hand, mentioned Fibonacci as his primary source, and Cardano describes a manuscript of the *Liber abaci* annotated by Pacioli.

Fibonacci must be regarded as superior to Chuquet where mathematical rigour is concerned. Chuquet, like many of his contemporaries, rarely gives a theoretical proof. Both writers tend to set out general rules for solving all problems in a particular class, rather than considering each example in isolation, and in his *rule of method* (cf *Problems* 124) Chuquet describes a way of deriving such rules, but this method is one of induction, not a abstract reasoning. Where Chuquet's work represents a considerable advance over Fibonacci's is in the generality of notation and terminology, and in the movement towards a symbolical algebra.

Pacioli's *Summa*

Pacioli, writing for the same generation as Chuquet, has made some, but not all, of the notational advances found in the latter's writings. Thus the same contractions are found for square root and cube root; but Pacioli does not generalize his notation to apply to roots of all orders, nor his terminology to apply to all powers. Instead of Chuquet's underlining for bracketing the terms in compound roots, he prefixes the expression whose root is to be taken by U or V; there is no indication of where the expression ends. Chuquet's use of a bracketing convention is the earliest recorded in Cajori's *History of Mathematical Notations* (1928, *I* p 385). Both writers use $\bar{p}$ and $\bar{m}$ for plus and minus in keeping with the standard practice in fifteenth-century Italy.

Like Chuquet, Pacioli accepts irrational solutions to problems in which only an answer in integers makes sense - for instance, the number of journeys undertaken by a merchant (cf *Problems* 95). Both authors attempt to solve problems of the form $n^x = c$ by first finding two integer values between which x must lie and then interpolating (*Problems* 94). Chuquet expresses his reservations with this solution, and does not present the most common type of problem leading to such an equation - that of finding the period for which a sum of money is lent at compound interest.

Pacioli does not generalize his remarks on equations in the same way as Chuquet. True, Pacioli considers specific cases in which quartic equations may be reduced to quadratics in x^2; but Chuquet expresses his laws for quadratic equations as canons for the solution of equations in three terms, evenly spaced, i.e. in x^{m+n}, x^m and x^{m-n} for arbitrary integer values of m and n.

Now that Chuquet's *Geometry* has been published, and the *Commercial Arithmetic* is shortly to be published, it is possible to compare Chuquet and Pacioli over a wider range of subjects. A number of commentators appear ignorant of the fact that Chuquet wrote on geometry, while Cantor refers to this work as part of the Appendix to the *Triparty*. Even Itard (1971) has little to say about the *Commercial Arithmetic*, and remarks that Chuquet used questions on the repayment of debts "only as pretexts for exercises in algebra".

In the realm of algebraic geometry, Chuquet and Pacioli treat a similar range of topics, and employ similar methods, with much stress on problems of inscribing one figure within another, and on solutions by algebra (usually by quadratic equations). L'Huillier (1979) has shown that a number of problem types are common to the two authors. (However, he found even more striking similarities between the list of problems in Chuquet's *Geometry* and in an anonymous manuscript located in Siena.) There was already a tradition of using algebraic methods for solving geometrical problems, and both men were aware of this tradition, even if they had not actually read the same manuscripts. L'Huillier was not able to trace back all the types of problem in Chuquet to other sources; in particular, some problems on solid bodies leading to cubic equations may be original to Chuquet, though L'Huillier inclines to the view that Chuquet took them from some manuscript unknown to us, just as many of his other problems were taken from contemporary Italian sources. (L'Huillier, 1979 p 64)

There is nothing in Chuquet's work to match Pacioli's *De divina proportione*, a book on polyhedra with illustrations by Leonardo da Vinci. Or at least, nothing of this nature has been discovered.

Pacioli's work is commonly described as "encyclopedic", and his *Summa* contains a great deal of material that is of utility to merchants - not just the applications of the rule of three which occupy the bulk of Chuquet's *Commercial Arithmetic*, but also a description of the many different types of barter and exchange which were practised, a merchants' compendium of relative weights, measures and currencies (copied without acknowledgment from Chiarini), and the classical treatise on double-entry book-keeping. This last is the only section which has been reprinted in extenso since the second edition of the *Summa* in 1523; it has also been translated into many languages, including a bilingual edition of Romanian and Italian in 1981.

Chuquet's examples on barter are sparse by comparison with those of Pacioli. The commodities bartered are often named in the most general terms - "cloth", "merchandise", etc. - and none of the variants such as dry barter by which merchants were wont to avoid the proscription on usury is mentioned. (De la Roche does include some of these topics; his treatment is based on Pacioli's, but it is not a direct translation.) In the rule of companies, Chuquet is more discursive; he was either familiar with conflicting arguments about the division of profits or debts or extremely inventive in advancing new arguments.

The one subject on which Chuquet has more to say than Pacioli is the minting of gold and silver coins. Here, his treatment is more detailed and more practical than in other parts of the work.

It could be argued that the differences in content between the two works can be explained in part by differences in culture and in the background of their students. The Italians were more advanced than the French in their commercial practices. Chuquet's pupils had a mathematical background in which the rule of three counted as an advanced method which some of them would prefer to avoid; questions involving algebra are not included in this part of the manuscript. The money-changer's table would have been a familiar sight to them, whereas financial transactions based on letters of credit were being introduced more gradually, through the Italian community. Even in the sixteenth century, the French merchants had difficulty in getting used to Italian practices (Brésard, 1914 p 285).

The question of originality of these two authors must now be considered. It has already been said that Pacioli copied one section of his work without acknowledgment, and it has also been claimed that he plagiarized Piero della Francesca. In the *Summa*, he does acknowledge several sources: Jordanus, Blasius of Parma, Prosdocimo de Beldomandi, and Thomas Bradwardine among others, and states that where he does not mention another source, he is following Fibonacci. Chuquet names only the unknown Berthelemy de Romanis, who is criticized in two places in the *Problems*, and Ramon Lull and Campanus once each, apart from the classical sources Euclid and Boethius. He also refers to "a certain wise man" (Archimedes), "the ancients", and "some" or "others".

Both authors had clearly read more widely than their works acknowledge, and it is inconceivable that they did not have common sources. (See, for example, L'Huillier's analysis of the problems in Chuquet's *Geometry* and a number of contemporary manuscripts.) Almost certainly, they both made notes of suitable problems as they came across them in manuscripts and later incorporated them into their own work. The copying of manuscripts was the mainstay of academic knowledge before the spread of printing, and copyright for life was unknown in the 15th and 16th centuries. A book would be published with a "privilege" which prevented other printers and booksellers from selling copies of the book while the privilege was in force (Schneider 1974); the privilege on the first edition of de la Roche (1520) ran for two years. To apply a modern notion of plagiarism in these circumstances would be anachronistic and unhelpful.

In spite of many similarities between the work of Pacioli and Chuquet, there is no proof that either knew of the other's writings. However, Chuquet's knowledge of contemporary mathematical literature argues that he probably made at least one visit to Italy, and it is not impossible that the two men may have met there.

If Chuquet had copied extensively from other sources, one might expect to find copying errors. There are a few places in the manuscript where a phrase has been repeated, or omitted, or the wrong word written down; but these are not so frequent as to suggest thoughtless copying or direct translations.

The contrast between Chuquet and Pacioli as writers and mathematicians has been overstated. A recent example is the following passage from Juschkewitsch (1961, p 433):

> "Although Chuquet's work is not as comprehensive as the *Summa* of Pacioli, although it is not as encyclopedic in its contents and presents significantly fewer examples, yet in terms of the ideas it contains it is essentially more abundant. Pacioli was merely the teacher who had thoroughly mastered all the mathematical knowledge to which he had access. Chuquet was an original thinker and creator of new generalized concepts which have driven algebra forward."

If "Chuquet's work is not as comprehensive as the *Summa*", it is nevertheless broader in scope than suggested by the *Triparty* alone, for the *Commercial Arithmetic* presents material more appropriate in range of subject and level of mathematical sophistication to the merchants and craftsmen of Lyon, while the *Geometry* covers the rudiments of surveying and gauging as well as questions in algebra drawn both from geometrical problems taken from daily life and from the pure geometry of regular polygons and circles.

If Pacioli indeed surpasses Chuquet as a teacher, it should be remembered that he had been teaching mathematics since his youth, first as a private tutor and then in universities, and had already written a mathematical manuscript sixteen years before the *Summa* was published. Chuquet, on the other hand, began his studies in medicine (though the Bachelor's degree he boasts would not have been sufficient to allow him to practise in Paris), and did not describe himself as an arithmetic master but as a copyist in the early years of his sojourn in Lyon.

(Considered as a textbook, de la Roche's *Larismethique* (1520) is better arranged and more clearly set out than the works of either Chuquet or Pacioli.)

Finally - and more controversially - if Chuquet produces the more original ideas and the more suggestive notation, it may be because he stood outside the Italian tradition of the followers of Fibonacci. His nationality, and the language in which he wrote, forced him into a pioneering role as the author of the first French algebra, and his situation in a town with no tradition of university education allowed him a greater freedom to experiment than Pacioli, whose teaching had to be consistent with that of other university professors.

A survey of Chuquet's sources

Chuquet's university studies would almost certainly have exposed him to Boethius' *Arithmetic*, the first six books of Euclid, and Sacrobosco on the sphere, for these were prescribed texts for the Master of Arts degree, the completion of which was a prerequisite for admission to the Faculty of Medicine in Paris. As mentioned above, he acknowledges Campanus, whose translation of Euclid was widely used in the Middle Ages. However, he misrepresents the author. Either he failed to understand the passage (perhaps because he read it in a corrupt form) or else he incorrectly remembered what it said. The citation of Ramon Lull is also strange, as the work he mentions, the *Ars magna*, has very little mathematical content.

Chuquet also mentions Berthelemy de Romanis. From what little he says about him, he would seem to have been a Dominican monk, presumably from the Abbey of Romans-sur-Isere near Vallence. Berthelemy is also mentioned in a manuscript written by Jehan Adam in 1473; since Adam uses the word "trimillion" where Chuquet uses "trillion", and since neither word has been found in other writings of the period, Berthelemy may have suggested such a terminology for large numbers. Chuquet mentions Berthelemy's "second book", so the work was probably divided into sections. It included a problem of the "strange journey" type requiring the sum of an arithmetic progression. This material could have been included in a gloss on Boethius, or in a textbook for monastery schools where the education was in Latin. In any case, it is unlikely that Berthelemy's book went as far as algebra.

Chuquet may have learned algebra either from an Arabic work in Latin translation, or from a Latin text, or from an Italian manuscript. The last seems more probable because of his inclusion of algebraic geometry and because of his adoption of the signs $\bar{p}$, $\bar{m}$ for plus and minus, which are found also in Pacioli. However, he refers to some previous writers on algebra as "the ancients" (see Chapter 5), which suggests that he had also consulted some older works on algebra.

There is nothing in the first part of the *Triparty* to suggest access to any particular writer, as almost all the material was a standard part of the mathematical corpus of knowledge, and the exceptions appear to be original to Chuquet. For the rest of the *Triparty*, he must have read about operations with irrational numbers including what he calls bound roots, and with algebraic expressions, and come across rules like "Plus and plus, minus and minus, add. Plus and minus, subtract". He proceeds to take these calculi far beyond the general level of fifteenth-century writings, and no source has been identified which might have led him in that direction.

An interesting, but unanswered, question is whether Chuquet knew Oresme's work on proportion. L'Huillier (1979) has established that Chuquet almost certainly copied one manuscript by Oresme, which is bound with an early version of the *Geometry*; but this is not a treatise on proportion. In fact, the differences are between the ways in which the two authors extend the range of exponents; they do so in totally different ways. Oresme used fractional indices; Chuquet used negative indices. Their approaches to mathematical argument and discovery are also completely different, with Oresme favouring the pattern of a scholastic disputation. Although Chuquet's R^n for nth roots is tidier to write or print than a notation based on fractional exponents, the latter is more pleasing from a point of view of generality, and if Chuquet had known Oresme's work he would at least have mentioned this possibility.

Most of the types of question treated in the *Problems* were widely known in the late fifteenth century, though

Chuquet is unusual in his emphasis on algebraic methods of solution and in the scope of the numerical answers he admits - not just fractional numbers of children but negative quantities of cloth and irrational numbers of deniers. Marre (1881, p 416) has traced some parallels with a Provencal manuscript written in Pamiers, which Sesiano (1984) dates at around 1430, and this manuscript also applies the rules for solving certain classes of problem even in cases where those rules lead to negative solutions. Battesti (1980) argues that Francesco Pellos was familiar with Chuquet's work; but although his *Compendion del abaco* was not published until 1492, Sesiano claims that it was written in about 1460, so any influence would have been in the other direction. In any case, it is clear that the data as well as the framework of such questions was frequently taken over without acknowledgment, so it cannot be concluded that Chuquet knew either of these works, only that he was well-read in practical mathematics as well as in the more ambitious algebraic manuscripts. It was almost certainly in such a work that he found the problem of the Christians and the Jews (*Problems* 146).

L'Huillier (1979) has studied the sources of the *Geometry* in more detail than this chapter can accommodate, and concludes that Chuquet had read a number of both medieval and Renaissance manuscripts; that he probably acquired much of his knowledge of geometry on a visit to Italy; and that it is not unlikely that he either saw an early manuscript by Pacioli or met the man himself. L'Huillier inclines to the belief that even those problems for which he has not identified a source were not original, and that Chuquet's contribution lay solely in the style and generality of his algebraic methods of solution. This study of the *Geometry* reveals Chuquet as more indebted to his contemporaries and less a lone voice in the wilderness than might be inferred from the *Triparty* and *Problems* alone. Sesiano's paper suggests, and Benoît's forthcoming edition of the *Commercial Arithmetic* will no doubt confirm, that here also a similar tradition was at work.

Estienne de la Roche and the transmission of Chuquet's ideas

Estienne de la Roche, described by LeFebvre (1897, p xxxiii) as a "vile plagiarist", and by Itard in scarcely less unfavourable terms, can probably be identified with a near neighbour of Chuquet's who was still a boy at the time when the *Triparty* was written and may well have been a pupil of Chuquet. Not only did he copy a great deal of material from the *Triparty*, but his handwriting can be found in the margins of the manuscript, which he probably inherited from its author. In defence of de la Roche, it should be stated that he made no secret of the debt he owed to Chuquet and Pacioli, and that his work is more suitable as an arithmetical handbook for merchants than those of his more eminent precursors. While the greater part of Chuquet's manuscript is arranged as an abstract treatise on arithmetic and algebra, with illustrative problems relegated to the appendix, two-thirds of de la Roche's *Larismetique* (1520) is set out as a commercial arithmetic, with problems classified by subject-matter (goods sold by weight, goods sold by quantity, etc.). Simple algebraic exercises from Chuquet's *Problems* are included under the appropriate heading, as well as the more mundane problems of the *Commercial Arithmetic* and a great deal of detail about the nature of goods sold or bartered. Besides Chuquet, he mentions two other sources, Pacioli, from whom he took a chapter on the non-mathematical properties of certain numbers (part of which comes from St. Augustine's *Civitas Dei*) as well as more worldly topics such as the different forms of barter and exchange, and Phillippe Friscobaldi of Florence whose work has not survived, though a certain Giovanni Frescobaldi, perhaps a member of the same mercantile family, wrote a handbook for merchants visiting England.

At worst, then, de la Roche cobbled together parts of the work of three authors, all of them inaccessible to the average French merchant, and turned them into an attractive volume. In its second edition (1538), the publisher, Gilles Huguetan, further added to its utility by including his own compilation of arithmetical tables, including ready reckoners for multiplication and division, tables of gold and silver more comprehensive than Chuquet's, special tables for booksellers and printers, and shorter tables for goods sold by length, weight, quantity or troy weight and for calculating the cost of a regular payment over various time periods. (There are no tables of compound interest.) If, as l'Huillier claims, the annotations on the *Triparty* manuscript which are not in Chuquet's hand may be attributed to de la Roche (with a very few exceptions), it is clear that he had worked through the *Triparty* and *Problems*. Marre (1880, 1881) has drawn attention to a number of places where the pupil has copied, either word for word or almost so, from the work of the master, concluding (1880, p 580):

> "Without being accused of injustice or exaggeration, one may say that he appropriated the work of Nicolas Chuquet, that he purely and simply copied the *Triparty* in a host of places, that he suppressed certain of the most important passages, especially in algebra, that he curtailed or prolonged others to make up his *Arismethique* vastly inferior to the *Triparty*, and finally, that if Nicolas Chuquet and his work have remained in obscurity for four hundred years, it is above all to him that the primary cause must be attributed."

For an evaluation of this judgement, it must be asked how de la Roche treated (or ignored) Chuquet's innovative passages. Another relevant question is how much difference a more faithful treatment would have made to the dissemination of Chuquet's ideas. The preceding arguments should have taken the sting out of the accusation of plagiarism and improper use of material, and justified de la Roche's book on its own terms, not as a scholarly algebra but as a manual for tradesmen.

Chuquet's first distinctive feature - counting up to *nonillion* - is reproduced in essence. De la Roche demonstrates some confusion in quoting 10 rather than 0 as one of the figures from which numbers are built up. Each writer makes a different slip of the pen in the example where a very large number is written out in words and in numbers. But de la Roche, having described in Chuquet's words how such numbers may be punctuated by the insertion of small marks after each group of 6, fails to insert these punctuation marks. This failure is repeated in the second edition, so the blaim cannot be shifted to the printer. In any case, the terminology did not survive; perhaps the world of vernacular arithmetic was not ready for the *trillion*.

Both Chuquet and de la Roche present the same quasi-algebraic rules. The rules of one and two positions are taken directly from Chuquet, except that de la Roche lets the more conventional name "rule of false position" slip in for the former. Of the rule of apposition and remotion, Chuquet takes a poor view, claiming that it should not be counted as a rule at all; de la Roche gives a more conventional treatment, including an example of a problem which cannot be solved in integers (and for this class of indeterminate equations, only integer solutions are required). Chuquet, on the other hand, ends his exposition by commending algebraic methods for these questions, and de la Roche omits this recommendation.

It will be remembered that the rule of intermediate numbers was claimed by Chuquet as an original discovery. De la Roche includes it under the heading of "rule of mediation between the greater and the less", but otherwise copies the section almost verbatim. This rule is not a very efficient procedure for approximation, but does have the advantage of producing fractions with small denominators. It is found in one late work the *Arithmetica* of Joan Perez de Moya (1590); Wertheim (1898) remarks:

> "The tedium of the process is a sufficient explanation of why Chuquet's method made no particular impact and soon passed into oblivion, at least I have not found it mentioned by any writer except those named."

Chuquet devotes more space than de la Roche to calculations with roots. The latter, though he presents Chuquet's unified notation for roots of any order, does not use it in his own working, and says very little about higher-order roots, regarding them as "not of great utility". He also states that bound roots should be linked with a line underneath, but fails to show them so linked. This invention of brackets, at least, was squandered by de la Roche, who might otherwise have communicated it to his readers; according to Cajori (1928, p 385) it would still have been only surpassed by Bombelli and others in the middle of the sixteenth century.

De la Roche follows Chuquet closely on the extraction of square and cube roots; for higher orders, he gives only some remarks about how roots of one order can sometimes be reduced to roots of a lower order. The table of the first

10 powers of numbers from 1 to 10 is not reproduced. There is a short note on the extraction of compound roots, followed by 14 sides of the calculus of surds; however, the sections on bound roots in the latter are rendered useless by de la Roche's failure to underscore the terms of the bound expressions.

The comments which have been made about the treatment of irrational numbers by Chuquet and de la Roche apply with similar force to the sections on algebra. Chuquet devotes sixty folios to the subject; de la Roche, twenty-eight. De la Roche normally uses the name "*regle de la chose*" (rule of the thing), derived from the Italian, to denote algebra, although he mentions that Chuquet calls the science the "rule of first terms". He introduces side by side Chuquet's notation, with its extension up to arbitrarily large powers of the unknown, and the notation which Chuquet attributes to "*les anciens*", going up to the fourth power only. However, in what follows, he uses the latter notation, succumbing, like many teachers, to the temptation to stick to the more familiar notation in spite of the advantages of a little-known alternative.

The much-praised table of powers of two with the exposition of the "secret which lies in proportional numbers" (*Triparty* fo. 87r) is given, but loses much of its force by the omission of 0 as the denomination of the number 1 and by the use of conventional symbols rather than index numbers for squares and cubes. After describing the operation of the four rules on algebraic expressions, de la Roche adds a section on proof and a chapter on fractions in algebra which are not found in Chuquet. The algebra is a reworking of Chuquet's work, or (which appears more likely) the conbination of Chuquet with some other author. But if Chuquet has the advantage in the algebraic notation, de la Roche is superior in his way of setting out the working; that is, he is in this respect one step closer to symbolical algebra.

The four canons, on the other hand, are set out more clearly by Chuquet. De la Roche interposes fresh material (on the rule of quantity, used for solving equations in more than one unknown, and on equations with more than one solution) between the first canon and the three canons for true quadratics equations, which occupy only six-and-a-half folios between them (nineteen in Chuquet). As far as problems on the first canon are concerned, several are adapted from the easier questions in the *Triparty* and the *Problems*, while the rest are mainly concerned with dividing a number into two or three parts with certain properties. Very few of his examples on this first canon use roots in the solution.

The passage on the fourth canon deserves special comment, because this is the one place where de la Roche was attacked by a contemporary, and he can be vindicated. The fourth canon deals with equations in which the first and third terms have the same sign, such as $x^2 + b = ax$. This is the only one of the three forms of quadratic equation admitted in the Middle Ages which is capable of having two positive solutions, and de la Roche follows Chuquet in saying so. Butéon, in his *Logistica* (1559), denies the possibility of two distinct roots, and claims that de la Roche is deceived because he incorrectly regards 1 as a root and 1 should not be considered as a number. In fact, de la Roche gives another example, with solutions 6 and 4: perhaps Butéon would have discounted this because the discriminant is 1. In the corresponding section of the *Triparty*, Chuquet does not present a single problem with two distinct integer solutions. His first example has no real roots; the second has a double root; and the third has solutions 5 ± R21. Pedagogically, de la Roche has the advantage; and the following delightful remark is not found in Chuquet:

> *Whence it follows that such calculations are impossible... And thus on every proposition, innumerable impossible questions may be posed to test out the masters and to know if they are expert in proportions.*

De la Roche's algebra is more limited in the complexity of its examples than Chuquet's, and is largely concerned with solving linear equations. The special techniques for such problems are often omitted from Chuquet's *Triparty*, being introduced only in the course of a solution in the *Problems*. Here again, de la Roche is more helpful to his readers.

Although de la Roche mentions geometry in the title of his book, very little space is devoted to it, and this is given over to simple problems and constructions, with no algebra. Thus de la Roche did nothing to disseminate Chuquet's work on algebraic geometry.

There is another important respect in which de la Roche fails to reflect Chuquet's insights; he avoids problems with negative solutions, where Chuquet, following Fibonacci, tries to provide an interpretation of the answer. But the general feeling, even as late as Cardano, was against negative numbers; if de la Roche had included these problems, they would have had little or no effect on this attitude.

To summarize: de la Roche, like many of his contemporaries, included only a small amount of algebra and geometry to supplement his mercantile arithmetic. Indeed, he included more algebra than was common at that time, particularly outside Italy. Within the limits of his scope, he presents most of Chuquet's ideas and notations, but his presentation is incomplete, and he reverts to an older notation and terminology for his own exposition. Far less inclined than his master to explore beyond the traditional domain of mathematics, he admits no negative solutions and has little to say about powers and radicals beyond the fourth degree.

Who were de la Roche's readers, and to what extent might they have been impressed by a more faithful treatment of Chuquet's work? Two French mathematicians of the sixteenth century, Gosselin and Butéon mention his name. Butéon cites de la Roche in the same sentence as Luca Pacioli; the two were regarded, at least within France, as worthy of comparison, though there are far more references to Pacioli. As stated above, in this case de la Roche was right, and Butéon guilty of wrong-headed conservatism in his rejection of 1 as a solution. The Italian mathematician Bonasoni points out this error and vindicates de la Roche, but he may have known of the latter only through Butéon's reference. If he did have a copy of de la Roche's book, it might have come to the attention of Rafaele Bombelli, who wrote one of the more substantial algebras of the mid-sixteenth century and adopted some notations similar to Chuquet's - but these notations are barely mentioned in de la Roche.

John Wallis's *Treatise on algebra both historical and practical* (1684-5) mentions de la Roche in the guise of "one Stephanus a Rupe Luodunensis, who had written in French", on the authority of Butéon, and two other seventeenth-century references are known, one in the preface to a collection of mathematical problems and one in the dedication of a Dutch book on arithmetic and book-keeping. After that, the work attracted the attention of bibliographers rather than mathematicians until its rediscovery of Michel Chasles (see below).

Apart from explicit references to de la Roche, there may have been some suggestions that other mathematicians may have had access to Chuquet's work. For instance, Battesti (1980), in a dissertation on Francesco Pellos whose *Compendion del Abaco* was published in 1492, argues on the strength of some similar passages that Pellos knew Chuquet's work. However, Sesiano (1984) argues that Pellos' work was actually written much earlier, so that the influence would be either in the opposite direction or through a common source - possibly the Pamiers manuscript.

A common source may also explain a parallel with Humphrey Baker's *The Well-Spring of Science* (1574), another commercial arithmetic with no algebra. Baker introduces the rule of three in exactly the same dry and theoretical terms as Chuquet and de la Roche, which is unusual in a work of this character.

To move into the realms of speculation, if Chuquet did indeed meet Pacioli on a visit to Italy, it is not impossible that some of the similarities in their choice of geometrical problems for algebraic solution are the result of Chuquet's influence on Pacioli rather than vice versa. However, Cantor (1901, p 347) remarks:

> "At that time, an Italian merchant would simply have regarded as ludicrous [the suggestion] that he should learn the principles of reckoning or the doctrine of equations from a German or a Frenchman."

Chuquet's inductive *rule of method* is very similar to a passage in which Cardano (1545) explains how he arrives at rules for solving certain kinds of problem. The present authors have found no other reference of this kind from which both Cardano and Chuquet might have drawn. However, Cardano writes as if this rule was an established procedure, and mentions Pacioli as someone who had applied it. Perhaps the rule of method can be seen as the first step towards algebra as a general method of solving problems rather than a notation for writing down the solution, but the empirical observations described by Chuquet and Cardano are a long way behind the precise science advocated by Viète.

The major influences on the development of algebra during the scientific renaissance (slightly later than the artistic and historical renaissance) were Pappus and Diophantus, whose works were rediscovered, reconstructed, and made available in Latin translation during the sixteenth century. Bombelli was the first modern writer to fall under the Diophantine influence: but Viète, Fermat and Descartes (all of them Frenchmen, and all of them law graduates) took upon themselves the task of rebuilding mathematics on the Greek model. They may have owed something to other sixteenth-century writers, notably Cardano and Ramus, but it is unlikely that Chuquet's work, with its drive towards invention rather than consolidation and rigour, could have made much of a contribution even if its had been better known. His notational improvements could have been helpful, but the lack of proof in his work would have hindered the adoption of his ideas.

However, it is certainly true that if there had been no algebra in the fifteenth century, then the history of the subject in the sixteenth century would have been very different. It was the shortcomings of their immediate predecessors that sent Bombelli and Viète back to the Greeks, but it was the achievement of these predecessors that they had sufficient knowledge to appreciate these shortcomings. A major achievement of Bombelli and Descartes was the invention and development of a suitable notation for the science of algebra; but the work of Chuquet shows that this quest was already well under way at the end of the fifteenth century. The details of this notational development will be discussed later; the next sections show how Chuquet's conceptual achievements can be set into the perspective of the history of algebra.

Michel Chasles on the history of algebra

After Wallis, the next writer on history of mathematics to discuss the work of de la Roche was Michel Chasles, a distinguished geometer and author of a history of geometry. He freely admitted that it was national pride that led him, in three papers presented to the Academie de Sciences (1841a-c) to enter into controversy with the Italian bibliophile and historian Libri:

> "for I have found in the course of my study that hasty and erroneous judgments have frequently sacrified even our finest geniuses...such as Viète, Descartes, Fermat, Pascal, to the glory of foreign names, some of them... scarcely known in the mathematical world." (1841c p616).

Libri, he claimed, had belittled Viète's achievement by stating that Fibonacci had already developed literal (i.e. symbolical) algebra, and consequently that the Frenchman's contribution was less important than that of the Italians, Ferro and Ferrari, who had solved the cubic and quartic equation respectively:

> "But suppress the discoveries of Viète, and science inevitably remains stationary in the state in which he found it; the discoveries of Descartes, Newton, Leibniz, Bernoulli, l'Hôpital, Euler, Lagrange, d'Alembert, Laplace Poisson, Legendre are no longer possible; they no longer have a foundation, they remain buried in nothingness. The illustrious analysts who are in the glory of our present century would likewise be unknown to us.
>
> On the other hand, suppress the discoveries of Ferro and Ferrari, and there would only be one page less in our mathematical history...science experiences no check in its triumphant progress. The discoveries of Ferro and Ferrari...are among those which unfortunately bear no seed, and are in no way fertile, because by their very nature they soon come up against frontiers which cannot be traversed. They could not even lead to the resolution of equations of the fifth degree." (1841b pp 518-9)

Viète's achievement, according to Chasles, lay in setting forth a calculus based on symbols, not merely using symbols as abbreviations for the names of variables but as entities on which arithmetical laws could be defined and arithmetical operations performed. Fibonacci and his sixteenth-century compatriots (including Cardano and Tartaglia) had restricted their calculations to numbers; even Cardano's rule of method was applied only to numeric data, and its application was recorded in ordinary language not in algebraic formula.

Chasles saw no anticipation of Viète's ideas among the Italians; but he instanced the German Stifel and the Frenchmen Peletier and Butéon because they used letters for unknowns and also a crude form of exponent notation. It is in this context that he first mentions de la Roche:

> "Notwithstanding a certain observation of Wallis in favour of Hariot... Descartes has remained in undisputed possession of his ingenious exponent notation, which has become in some measure a scientific concept by the extension it has undergone. But it has not been known until now that this notation is much older, and that it is found in a work which appeared in 1520 and which was reprinted in 1538, *Larismethique (sic) nouvellement composee par maistre Estienne de la Roche dict Villefrache, natif de Lyon*... The author represents the second, third, fourth, etc. powers of a number, of 12 for example, thus: 12^2, 12^3, 12^4, etc... Further, he applies the same exponents to the expression of roots, using the sign R in place of $\sqrt{\ }$. Thus he writes: $R^2 12$, $R^3 12$, $R^4 12$, etc..."
> (1841a p752n)

It has been observed above that de la Roche makes relatively little use of the former notation, which makes it the more creditable that Chasles should have noticed it, and the more excusable that he misinterpreted 12^3 as having its modern sense rather than that of $12x^3$ as Chuquet and de la Roche intended. However, his point remains. Chasles concludes his discussion of de la Roche with the remark:

> "The author cites the Treatise on Algebra of master Nicolas Chuquet, Parisian, another work by a French author before 1520. Perhaps the exponent notation is

already to be found there. It is to be hoped, in the interests of history, that this work has not been completely lost."

It is ironic that the publication of the first part of the *Triparty* in 1880 should have ended immediately opposite the necrology on Michel Chasles, who would have been able to give the whole manuscript an evaluation in which national pride was finely balanced by mathematical insight. Other nineteenth-century writers mocked or castigated de la Roche; Chasles recognized the passeges where he followed Chuquet in the direction of modern algebra. There is one other reference in these papers to *Larismethique*, to de la Roche's recognition of the double root in certain quadratic equations. Chasles does not appear to notice what little remains of Chuquet's terminology for the first, second, third...powers, which he would certainly have regarded as a great advance on the terminology of Hindu and Arab algebra:

"This most imperfect system was that adopted by the European Christians in the Middle Ages, in the Renaissance, and still employed in all their works of the sixteenth century. The most celebrated analysts of that epoch, Cardan, Tartaglia and Bombelli, only knew this imperfect nomenclature which halted the development of algebra."

Chuquet's mathematical achievements

The most extreme panegyric lavished on Chuquet following the publication of Marre's transcription of the *Triparty* occurs in a little-known sketch of the history of algebra by B. LeFèbvre [1897, p xxxiii]:

"In the author of this treatise on arithmetic and algebra, the wisest and most profoundly original algebraist who had written since Leonardo of Pisa [Fibonacci] is revealed. Anticipating Descartes, he expounds the true conception of negative quantity, and he invents and utilizes the notation of exponents; before Wallis, he formulates the rules of the calculus of exponents, without fear either of the negative exponent or of the exponent zero; before Napier, he discovers logarithms and realizes the value of this

> future tool of mathematicians; the theory of the equation of the first degree in one unknown is excellent and foresees *infinitely many solutions* and *impossible solutions;* the theory of the equation of the second degree is treated in full and in generality by the exposition of four rules or *general canons*"

And in a footnote on the following page, he writes

> "It is a literary and scientific delight to read through these 3000 quarto pages [of Marre's version], in which the ancient author combines with an arithmetical and algebraic science ahead of his time an inimitable naivety of style and an admirable clarity of exposition."

Alas, it is all too easy, in reading old books in a recognizable but imperfect version of one's own language to be charmed by all that appears strange and impressed by the clarity of all that does not. There is little naivety in Chuquet, and the term may be a euphemism for what in a modern writer would be regarded as imprecision or uncritical acceptance of the result of applying an inappropriate method. Compared to his contemporaries, however, it must be admitted that Chuquet is succinct, usually lucid, and admirably free from digressions into the mystical aspects of number.

It is LeFèbvre's mathematical claims that represent an obvious travesty of historical record. Chuquet did not discover logarithms. He had sufficient knowledge of the index laws to have done so; but so did Nicole Oresme, over a century earlier, and the additive law is even hinted at, in a special case, by Boethius. True, Chuquet discusses a problem whose accurate solution calls for logarithms, but he expresses only a grumble of dissatisfaction with his approximate solution by linear interpolation. Again, under the heading of the compound rule of three in the *Commercial Arithmetic*, he solves a problem in compound interest by finding mean proportionals; but he avoids, as Pacioli does not, a general discussion of compound interest which would have led him to ask for the time in which the sum invested would double itself at a given rate of interest, and thereby lost the best opportunity he had of realizing the value of logarithms for mathematics. (Astronomy and navigation, which motivated Napier's discovery, did not concern Chuquet - or at least, not in the only manuscripts which have survived in his name.)

Charles Naux, in his study of the history of logarithms [1966, I, 16], gives a fairer evaluation of Chuquet's remarks on the problem of the barrel (*Problems*, 94) when he describes Chuquet as "catching a glimpse of the phantom of the logarithmic function"; but he over-rates the originality and achievement of the *Triparty* in his earlier remarks on the same page:

> "This treatise, unique of its kind among those of his period, shows us a mathematics not enslaved to that of Euclid, and most often of a higher level than that which would prevail in the following century. For example, Chuquet handles the question and usage of exponents with plenty of confidence and accuracy; he considers, without hesitations or false steps, the introduction of fractional, and even negative exponents in the study of radicals and irrational numbers."

Negative exponents, yes; but Chuquet dealt in *n*th roots rather than in fractional exponents, and is further from the latter than Oresme. Nor could Chuquet's compact notation easily have accommodated fractions as exponents; it is hard enough to distinguish visually between his superscripts representing exponents and the numerators of his fractions. It may be true that Chuquet was no slavish follower of Euclid; but neither were his contemporaries in the realm of vernacular arithmetic, and through their liberation they all, Chuquet included, lost much of the glory and rigour of mathematical proof, leaving it for the mathematicians of the next century, inspired by the rediscovery of Pappus and Diophantus, to refurbish the science of algebra.

Naux is more perceptive in his critique of Chuquet's index notation, in which he did indeed anticipate Descartes, though not as precisely as Chasles and LeFèbvre believed:

> "Manifestly Chuquet did not succeed in perfecting his new algebra, and did not know how to give it that unity which would constitute its strength. He wrote the powers of 2 in their common forms 2, 4, 8, 16, etc., although his theory of exponents would have allowed him to adopt the more learned and effective notation 2, 2^2, 2^3, etc.; this lack of insight distanced him

from the theory of logarithms. The kernel of this theory resides in the correlation between the operations of multiplication and addition, whose existence he had glimpsed; but the unfortunate choice of rank, of a qualitative rather than a quantitative order, concealed from him the real possibility of addition, and rendered him deaf to the appeal of the logarithmic method."

And yet Chuquet possessed an index notation; the only problem was that he used it exclusively for powers of the unknown, and this was forced on him because he had no other way of denoting the presence of the unknown. If there is no symbol for x, 12^2 cannot stand both for 12×12 and for $12x^2$, the second power of the unknown. Since his notation was forgotten, future generations were able to rediscover indices for themselves, and to apply them in a form better capable of advancing arithmetic and algebra simultaneously.

Chuquet's use of notation

Chuquet's use of notation is a valuable source for the study of the movement towards symbolic algebra. Although the odds must be against Chuquet's use of a particular symbol influencing anybody else, it affords evidence of the evolution and diffusion of that symbol, and of what an ingenious mathematician of the day could make out of it. For instance, the use of a crossed R for roots was common in the later Middle Ages [Cajori, 1928 I p 361], but Chuquet extended it to provide a common notation for square roots, cube roots, and roots of higher order, far beyond those usually investigated. Chuquet also broke new ground in using an underline as a bracketing convention, and adopted the abbreviations $\bar{p}$ and $\bar{m}$ for plus and minus which were then becoming popular in Italy. [Cajori, *op cit*, 100-103].

The history of Chuquet's first innovation, his terminology for large numbers, is mysterious. *Million* was a well-established Italian word, and perhaps *billion* also, but the words for the higher powers of a million are known only from Chuquet and Jehan Adam until they were rediscovered in the late seventeenth century. Adam needed these numbers for some calculations in which Chuquet would have used mixed fractions, but Chuquet himself never names such a large number after the initial section, nor does he use the system of dots to mark off every sixth digit (the common practice then, as now, was to mark off in threes).

His notations for powers and roots, on the other hand, can be compared with similar notations being developed in his day and over the following century and a half. In what follows, the dates are given for interest only; for bibliographic references, the reader is referred to (Cajori, 1928 I).

Index notation in its current form (e.g. x^2-2x+5) was not established until the work of Descartes (1637). Most writers in the sixteenth century preferred to use a letter or letters, or special symbols, to denote the power concerned, which in practice was almost always a square or a cube. Pacioli did have a numeric notation, using R 2^a for x, R 3^a for x^2, and so on; but apart from the confusion between the use of R for roots and for powers, the assignment of 1 (*prima*) instead of 0 to the constant term hindered him from perceiving the index law of addition ($x^m.x^n=x^{m+n}$), and he did not use the notation to simplify his calculations, merely to record the answers. Cardano (1545) used no numerical index notation at all, writing the equation $x^3+3x^2=21$ as "the cube & 3 squares are equal to 21".

The first writer after Chuquet to use anything similar to his notation was Rafaele Bombelli, who wrote $4x^2+20x+16$ as $4\overset{2}{\smile}p.20\overset{1}{\smile}p.16$; the indices with the arcs also appeared in his working, written directly over the coefficient. This notation appears in a manuscript dating from about 1550 and in his printed *Algebra* of 1572. Stevin (1585) replaced Bombelli's arc by a full circle, and used the notation not only in algebra but also in his decimal fractions, where 5 ② 4 ③ stood for $5.10^{-2} + 4.10^{-3}$ or 0.054. He also suggested that a fraction such as $\frac{3}{2}$ could be inscribed in a circle to represent the square root of the cube, and used his indices to apply to both constants and variables; in the former case, they were written to the left of the number.

Chuquet's terminology of first, second and third terms, which also encourages the use of index laws by making explicit the relation between numbers or powers, became used as a notation in the work of Schreiber or Grammateus

(1518), who wrote 30 se. - 2 pri. -56 N for $30x^2 - 2x - 56$. This was almost certainly an independent development, and was adopted by other German writers in the sixteenth century, though not by all of them.

Chuquet also used the notation 1^2 to refer to the second unknown in an equation. This is less satisfactory, as it is indistinguishable from his notation for x^2, but in fact the meaning is always clear from the context. Pacioli used the terms *cosa* and *quantita* for the first and second variables; the latter is reflected in de la Roche's marginal note about the rule of quantity in the *Triparty* manuscript. Cardano used contractions of these terms, and the first recognizable approach to the modern notation is Butéon's use of A, B and C for three variables in 1559.

While the radix notation R (crossed) was common enough, very few writers before the seventeenth century used numbers to denote the order of the root, though R^3 is sometimes found. Bombelli used R.q and R.c for square and cube roots. Stevin used √③ for cube roots but √√ for the fourth root, or square root taken twice. Oughtred, in the middle of the seventeenth century, used both √cccc and √[12] for twelfth roots. Descartes joined together the radical sign √ and the vinculum or overscore (see the next paragraph), and used both √C and √3 for cube roots. Wallis (1655) wrote $\sqrt{}^3$, which became $\sqrt[3]{}$ in the eighteenth century. Cajori (1928, I p 375) remarks that Descartes "missed a splendid opportunity of rendering a still greater service" by improving the radix notation when he might have discarded it in favour of fractional exponents.

The history of bracketing is closely linked with that of radicals. Chuquet's use of an underscore to denote bound roots is the earliest bracketing convention noted by Cajori. Pacioli used V (for *universale*) in front of the expression whose root was to be extracted, but no special symbol marked the end of the expression. German writers, from Rudolff (1525) onwards, used a dot at the beginning, and Stifel (1544) sometimes used a dot at the end as well. Bombelli's manuscript (c.1550) has underscores with two vertical lines, one at either end, but in the printed version he enclosed the expression between two bold-faced Ls, the second being reflected: **L** ... **⅃** Hariot (1631) put a long horizontal brace over the expression, and this became the vinculum to which Descartes joined the radical sign to make the familiar $\sqrt{\ }$. Parentheses were used in some editions of Viète's works, but did not become common until the eighteenth century (Cajori, 1928 I p 303).

In conclusion, Chuquet's major notational innovations were typical of the quest for a rational symbolism which was to evolve slowly over more than 150 years before stabilizing in the writings of Descartes and his successors. Chuquet stands alongside Bombelli in his development of index notation and use of bracketing, and surpasses him in his unified notation for radicals. However, Bombelli's notation found no immediate followers apart from Simon Stevin, a self-taught mathematician born in Bruges. It took a mathematician with the international stature of Descartes to establish the language of algebra as we know it today. It the language of algebra as we know it today. It was quite natural for an algebraist of the late fifteenth or sixteenth century to develop his own notation, and Chuquet did so to better effect than most.

The limitations of Nicolas Chuquet

To conclude this survey of the work of a remarkable figure in the history of mathematics, the limitations and weaknesses revealed by Chuquet's work must also be indicated. One of the most critical reviews (though by no means unfavourable) is Jean Itard's article in the *Dictionary of Scientific Biography* (1971). His main adverse comments are discussed below, together with some other points which have arisen in the foregoing study of Chuquet's text.

While Chuquet's concept of number was broad enough to embrace negative and zero solutions and to manipulate negative and zero coefficients, he still adopted the medieval division of quadratic equations into the three forms capable of expression in terms of positive coefficients. He also used the classical rules of solution, mentioning the possibility of a second solution only in the case of $ax^2 + c = bx$ which can have two positive solutions. He states that the equation $ax^n = bx^n$ has no solution if $a = b$, quoting the received opinion and overlooking the solution $x = 0$. On the other hand, he claims to find a zero solution to the impossible equation $R\ \underline{(12x)/(3x)} = R\ 5$. Although he uses zero solutions correctly and confidently in linear problems, they have not been assimilated into his more advanced algebra.

Itard reproaches Chuquet with failing to appreciate the difference between irrational solutions, rational solutions and integers. Thus he misunderstands a remark by Campanus about the impossibility of finding two numbers in a certain proportion, and claims that he has refuted Campanus by finding an irrational solution to the equation concerned. Since the whole book of Euclid on which that section of the commentary is based is devoted to proportions of numbers, it would appear that Chuquet misunderstood far more than the one remark on which he comments. However, it is not known how good a text of Campanus he had consulted; it is possible that he did not have the manuscript before him as he wrote, but only a reference in another author or incomplete notes from his earlier reading.

There are several facets of Chuquet's lack of respect for integer solutions. The proposal to divide an estate between 2 5/6 children (with the 5/6 child getting a 5/6 share) may evoke no more than amusement, though his pupils must have wondered what sort of merchant would sell fractions of an apple. His application of the formula for the sum of an arithmetic progression to a progression with 8 1/3 terms raises some questions about the accuracy of the model, but the problem (105) is sufficiently bizarre to admit this interpretation. It must be admitted that Chuquet's attitude blinded him to the joys of Diophantine equations, and led him to consider the rule of apposition and remotion, whose purpose was to find integer solutions, as "a science of small recommendation"; but the problems to which this rule was usually applied in the Middle Ages were not so different from those traditionally solved by the rule of three or the rule of false position to make it obvious that an algebraic solution was inappropriate. Chuquet had no access to the work of Diophantus, so he cannot be blamed for failing to appreciate it.

It is, however, a shortcoming in a work intended for the instruction of merchants that the answers to exercises should be left as complicated surds or fractions with no integer approximation. Chuquet could have used his rule of intermediate terms to find a rational approximation, or at least given the nearest integer, but he very rarely gives any approximation whatsoever. Some of his examples involve long calculations in order to simplify the answer to a recognizable form, and he makes no effort to grade the problems so that the early ones work out neatly.

Chuquet is sometimes misled by his method of induction into presenting a short cut which gives incorrect answers. One example is in his discussion on perfect numbers, where he states correctly the Euclidean rule that $2^n(2^{n+1} - 1)$ is perfect if $2^{n+1} - 1$ is prime (which implies that n is even unless $n = 1$), but then claims that any product of the form $2^{2k}(2^{2k+1} - 1)$ is perfect, without testing for primacy. The first few examples are indeed perfect, but two of the other numbers that he classes as perfect are not, because the second factor is composite in each case. He also claims, whether by induction on the known perfect numbers or as a consequence of his adopted rule, that alternate perfect numbers terminate in 6 and 8; this was refuted in the seventeenth century.

The novelty of his notation is also a source of confusion. Chuquet's use of bound roots makes it possible to write down complex surds without ambiguity, but his discussion in Book 2 of the *Triparty* on bound roots of first and second intention makes it clear that he has not completely assimilated this fact. In one of the problems, he falls into error by applying the operation of cube root to only the first term of the two bound together by the underscore. Had he not made this mistake, the algorithm would have led him to take the square root of an expression with a negative value and invent complex numbers by accident. However, the marginal note in which he shows that the question is impossible does not comment on or correct the errors in the solution, or explain how a correct application of the algorithm could be reconciled with his proof that there is no solution.

Excellent as Chuquet's terminology is in many respects, the words used to describe the various orders of roots in Book 2 of the *Triparty* are rather too close to those used for powers of the unknown in Book 3. Ordinal adjectives (*premier*, *second*, etc.) are used in both cases, with no noun. However, this may be more of a problem for the modern reader or translator than it was in his own day.

The details of Chuquet's working are frequently ponderous, in the algebraic parts of the *Geometry* as well as in the *Triparty* and its appendix. His notation, excellent though it was in comparison with that of his contemporaries, was still a long way from a full symbolical algebra. It is marvellous that, faced with the task of developing his own language of both French words and symbols to describe his calculations, and working outside any recognized mathematical community, Chuquet can nevertheless be placed among the leading mathematicians of his day.

Appendix

TABLE OF CONTENTS FOR CHUQUET'S MATHEMATICAL MANUSCRIPTS

The detailed table of contents for Chuquet's mathematical manuscripts collates the pagination of the manuscript with that of various printed editions of parts of Chuquet's manuscript. For the *Triparty*, the manuscript pagination is collated with Marre's publication in the Bulletino (1880 and 1881). For the *Geometry*, the pagination is collated with L'Huillier's edition (1979). The table of contents for the *Geometry* includes the detailed contents for the third part of the work added by L'Huillier. L'Huillier's additions to Chuquet's headings provide a conspectus of the contents of this central section of the *Geometry*. (The paginations are headed MS, and BB or L'H. The column headed Page refers to this book.)

		MS	L'H	Page
IC.	Concerning objects which are measured in three dimensions	216v	142	245
	Of the sphere	216v	142	245
	Of pyramidal bodies	216v	145	245
	Of cubical and all other bodies	217v	151	-
	Of misshapen and crooked bodies	219r	160	246
	Conversion of bodies	219v	162	247
II.	There follows the construction of the quadrant and its use	219v	165	248
	Description	219v	165	248
	The first chapter which is about the length of any object	220r	168	250
	The second chapter which is about the width of objects	221v	174	-
	The third chapter which is about the depth of objects	222r	178	250
	The fourth and last chapter which deals with the height of objects	223r	185	-
	Another way of measuring the height of something without the aforesaid instrument	225r	198	253
III.	Application of the aforesaid rules	226r	204	255
IIIA.	Little problems	226r	204	255
IIIB.	Problems dealing with each particular figure	228v	216	257
	Area of a figure formed from the circumferences of two others	228v	216	257
	Problems on the circle	229v	218	257
	Problems on triangles	230v	223	258
	Equilateral	230v	223	-
	Scalene	231v	226	-
	Isosceles	232v	230	258
	Right-angled	233r	233	-
	Two juxtaposed right-angled triangles	234v	242	-

BIBLIOGRAPHY

Anon
(1963) *Journal d'un Bourgeois de Paris à la fin de la guerre de cent ans.* Edited by Jean Thielly from a 15th-century manuscript.
Union Générale, Paris

Anon
1501? *Livre de chiffres et de getz.* Copy in Kreisbibliothek, Augsburg.
Mareschal, Lyon

Adam, Jehan
1475 *Traicte d'arismetique pour la pratique par gectoners.* Manuscript in Bib. Sainte Geneviève, Paris, MS Français 3143.

Bachet, C.B., Sieur de Méziriac
1612 *Problems plaisans et delectables qui se font par les nombres.*
Rigaud, Lyon

Baker, Humphrey
1562? *The Well-Spring of Sciences. Which teacheth the perfect worke and practise of Arithmeticke.* Edition consulted: 1574.
London

Ball, W. W. R.
1892 *Mathematical recreations and problems.* (Many later editions under title *Mathematical recreations and essays.*)
Macmillan, London

Battesti, François
1980 *Le "Compension de l'Abaco" de Frances Pellos.* Dissertation, Université de Nice.

Baxendall, Michael
1972 *Painting and experience in fifteenth century Italy.*
Oxford, UP, Oxford

Bortolotti, E.
1925 *'Primordi della Geometria Analitica. L'Algebra Geometria di Paolo Bonasoni ...'* in his *Studi e ricerche ...*, 1928
Zanichelli, Bologna

Boyer, Carl Benjamin
1968 *A History of Mathematics.*
Wiley, New York

Brésard, Marc
1914 *Les foires de Lyon aux VXe et XVIe siècles*
Picard, Paris

Butéon, Jean
1559 *Logistica, quae & Arithmetica vulgo dicitur in libros quinque digesta.*
Lyon

Cajori, Florian
1928 *A History of Mathematical Notations. Volume I. Notations in Elementary Mathematics.*
Open Court, La Salle, Ill.

Cantor, Moritz
1892 *Vorlesungen über Geshichte der Mathematik. II.Vom Jahre 1200 bis zum Jahre 1668.* Originally issued in parts. 2nd ed. 1900
Teubner, Leipzig.

Cardan, Girolamo
1545, 1968 *The Great Art or the Rules of Algebra.* Tr. T. R. Witmer, from *Ars Magna sive de regulis algebraicis* (Nuremberg).
MIT Press, Cambridge, Mass

Cardan(o) Girolamo
1603, 1966 *Opera omnia* Facsimile editions by Frommann, Stuttgart and Johnson Reprint Corp, New York

Chasles, Michel
1841a 'Note sur la nature des opérations algèbriques ...' *Comptes Rendus de l'Académie des Sciences*, 12, 741-756
Paris.

Chasles, Michel
1841b 'Sur l'époque où l'Algèbre a été introduite en Europe'. *ibid.*, 13, 497-524.
Paris

Chasles, Michel
1841c 'Note de M. Libri en réponse au derneir Memoire de M. Chasles. Reponse de M. Chasles'. *ibid.*, 13, 560-563.
Paris

Chasles, Michel
1841d 'Sur les expressions *res* et *census*. Et sur le nom de la science, *Algebra et Almuchabala*'. *ibid.*, 13, 601-628
Paris

Chuquet, Nicolas
see Marre, Aristide, 1880 and 1881 and l'Huillier, Hervé, 1979 for printed editions of his work.

Chuquet Nicolas
1484 *Triparty en la science des nombres* ... Manuscript in Bibliothèque Nationale, Paris, fonds français 1346.
Lyon

Chuquet, Nicolas
14?? *Ung petit traicte de la pratique de geometrie* ... Manuscript in Bibliothèque Nationale, Paris, fonds français 1052 fo. 39r-67v

Clagett, Marshall (ed.)
1968 *Nicole Oresme and the Medieval Geometry of Qualities and Motions*
U. Wisconsin Pr, Madison

Cotgrave, Randle
1611, 1950 *A dictionarie of the French and English Tongues*, Reprinted in facsimile, with an introduction.
U. South Carolina P, Columbia

Davis, Natalie Zemon
1960 'Sixteenth century French arithmetics on the business life', in *Journal of History of Ideas* 21, 18-48.

de la Roche, Estienne
1520 *Larismethique nouellement composee.* (Second edition (1538) under title *Larismethique & Geometrie* published by Huguetan.)
Lyon

De Morgan, Augustus
1847 *Arithmetical Books from the invention of printing to the present time.* Reprinted in 4th ed (1970) of D.E. Smith (1908).
Taylor & Walton, London

Dudeney, Henry
1926 *Modern problems and how to solve them.*
Pearson, London

Egmond, Warren van
see Van Egmond, Warren

Eves, Howard
1953 *An Introduction to the History of Mathematics.* (The fourth edition (1976) was published by Holt, Rinehart and Winston.)
Rinehart, New York

Fibonacci, Leonardo
see Ver Eecke (1952) for translation of his *Liber quadratorum.*

Fibonacci, Leonardo
(1857-62) *Scritti.* Edited by B. Boncompagni. Volume I = *Liber abaci.* Volume II = other works.

Gosselin, Guillaume
1577 *De Arte Magna seu de occulta parte numerorum, quae & Algebra & Almuchabala vulgo dicitur.*
Paris

Gosselin, Guillaume
1578 *L'arithmêtique de Nicolas Tartaglia ... avec ... plusieurs inventions dudit Gosselin.*

Grant, Edward (ed.)
1966 *Nicole Oresme: ad pauca respicientes.*
U. Wisconsin Pr, Madison

Grant, Edward (ed.)
1971 *Nicole Oresme: De proportionibus proportionum.*
U. Wisconsin Pr, Madison

Hughes, Barnabas B. (ed.)
1983 *Jordanus de Nemore, De Numeris Datis.*
Publications of the Center for Medieval and Renaissance Mathematics 14. Los Angeles.
U. California Press

Itard, Jean
1971 'Chuquet, Nicolas' in C. C. Gillispie (ed.), *Dictionary of Scientific Biography*, 3, 272-278.
Scribner, New York

Johnson, Charles (ed.)
1956 *The De Moneta of Nicole Oresme and English Mint Documents.*
Nelson, Edinburgh

Jordanus de Nemore
see Hughes, Barnabas B. (ed.)

Josephus, Flavius
(1959) *The Jewish War.* Edited and translated by G. A. Williamson. (Written in 1st century).
Penguin Books, London

Juschkewitsch, A.P.
1964 *Geschichte der Mathematik im Mittelalter.*
Tr. from Russian, 1961. (For French translation see Youschkevitsch.)
Teubner, Leipzig

L'Huillier, Hervé (ed.)
1979 *Nicolas Chuquet, La géométrie. Premiere Géométrie algèbrique en langue française(1484).*
Vrin, Paris

la Roche, Estienne de
see de la Roche

LeFèbvre, B.
1897 'Aperçu historique sue les origines des mathêmatique elementarires' in his *Cours dêveloppê d'algèbre êlêmentaire I*
Namur

Mahoney, Michael S.
1978 'Mathematics', in David C. Lindberg (ed.), *Science in the Middle Ages.*
U. Chicago Press, Chicago

Marre, Aristide
1880 'Notice sur Nicolas Chuquet et son Triparty en la Science des Nombres', *Bull di Bibliog e di Storia d Scienze* 13, 555-592.
Rome

Marre, Aristide (ed.)
1880 'Le Triparty en la Science des Nombres par Maistre Nicolas Chuquet, Parisien, *ibid.*, 593-658, 693-814.
Rome

Marre, Aristide (ed.)
1881 'Appendice au Triparty en la Science des Nombres de Nicolas Chuquet, Parisien', *ibid.*, 14, 413-460.
Rome

Naux, Charles
1966 *Histoire des Logarithmes de Neper à Euler. I. La dêcouverte des logarithmes et le calcul des premieres tables.*
Blanshard, Paris

Oresme, Nicole
see Johnson, Charles (ed.), 1956; Grant, Edward (ed.), 1966, 1971; Claggett, Marshall (ed.), 1968.

Pacioli, Luca
1494 *Summa de arithmetica, geometrica, proportioni et proportionalita.* Second edition (1523) at Toscolano.
Venice

Pelloa, Francesco
1492 *Compendion de'l abaco.* First edition published in Turin, reprinted 1967, *Editions de la Revue des Langues Romanes.*

Rashdall, Hastings
1895, 1936 *The Universities of Europe in the Middle Ages.* New edition (1936) edited by F. M. Powicke and A. B. Emden. 3 volumes.
Oxford UP, Oxford

Record(e), Robert
1552 *The Ground of Artes teachying the worke and practise of Arithmetike.* Second edition.
Wolff, London

Roche, Estienne de la
see de la Roche

Schneider, Ivo
1974 'Urheberrechtliche Sicherung im naturwissenschaftlichen Schrifttum des 16. Jahrh.' in *Das Antiquariat* 5, 145-151

Sesiano, Jacques
1984 'Une arithmétique médiévale en langue provençale' in *Centaurus* 1984, 26-75.

Smith, David Eugene
1908 *Rara Arithmetica. A catalogue of the arithmetics written before the year MDCI ...*
Fourth edition (1970) pub. by Chelsea, New York

Sullivan, J.W.N.
1924 *The history of mathematics in Europe. From the fall of Greek science to the rise of the concept of mathematical rigour.*
Oxford UP, London

Tartaglia, N.
1556 *General trattato di numeri et misure*
Venice

Terquem, Olry
1847 'Note historique sur la notation cartésienne des exposants. Estienne de la Roche.' *Nouvelles Ann. Math.*, ser 6, 44.
Paris

Thorndike, Lynn
1926 'Jehan Adam' in *American Mathematical Monthly* 33, 24-28

Thorndike, Lynn
1929 *Science and thought in the XVth century. Studies in the history of medicine and surgery, natural and mathematical science, philosophy and politics.*
Columbia UP, New York

Trenchant, Jean
1558? *L'Arithmétique départie en trois livres. Ensemble un petit discours des changes. Avec l'art de calculer avec Getons.*
Lyon

Tropfke, Johannes
(1980) *Geschichte der Elementarmathematik. I. Arithmetik und Algebra.* 4th edition, completely revised by Kurt Vogel, Karin Reich and Helmuth Gericke.
de Gruyter, Berlin

Van Egmond, Warren
1976 *The Commercial Revolution and the Beginnings of Western Mathematics in Renaissance Florence, 1300-1500.*
Ph.D. thesis, Indiana U.

Ver Eecke, Paul
1952 *Léonard de Pise. Le Livre des nombres carrés.* Translated with introduction and notes.
Desclée &c., Bruges

Victor, Stephen K.
1979 *Practical Geometry in the High Middle Ages.* An edition of *Ars cuiuslibet consummatio* and *Pratike de Geometrie.*
Amer Phil S, Philadelphia

Vogel, Kurt
1954 *Die Practica des Algorismus Ratisbonensis. Ein Rechenbuch ... aus der Mitte des 15. Jahrhunderts.*
Beck, Munich

Wallis, John
1685 *A treatise of algebra both historical and practical. Shewing the original, progress, and advancement thereof ...*
Playford, London

Wetheim, G.
1898 'Die Berechnung der irrationalen Quadratwerzeln und die Erfindung der Kettenbrüche' in *Abh. Gesch. Mathematik* 8, 147-160.

Wickersheimer, C.A.E.
1906 *La médecine et les médecins en France à l'époque de la Renaissance.*
Maloine, Paris

Wickersheimer, C.A.E. (ed.)
1915 *Commentaires de la faculté de médecine de l'Université de Paris 1395-1516.*
Imp. Nationale, Paris

Youschkevitsch, A.P.
1976 *Les Mathématiques Arabes (VIIIe-XVe siècles).* Translated by M. Cazenave and K. Jaouiche from part 3 of Juschkewitsch (1964).
Vrin, Paris

INDEX OF NAMES AND WORKS